43 wm 100
lbj 30005

Ausgeschieden im Jahr 20 25

D1731481

PRINCIPLES OF
ELECTRONIC CERAMICS

PRINCIPLES OF ELECTRONIC CERAMICS

L. L. HENCH
Advanced Materials Research Center
University of Florida
Gainesville, Florida

and

J. K. WEST
Advanced Materials Research Center
University of Florida
Gainesville, Florida

WILEY

A WILEY-INTERSCIENCE PUBLICATION
JOHN WILEY & SONS
New York / Chichester / Brisbane / Toronto / Singapore

A NOTE TO THE READER:
This book has been electronically reproduced from digital information stored at John Wiley & Sons, Inc. We are pleased that the use of this new technology will enable us to keep works of enduring scholarly value in print as long as there is a reasonable demand for them. The content of this book is identical to previous printings.

Copyright © 1990 by John Wiley & Sons, Inc.

All rights reserved. Published simultaneously in Canada.

Reproduction or translation of any part of this work
beyond that permitted by Section 107 or 108 of the
1976 United States Copyright Act without the permission
of the copyright owner is unlawful. Requests for
permission or further information should be addressed to
the Permissions Department, John Wiley & Sons, Inc.

Library of Congress Cataloging in Publication Data:
Hench, L. L.
    Principles of electronic ceramics/L. L. Hench and J. K. West.
      p.   cm.
    Includes bibliographies and index.
    "A Wiley-Interscience publication."
    1. Electronic ceramics. I. West, J. K. II. Title.
    TK7871.15.C4H38   1989   89-30955
    621.381—dc19   CIP
    ISBN 0-471-61821-7

Printed in the United States of America

10 9 8 7 6 5 4

To June and Donna

For their patience, understanding, and support without which this book would still be a set of unedited, coffee-stained notes

CONTENTS

Preface

CHAPTER 1   INTRODUCTION   1

    1.1   Overview / 1
    1.2   The Materials Spectrum / 2
    1.3   What are Electronic Ceramics? / 4

        1.3.1   Crystal Compounds / 4
        1.3.2   High Temperature Processing / 5
        1.3.3   Crystals and Glasses / 5

    1.4   Structure of Crystal Compounds / 6

        1.4.1   Structure Rules / 6
        1.4.2   Radius Ratio Concept / 6
        1.4.3   Electrical Neutrality / 8
        1.4.4   Ionic Bonding / 14

    1.5   Thermal Expansion / 17
    1.6   Glass / 19
    1.7   Processing and Microstructures / 20
    1.8   Range of Electronic Properties / 27
    1.9   Electronic Ceramics Market / 29
        Problems / 30
        Reading List / 31

CHAPTER 2 QUANTUM MECHANICS AND THE BAND THEORY OF SOLIDS — 33

- 2.1 General Applications of Quantum Mechanics / 33
- 2.2 Postulate the Five Rules of Quantum Mechanics / 36
- 2.3 De Broglie's Wavelength / 38
- 2.4 Electron Scattering / 39
- 2.5 Simple Harmonic Oscillator (SHO) / 40
- 2.6 Probability / 42
- 2.7 Superposition / 43
- 2.8 Heisenberg Uncertainty / 44
- 2.9 Quantization of Wavefunctions and Energies / 45
- 2.10 Probability of Locating a Particle / 47
- 2.11 Bohr Model of the Hydrogen Atom / 51
- 2.12 Quantum Numbers / 54
- 2.13 Molecular Orbitals / 55
- 2.14 *Ab-initio* Calculations / 57
- 2.15 Semi-Empirical Calculations / 58
- 2.16 Huckel Molecular Orbital Theory / 59
- 2.17 Band Theory of Solids / 64
- 2.18 Silicon Carbide High-Band-Gap Semiconductor / 69
  Problems / 71
  Reading List / 73

CHAPTER 3 SEMICONDUCTORS — 74

- 3.1 Introduction / 74
- 3.2 Theory of Superposition / 75
- 3.3 Hall Effect / 81
- 3.4 Electronic Conductivity / 83
- 3.5 Boltzmann Energy Distribution / 84
- 3.6 Fermi–Dirac Distribution / 86
- 3.7 Fermi Level / 90
- 3.8 Extrinsic Semiconduction / 93
- 3.9 Concentration of Carriers / 96
- 3.10 Polarons / 96
- 3.11 Excitons / 96
- 3.12 Semiconductor Surfaces / 97
- 3.13 Semiconductor Contacts / 99
- 3.14 P–N Junctions / 104
- 3.15 Diodes / 105
- 3.16 Junction Transistor / 109
- 3.17 Gallium Arsenide / 111
- 3.18 Silicon Carbide Semiconductors / 112
- 3.19 Amorphous Semiconductors / 117

3.20 Radial Distribution Function / 120
3.21 Amorphous Germanium / 121
3.22 One-Dimensional Model / 121
3.23 Transport / 124
3.24 Materials / 130
   Problems / 133
   Reading List / 134

CHAPTER 4 IONIC AND DEFECT CONDUCTORS   136

4.1 Introduction / 136
4.2 Theory of Ionic Conductivity / 139
4.3 Ionic Conductivity / 144
   4.3.1 Electrode Polarization / 144
   4.3.2 Glass Composition and Structure Effects / 147
   4.3.3 Molten Silicates / 150
4.4 Solid Electrolytes and Fast Ion Conductors / 154
4.5 Criterion for Fast Ion Conduction / 157
4.6 β-Alumina and β''-Alumina / 159
4.7 Grain Boundary Effects / 163
4.8 Defect Semiconductors / 163
4.9 Frenkel Disorder in a Stoichiometric Crystal / 165
4.10 Example of a Defect Semiconductor: Lead Sulfide / 174
4.11 Protonic Conduction in Glasses / 180
   Problems / 183
   Reading List / 184

CHAPTER 5 LINEAR DIELECTRICS   185

5.1 Importance / 185
5.2 Theory / 185
5.3 Circuit Description of a Linear Dielectric / 191
5.4 Relation of Dielectric Constant to Polarization / 193
5.5 Dipolar Polarization Theory / 195
5.6 Time Dependence of Dipolar Polarization / 198
5.7 Cole–Cole Distributions / 203
5.8 Temperature Dependence of Dipolar Polarization / 205
5.9 Interfacial Polarization / 208
5.10 Composition and Structural Effects of Glasses / 212
5.11 Sol–Gel Silica Dielectric Spectra / 216
5.12 Dielectric Breakdown / 222
5.13 Intrinsic Breakdown Mechanisms / 224

5.14 Thermal Breakdown Mechanics / 227
5.15 Avalanche Breakdown Mechanisms / 229
5.16 Experimental Variables / 231
Problems / 233
Reading List / 235

CHAPTER 6 NONLINEAR DIELECTRICS 237

6.1 Introduction / 237
6.2 Crystallographic Considerations / 237
6.3 Ferroelectric Theory / 240
6.4 Structural Origin of the Ferroelectric State / 244
6.5 Hysteresis / 250
6.6 Description of Ferroelectricity Based on Local Fields / 252
6.7 Effect of Temperature on Polarizability / 259
6.8 Ferroelectric Domains / 259
6.9 Effects of Environment on Switching and Transitions / 266
6.10 Effect of an Electric Field on T_c / 267
6.11 Compositional Factors / 270
6.12 Effect of Grain Size on Ferroelectric Behavior / 272
6.13 Antiferroelectric–Ferroelectric Transition / 272
6.14 PLZT Ceramics / 276
Problems / 278
Reading List / 280

CHAPTER 7 MAGNETIC CERAMICS 282

7.1 Introduction / 282
7.2 Basic Theory / 282
7.3 Diamagnetism / 289
7.4 Paramagnetism / 291
7.5 Ferromagnetism / 292
7.6 Antiferromagnetism / 295
7.7 Ferrimagnetism / 296
7.8 Exchange and Indirect Exchange Interactions / 297
7.9 Spin Order / 303
7.10 Lattice Interactions / 304
7.11 Ferrimagnetic and Ferromagnetic Domains and Domain Motion / 305
7.12 Classes of Magnetic Ceramics / 309
7.13 Spinel Ferrites / 311
7.14 Fe_3O_4 / 313
7.15 Structure of Spinel Ferrites / 315

7.16 Spinel Lattice Interactions / 314
7.17 Effect of Composition in Ferrites / 315
7.18 Effect of Thermal Treatment / 316
7.19 Oxidation States / 317
7.20 Manganese and Nickel Zinc Ferrites / 317
7.21 Hexagonal Ferrites / 321
7.22 Garnet Ferrites / 327
   Problems / 333
   Reading List / 334

CHAPTER 8   PHOTONIC CERAMICS   335

8.1 Introduction / 335
8.2 Radiation / 337
8.3 Reflection / 338
8.4 Refraction / 341
8.5 Scattering / 351
8.6 Absorption / 352
8.7 Absorption in the Visible / 358
8.8 Ions in Ligand Fields / 358
8.9 Intraionic Absorption and Ligand Field Theory / 359
8.10 Optical Filters / 376
8.11 Polarization / 378
8.12 Ionic Polarizability / 385
   Problems / 389
   Reading List / 391

CHAPTER 9   OPTICAL WAVEGUIDES   392

9.1 Introduction / 392
9.2 Optical Fibers / 393
9.3 Generations of Optical Communications / 395
9.4 Hollow Conducting Waveguides / 396
9.5 TEM Waves in a Metallic Waveguide / 402
9.6 Hollow Rectangular Waveguide for a TE Wave $(E_{oz} = 0)$ / 407
9.7 Refraction and Reflection / 410
9.8 Reflectance / 418
9.9 Dielectric Waveguides / 419
9.10 Goos–Haenchen Shift / 420
9.11 Graded-Index Waveguides / 420
9.12 Ray Deviations from Central Core / 422
9.13 Planar Waveguides / 425
9.14 Laser Fabrication of Gel-Silica Waveguides / 426

9.15 Application of the Fresnel Equations / 427
9.16 Infrared Waveguides / 429
9.17 Waveguide Integrated Optics / 429
Problems / 432
Reading List / 433

CHAPTER 10 ELECTRO-OPTICAL CERAMICS   434

10.1 Introduction / 434
10.2 Birefringence / 434
10.3 Optical Principal Axis / 436
10.4 Isometric Optical Indicatrix / 437
10.5 Uniaxial Optical Indicatrix / 437
10.6 Biaxial Optical Indicatrix / 438
10.7 Electric Field Dependence of Index of Refraction / 439
10.8 Pockels (Linear) Effect / 440
10.9 Kerr (Quadratic) Effect / 440
10.10 Optical Activity / 441
10.11 Faraday Effect / 441
10.12 Materials / 442
10.13 Memory Effect / 447
10.14 Parametric Amplification Oscillation / 448
10.15 Fluorescence / 449
10.16 Light Emitting Diodes (LED) / 451
10.17 Blue Light Emitting Diodes / 452
10.18 Avalanche Photodiodes (APDs) / 454
10.19 Excitons / 455
10.20 Quantum Confinement / 457
10.21 Photoconductive Semiconductors / 459
10.22 Nonlinear Optics / 460
Problems / 461
Reading List / 462

CHAPTER 11 GLASS AND CRYSTALLINE LASERS   463

11.1 Introduction / 463
11.2 Optical Cavities: Feedback and Lasers / 464
11.3 Metastable States and Stimulated Emission / 467
11.4 Three Level Lasers (Cr—Al_2O_3) / 468
11.5 Four Level Lasers (Nd-YAG, Nd-Glass) / 470
11.6 Emission Cross Section / 476
11.7 Laser Modes and Focusing / 477
11.8 Etalons: Fabry–Perot Interferometer / 479
11.9 Brewster Angle Windows / 485

CONTENTS   xiii

11.10 Q-Switching / 485
11.11 GaAs Lasers / 485
11.12 Quantum Well Lasers / 487
11.13 Limitations in Laser Power / 487
11.14 Fluorescence Lifetimes / 488
11.15 Parameters of Various Lasers / 490
Problems / 491
Reading List / 492

CHAPTER 12   CERAMIC SUPERCONDUCTORS   493

12.1 High-Temperature Superconducting Ceramics / 493
12.2 Background / 495
12.3 Meissner Effect / 496
12.4 The Critical Field / 498
12.5 Early Theories on Superconductivity / 498
12.6 The BCS Theory / 499
12.7 Electron Pairs / 499
12.8 Energy of the Cooper Pair / 504
12.9 The Isotope Effect / 505
12.10 Debye Theory of Specific Heat / 507
12.11 BCS Phonon–Electron Coupling / 508
12.12 Type I Superconductors / 508
12.13 Magnetic Field Penetration Depth / 510
12.14 Type II Superconductors / 513
12.15 Structure of High T_c Superconducting Ceramics / 515
12.16 High-T_c Electron Pairs and Coupling Mechanisms / 516
12.17 Resonating Valence Bonds / 517
12.18 Spin Bag Model / 517
12.19 Phase Diagrams of High-T_c Superconducting Ceramics / 518
12.20 Crystalline Structure / 521
12.21 Defect Chemistry of Superconducting Ceramics / 523
12.22 Conclusions / 525
Problems / 526
Reading List / 528

APPENDIX I   **537**

INDEX   **539**

PREFACE

Many thousands of years ago, the conversion of molded, naturally occurring clay into fired, insoluble ceramic pottery began mankind's technical mastery over nature. During the following millennia (see Figure 1), an ever-increasing ability to manipulate and control the properties of ceramic materials was achieved. These developments led to the refractories essential for production of high-melting metallic alloys. The transition from the Ages of Pottery to Bronze to Iron and Steel has culminated in our modern worldwide industrial society. Figure 1 also illustrates that the information content in these progressive technological developments has been additive, (i.e., the knowledge inherent in one advance has been incorporated in each succeeding development).

In past ages, the rate of change of technology development has been relatively slow, taking many decades to complete and be incorporated into worldwide society. Figure 1 shows, however, that during the last 40 years, the rate of technological change has increased dramatically. The result of this almost explosive rate of change is that new technological developments are introduced and, in some cases, replaced within a few decades with even more advanced concepts.

This rapid change in technology coincides with the development and growth of the electronic ceramics industry. In fact, man's ability to manipulate and control the electronic properties of ceramic materials has played a major role in our transformation from the Steel Age to the Information Age and our transformation from local economies to a world economy. Nearly instantaneous satellite communications from coast to coast and around the world would not have been possible without simultaneous advances in semiconductors, dielectrics, ferroelectrics, ferrites, and opto-

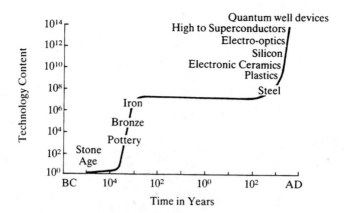

FIGURE 1. The evolution of materials.

electronics. The continued developments in these fields are being matched, and perhaps eventually superseded, with advances in superconducting ceramics, photonics, optical waveguides, and potentially even optical computing and information processing.

It is difficult to establish a realistic historical date for the origin of electronic ceramics as a field. Development of the technology for evacuation and sealing of glass envelopes for incandescent lighting and vacuum tubes dates back a century. Porcelain ceramics have been used for electrical insulation from the beginning of the electrical industry, also a century ago. The discovery of ferrimagnetic behavior in iron oxide spinels by Snoek occurred as long ago as 1919, and ferroelectricity was observed in Rochelle salt as early as 1921.

However, the field of modern electronic ceramics, as presently known, is primarily based upon the specialty electronic materials developed since the 1940s. These materials, such as barium titanate, developed by von Hippel in 1945, were optimized for specific electronic characteristics by controlling their composition and microstructure. Equally important, the successful engineering of new electronic materials was accompanied by the development of solid-state physics and an atomistic level of understanding of the physical principles responsible for the improved performance of the materials.

The objective of this book is to describe the physical principles of this broad class of electronic materials. It is an exciting challenge, since the range of properties of electronic ceramics is the largest known, from the highest electrical resistance to the lowest. The origin of their differences derives from variations in the electronic bonding and the atomic structure of the materials. Consequently, the book begins with a review of the structural concepts of ionic compounds and amorphous solids (Chapter 1) followed with a development of quantum mechanics and electronic bonding which leads to energy band theory (Chapter 2).

The description of semiconductor properties (Chapter 3) is based upon energy band theory and electron transport. Comparisons are developed between low-band-gap Si and GaAs semiconductors with a high-band-gap semiconductor, SiC. Also, the theory of amorphous semiconductors is compared with that of crystalline semiconductors.

Ionic conductivity is described in Chapter 4, with application to both insulators and fast ionic conductors. Crystal compound semiconductors that possess both electronic and ionic defect structures are also treated in Chapter 4.

The theory of linear and nonlinear dielectrics is presented in Chapters 5 and 6, with emphasis on the structural mechanisms of the electric field dependence of these materials. Applications to several important commercial ceramic insulators and ferroelectrics are described.

Magnetic ceramics (i.e., those exhibiting ferrimagnetism) are discussed in Chapter 7. The optical properties of glasses and crystalline ceramics are described in Chapter 8 in terms of the interaction of photons of a characteristic energy, or wavelength, with the electronic and ionic structure of the material. Use of ceramics and glasses to guide light is the subject of Chapter 9. The switching of light by various optically active ceramics is the subject of Chapter 10, and light amplification by simulated emission of radiation, or lasers, is developed in Chapter 11. Finally, the theoretical and structural basis of the newest class of electronic ceramics, the high-T_c superconductors, is presented in Chapter 12.

Throughout this volume we have attempted to describe the theory in structural terms, using quantum mechanics or statistical mechanics as an aid, but not an end in itself. Thus, this text is designed to fit in the gap between a solid-state physics text, where the emphasis is entirely on theory, and a ceramics or materials science text, where the emphasis is largely on processing and microstructure. Nearly all equations are derived, and where possible from first principles. Some applications of the theory are given in each chapter with a description of the relevant properties of a range of materials. However, with no apologies, this volume is not a handbook of electronic ceramics. There will obviously be many useful materials excluded from the various tables and figures. An ample reading list is provided though, for the serious student, scientist, or engineer who wishes to explore the subject more deeply or find additional property data.

Throughout the book we adopt the convention of the American Ceramic Society, where glasses and amorphous solids are considered as a subset of the generic field of ceramics, defined broadly as inorganic-nonmetallic materials.

For the student and serious reader, we have attempted to present a useful set of calculational problems for each chapter. We hope to advance the depth of understanding through some of the problems in each set. Consequently, several are quite difficult. Also, at least one problem per set requires use of a computer to do it properly. This is an effort to aid the student in preparation for the "real world," where compositional tailoring of the properties of electronic ceramics are commonplace and one of the great strengths of the field. An answer key is available for instructors or nonstudent readers.

The level of the text is for a senior in an engineering or science curriculum or first-year graduate student. It is assumed that the reader has had at least an introductory course in materials science, knows the rudiments of crystallography, and understands the difference between single crystal, polycrystalline and amorphous materials. For the reader without this background, Chapter 1 provides a brief review. We also assume the reader has some knowledge of thermodynamics, phase equilibria, and phase transformations. The reader also should understand calculus with at least a modest background in differential equations. Our goal is to prepare the reader to explore the literature in this exciting and enormously rewarding field. There are thousands of research papers published annually concerning the materials covered in this volume. They all assume an understanding of the concepts presented herein. We hope this book will open the door to this rapidly growing library of information.

In order to enhance clarity we have chosen to minimize reference citations within the text. The reading lists have been chosen to satisfy most needs for additional references. We hope the many researchers who have contributed to the scientific development of electronic ceramics will forgive us for not mentioning them by name.

Finally, we wish to acknowledge the continued spirit and enthusiasm for research and teaching in the University of Florida Department of Materials Science and Engineering and a great group of graduate students over the years, which provided the incentive for producing a text such as this. The assistance of Frank Cerra and Carol Beasley at John Wiley has also been invaluable. The continued intellectual and research support of Don Ulrich and the Air Force Office of Scientific Research has also been highly important to us for a long period of time. A very special acknowledgment is noted for the assistance of Alice Holt throughout this project for her management and secretarial leadership, and Minnie Stalvey for her typing.

L. L. HENCH
J. K. WEST

Gainesville, Florida
August 1989

PRINCIPLES OF ELECTRONIC CERAMICS

1
INTRODUCTION

1.1 OVERVIEW

Electronic ceramics originated as speciality electrical insulators made by the ceramic whitewares industry. The need for progressively higher performance sparkplugs for automotive and aircraft engines during World War II provided the motivation to improve the traditional clay–feldspar–flint triaxial porcelain body widely used in housewares and for low-voltage insulation. Removal of the ionically conducting potassium ions from the insulator composition was achieved, leading eventually to an all-alumina body and the beginnings of high performance molecularly tailored electronic ceramics.

Development of ferroelectric and piezoelectric components occurred at the same time, also utilizing the new concept of solid-state sintering for densification of the materials. Solid-state sintering provided the advantage of eliminating the ionically conducting, glassy grain boundary phase which results from liquid phase sintering or vitrification. Consequently, the physical properties of the new electronic ceramics were dependent solely on an assemblage of polycrystalline phases and a very small amount of residual porosity.

However, in order to achieve a high-density sintered compact without use of liquid phase sintering, it is essential to control the purity and size of the starting powders, their size distribution, and extent of agglomeration. Without such processing controls the fired ceramic can possess a broad range of grain sizes and porosity, and uncontrolled grain boundary phases. These can degrade the physical properties and increase the environmental sensitivity of the ceramic component. Variability in properties and environmental resistance must be minimized because electronic ceramics are often selected for high-performance applications over organic-based electronic components, due to greater stability and reliability. Therefore, it is essential that processing controls be sufficient to produce uniform and reliable microstructures. The first step in controlling reliability is understanding the relationships between chemical composition, crystal structure, and microstructure and the electronic properties of electronic ceramics. Consequently, the emphasis in this volume will be on the physical principles underlying the various types of electronic ceramics. Some attention will also be devoted to the effects of microstructure of the electronic ceramic on its properties and performance.

For the reader unfamiliar with the processing of ceramics, their characterization, or microstructural features, a brief overview follows. For detailed understanding of processing or characterization of electronic ceramics, the reader is referred to a supplementary reading list at the end of this chapter.

1.2 THE MATERIALS SPECTRUM

All engineering materials can generally be identified as part of a materials spectrum consisting of three major classes of materials: metals, plastics, and ceramics (see Table 1.1).

Metals, plastics, and ceramics each have distinctive properties. The properties are different because of a difference in the nature of the electronic bonding in the materials. Bonding in plastics basically consists of covalent bonds within hydrocarbon chains and cross-linking between chains. Metals can be considered to be held together by mutually shared free-electron bonding. The chemical bonding in ceramics generally consists of a combination of ionic and covalent bonds.

The high strength of the ceramic bond is responsible for the generally high melting point of ceramics, their brittleness, good corrosion resistance, low thermal conductivity, and high compressive strength. Variations in the chemical bonding within differing crystal structures give rise to the enormous range of electronic and magnetic properties of ceramics.

A relative comparison of the various classes of materials and their properties is shown in the materials spectrum of Table 1.1. It must be cautioned that much like in an optical spectrum, the class distinction between various materials is at times nearly indistinguishable. For example, at room temperature the metal tungsten is brittle, has a poor electrical conductivity, and a very high melting point. It behaves much like a ceramic in these respects, but it differs markedly in that it is an elemental metal and oxidizes readily. Therefore, the decision as to which class of materials a particular material belongs is at best a relative one.

TABLE 1.1 Materials Spectrum

Relative property comparisons	Plastics	Metals	Semi-conductors, cermets	Ceramics[†]
		←——— Electronic bond strength ———→		
Bond type	Hydrocarbon	Metallic	Partial covalent	Ionic-covalent
Materials example	Polyethylene	Fe, Ag, Ti (metallic elements)	Si, Ge, $Ni_{1-x}O$	SiO_2, Al_2O_3, SiC
Mechanical deformation	Plastic	Ductile	Brittle	Brittle
Melting point	Low	Intermediate	Intermediate	High
Electrical conductivity (σ)	Low	High	Intermediate	Very low to very high
Temperature dependence of σ	Positive?	Negative	Positive	Positive
Thermal conductivity	Intermediate	High	Intermediate	Low
Thermal expansion	Intermediate	High	Intermediate	Low
Corrosion resistance	Good	Poor	Poor	Very good

[†] Ceramics include glasses, glass-ceramics, polycrystalline ceramics, and single crystals.

1.3 WHAT ARE ELECTRONIC CERAMICS?

Electronic ceramics are ceramic materials that are specially formulated for specific electrical, magnetic, or optical circuit applications. The features that generally describe ceramics, and consequently electronic ceramics are:

1. The presence of ionic-covalent bonding.
2. Microstructures comprising inorganic crystal compounds and/or amorphous glass in varying proportions.
3. Thermal processing conducted at elevated temperatures.

Let us examine each of these characteristics in the following sections.

1.3.1 Crystal Compounds

Ceramics can generally be designated by the chemical compound formulae, $M_a X_c$, $M_a N_b X_c$ or in some cases combinations of several compounds. M and N can represent any metal and X represents any nonmetallic element that can form a stable compound with a metal. X most commonly represents O (oxygen) in commercial electronic ceramics, but also includes: Cl (chlorine and other halides), C (carbon), N (nitrogen), S (sulfur and other chalcogenides). For example, the most common solid chemical compound on Earth's surface is SiO_2 (silica). In this case $M = Si$, $X = O$, $a = 1$, and $c = 2$. Other ceramic compounds include: Al_2O_3, CaO, Na_2O, TiC, UO_2, PbS, $MgSiO_3 \cdot 2H_2O$, $PbZrTiO_6$, and many others.

In each of these compounds, the outer valence electrons of the metal ions (cations) are shared with the outer valence shell of the nonmetal ions (anions). A cation is the result of an atom giving up an electron, thus producing a net unbalanced positive charge on the nucleus (Figure 1.1). An anion possesses a net negative charge because it has gained an extra electron from the cation. The electron transfer occurs because it is a lower energy state for each atom to possess a closed shell of electrons. The shells of X are filled when it accepts the extra electron from the M atom. Giving up its outer electrons results in a closed shell for M. Figure 1.1 illustrates the formation of Mg^{2+} and O^{2-} ions by electron transfer.

Complete transfer of charge between atoms does not occur in most crystal compounds. Instead, the transferring electrons are shared between neighboring ions providing a partial covalent contribution to the chemical bond. The high concentration of electrons between the atoms in the crystal results in a very strong electron bond. This type of mutually shared, partly directional bond is termed an ionic-covalent bond and is characteristic of most ceramic materials. Chapter 2 discusses the quantum nature of chemical bonding and how shared bonds result in electronic band theory.

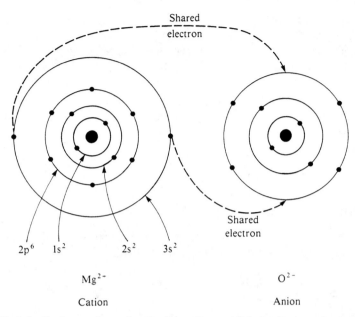

FIGURE 1.1 Ionic-covalent chemical bonding of MgO due to sharing of two electrons.

1.3.2 High Temperature Processing

Ceramic products are usually given a thermal treatment at temperatures in excess of 700°C for two major reasons. First, chemical reactions are accelerated at high temperatures. This is important, because many raw material constituents of ceramic bodies decompose at high temperatures and form more stable compounds. High temperatures are also necessary to produce new crystal compounds and form homogeneous solid solutions in the crystals. Reactions that produce a liquid phase are often desirable and such dissolution or melting reactions become more rapid at elevated temperatures. Also, quenching of a liquid phase from high temperature is necessary to produce most glasses. Second, thermal processing increases the density of ceramic objects. Before firing, the ware is very weak because there is a large percentage of porosity and only weak bonding between neighboring particles. During thermal treatment, the porosity is gradually eliminated and strong intergranular bonds are developed. The density and strength of the object increases as a result.

1.3.3 Crystals and Glasses

Some ceramic compositions possess the unusual property of being able to retain a random, nonperiodic structure (glass) when cooled from the liquid state. The nature of the glassy structure produced will be discussed later. Since many ceramics consist of a glassy phase as well as crystalline phases, a wide variety of microstructures is possible. Some products may be completely glassy, others completely crystalline, and others a combination of glassy and crystalline phases. As discussed later, the particular type of microstructure that is produced during the thermal processing is a result of the composition, temperature, and duration of the thermal treatment employed.

1.4 STRUCTURE OF CRYSTAL COMPOUNDS

1.4.1 Structure Rules

The principles that underlie the structural characteristics of crystal compounds can be generalized into two structural rules:

Rule 1: Achieve a maximum packing of oppositely charged ions around each reference ion.
Rule 2: Preserve electrical neutrality.

Rule 1 arises as a consequence of all physical systems tending towards a state of lowest energy. In other words, a crystal compound is bonded chemically so that there are as many negative ions as possible around each positive ion and vice versa. This arrangement minimizes the electrostatic energy and maximizes the electronic bonding. However, there is a spatial limitation on the arrangement of ionic charges in a crystal. Ions have a finite size (see Table 1.2), and consequently there is a spatial limit to the number of ions which can be arranged around each reference ion.

1.4.2 Radius Ratio Concept

An understanding of the packing of ions can be obtained through the radius ratio concept. The ratio of the size of the small ion r to the size of the large ion R determines the number of large ions which can be packed around the small ion. To see this, consider the geometrical construction of Figure 1.2. The maximum radius that a small sphere can possess and fit in the interstice of three large spheres is 0.155 times the radius of the large sphere.

To apply this geometrical packing argument to ions, we must make two assumptions. First, we must assume that the ions are spherical in shape. This is a good assumption, because ions generally have closed electron shells which are spherical in shape (see Figure 1.1).

Second, it must be assumed that the electron shells of ions of the same sign will not overlap. This is reasonable, since like charges repel. A consequence of this assumption is that the radius of the small ion will always

TABLE 1.2 Table of Ionic Radii, Å

Ion	Valence	Radius	Ion	Valence	Radius	Ion	Valence	Radius
Ac	+3	1.15	I	+7	0.5	Ru	+4	0.64
Ag	+1	1.22	I	−1	2.18	S	+4	0.37
Ag	+2	0.89	In	+3	0.85	S	+6	0.31
Al	+3	0.51	Ir	+4	0.66	S	−2	1.83
Am	+3	1.03	K	+1	1.33	Sb	+3	0.76
Am	+4	0.91	La	+3	1.14	Sb	+5	0.62
As	+3	0.58	Li	+1	0.69	Sc	+3	0.78
As	+5	0.47	Lu	+3	0.92	Se	+4	0.30
At	+7	0.62	Mg	+2	0.69	Se	+6	0.43
At	−1	2.27	Mn	+2	0.84	Se	−2	1.97
Au	+1	1.37	Mn	+3	0.56	Si	+4	0.45
Au	+3	0.85	Mn	+4	0.53	Sm	+3	1.00
B	+3	0.20	Mn	+7	0.46	Sn	+2	0.93
Ba	+2	1.35	Mo	+4	0.68	Sn	+4	0.72
Be	+2	0.43	Mo	+6	0.62	Sr	+2	1.16
Bi	+3	0.93	N	+3	0.16	Ta	+5	0.68
Bi	+5	0.74	N	+5	0.14	Tb	+3	0.93
Br	−1	1.96	NH_2	+1	1.43	Tb	+4	0.81
Br	+7	0.39	Na	+1	0.96	Tc	+7	0.57
C	+4	0.18	Nb	+4	0.74	Te	+4	0.80
Ca	+2	1.00	Nb	+5	0.68	Te	+6	0.56
Cd	+2	0.99	Nd	+3	1.04	Te	−2	2.18
Ce	+3	1.13	Ni	+2	0.72	Th	+4	1.03
Ce	+4	0.97	Np	+3	1.06	Ti	+3	0.76
Cl	+7	0.27	Np	+4	0.93	Ti	+4	0.65
Cl	−1	1.81	Np	+7	0.71	Tl	+1	1.47
Co	+2	0.76	O	+6	0.10	Tl	+3	0.98
Co	+3	0.63	O	−2	1.39	Tm	+3	0.96
Cr	+3	0.64	Os	+4	0.67	U	+4	0.98
Cr	+6	0.45	P	+3	0.44	U	+6	0.82
Cs	+1	1.61	P	+5	0.35	V	+2	0.95
Cu	+1	0.96	Pa	+3	1.12	V	+3	0.74
Cu	+2	0.72	Pa	+4	0.96	V	+4	0.61
Dy	+3	0.92	Pa	+5	0.90	V	+5	0.53
Er	+3	0.97	Pb	+2	1.24	W	+4	0.68
Eu	+3	0.98	Pb	+4	0.84	W	+6	0.62
F	−7	0.08	Pd	+2	0.80	Y	+3	0.95
F	−1	1.34	Pd	+4	0.65	Yb	+3	0.93
Fe	+2	0.77	Pm	+3	1.06	Zn	+2	0.77
Fe	+3	0.65	Po	+6	0.67	Zr	+4	0.81
Fr	+1	1.76	Pr	+3	1.11	BO_3	−3	1.35
Ga	+3	0.62	Pr	+4	0.95	CO_3	−2	1.31
Gd	+3	1.04	Pt	+2	0.80	NO_3	−1	1.23
Ge	+2	0.73	Pt	+4	0.65	OH	−1	1.33
Ge	+4	0.50	Pu	+3	1.04	AsO_3	−1	2.01
Hf	+4	0.78	Pu	+4	0.92	BrO_3	−1	1.68
Hg	+3	1.11	Ra	+2	1.40	SbO_8	−1	2.22
Ho	+3	0.91	Rb	+1	1.48	NH_4	−1	1.74
I	+1	0.62	Re	+7	0.56			
			Rh	+3	0.69			

8 CHAPTER 1: INTRODUCTION

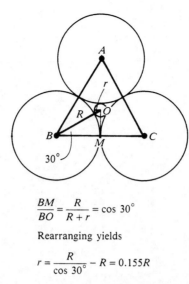

$$\frac{BM}{BO} = \frac{R}{R+r} = \cos 30°$$

Rearranging yields

$$r = \frac{R}{\cos 30°} - R = 0.155R$$

FIGURE 1.2 Geometry of triangular coordination of ions of small radius (r) with large ions (radius = R).

be larger than the maximum calculated for a particular packing. In other words, the small ion will always be sufficiently big to keep the electron shells of the large ions from overlapping. When this condition is met, the packing arrangement will be stable.

If the radius of the small ion is much larger than $0.155R$, it will be possible for it to keep four large ions from overlapping, as shown in Figure 1.3(d). In this case, the stable coordination of large ions will be four, in order to maximize the concentration of oppositely charged ions. A similar geometrical analysis can be performed for tetrahedral, octahedral and cubic packing of ions. Figure 1.4 relates the geometrical packing of ions (coordination number) to their radius ratio. Larger ions require more ions around them to maximize ionic-covalent bonding.

1.4.3 Electrical Neutrality

It is easiest to see the influence of electrical charge balance in establishing crystal structure by considering examples, such as MgO and CaF_2. The radius of Mg^{2+} (r_{Mg}) = 0.69 Å, and O^{2-} (R_O) = 1.39 Å (see Table 1.2). Computing the radius ratio yields a value of $r/R = 0.467$, which predicts octahedral coordination of oxygen ions around magnesium ions (Figure 1.4). The chemical formula of MgO tells us that there must be one Mg^{2+} for each O^{2-} and vice versa, therefore, there must be octahedral coordination of magnesium cations around oxygen anions as well. This configuration,

1.4 STRUCTURE OF CRYSTAL COMPOUNDS

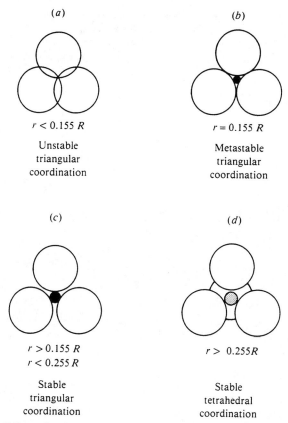

FIGURE 1.3 Effect of radius of small ion (r) on pacing of large ions of radius R.

when extended in three dimensions, results in a crystal structure, as shown in Figure 1.5.

This particular structure is called the NaCl or rock salt structure and is shown in Figure 1.5. NaCl (table salt) crystallizes in this cubic form. NaCl possesses completely different properties than MgO even though the crystal structure is the same. For example, the melting point of MgO is 2800°C, whereas NaCl melts at 801°C; NaCl readily dissolves and ionizes in water, while MgO is nearly insoluble. The difference in properties is due to the number of electrons involved in the bonding between the ions in the crystals. In MgO there are two electrons available for bonding in each Mg—O bond, giving a relative bond strength for each coordination octahedron of $2e^-$ as shown in Figure 1.6. For NaCl there is only one bonding electron per coordination octahedron, so the relative bond strength is much less and the physical and chemical properties reflect this difference. The lesson to be remembered is that not only crystal structure, but crystal bond strength, determines the properties of crystal compounds.

Coordination number		Disposition of ions about central ion	Radius ratio (r/R)
	2	Linear	0.155 – 0
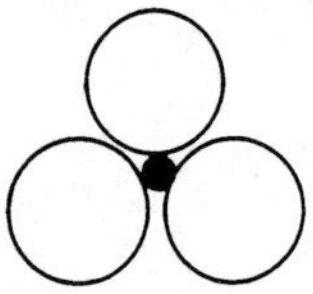	3	Corners of triangle	0.225 – 0.155
	4	Corners of tetrahedron	0.414 – 0.225
	6	Corners of octahedron	0.732 – 0.414
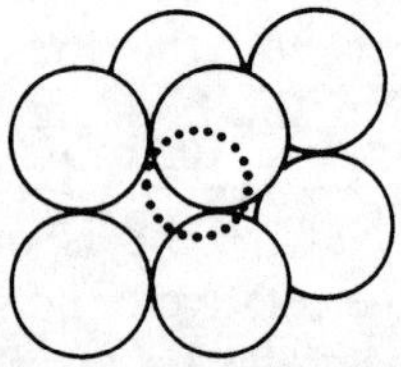	8	Corners of cube	1.0 – 0.732

FIGURE 1.4 Coordination of ions as a function of ionic radius ratios.

1.4 STRUCTURE OF CRYSTAL COMPOUNDS

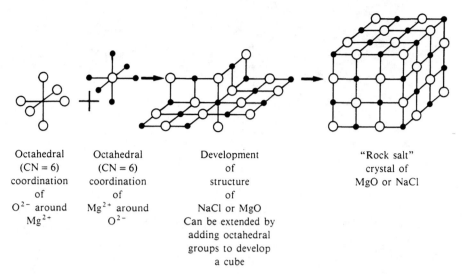

| Octahedral (CN = 6) coordination of O^{2-} around Mg^{2+} | Octahedral (CN = 6) coordination of Mg^{2+} around O^{2-} | Development of structure of NaCl or MgO Can be extended by adding octahedral groups to develop a cube | "Rock salt" crystal of MgO or NaCl |

FIGURE 1.5 Construction of cubic lattice from octahedral configurations of cations and anions.

Another feature of the NaCl structure (Figure 1.7) is that the anions form a close-packed layer and the cations occupy octahedral interstitial positions. This is best seen by looking at a (111) plane in the lattice. The Miller indices (ijk) represent the reciprocals of the x, y, and z axis intercepts as measured in units of the crystal cell dimensions a, b, and c respectively.

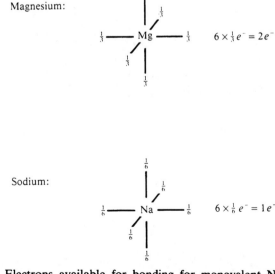

FIGURE 1.6 Electrons available for bonding for monovalent Na^+ and divalent Ca^{2+} in octahedral coordination.

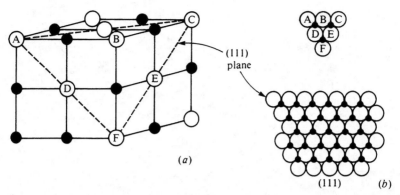

FIGURE 1.7 (*a*) NaCl lattice. (*b*) Close-packed plane of oxygen ions. Mg^{2+} ions in octahedral intersices. Three more O^{2-} are provided by the close-packed oxygen layer above.

(See Van Vlack, 1985 for a discussion of crystal planes and notation.) The same feature of close-packed anions and interstitial cations occurs throughout crystal chemistry. In Al_2O_3 for example, only two-thirds of the octahedral positions in the close-packed oxygen lattice are filled with Al^{3+} ions, because of electrical neutrality requirements. Note the hexagonal symmetry due to the Al^{3+} ions occupying only two-thirds of the octahedral interstices of the O^{2-} lattice in Figure 1.8.

Now consider a more complex crystal compound structure, CaF_2. It is more complex because the chemical formula requires that there be two F^- ions for each Ca^{2+} ion in the crystal in order to preserve electrical neutrality. The radius ratio is:

$$\frac{rCa^{2+}}{R_{F^-}} = \frac{1.00 \text{ Å}}{1.34 \text{ Å}} = 0.75 \tag{1.1}$$

which corresponds to a cubic coordination of monovalent fluorine ions around divalent calcium ions. (Confirm from Figure 1.4.) Based on this

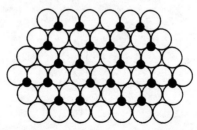

FIGURE 1.8 Al_2O_3 close-packed oxygen anion lattice with Al^{3+} cations in two-thirds of the interstitial positions.

1.4 STRUCTURE OF CRYSTAL COMPOUNDS

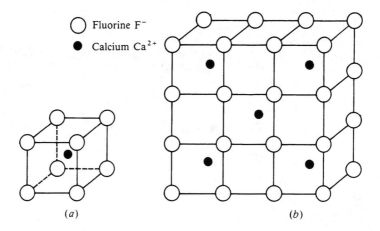

FIGURE 1.9 Calcium fluorite structure (*a*) involving cubic coordination of fluorine anions around calcium cations (*b*).

prediction, we can construct a lattice as shown in Figure 1.9. Only every other F^- cube contains a Ca^{2+} ion. This is necessary in order to retain charge neutrality, as can be seen by doing a little bookkeeping of electrical charges. In the cube (*a*) there are eight corner F^- ions, each ion having one minus charge, and this charge is shared with eight neighboring cubes. Therefore,

$$(8F^-)(-1e^-)\left(\frac{1}{8}\right) = -1e^- \qquad (1.2)$$

The doubly charged calcium ion is not shared; therefore,

$$(1Ca^{2+})(+2e^-)(1) = +2e^- \qquad (1.3)$$

so there is an excess of one charge in this cube:

$$-1e^- + 2e^- = +1e^- \qquad (1.4)$$

However, by Ca^{2+} occupying only every other cell, the net charges on the cells balance,

$$-1 + 1 - 1 + 1 - 1 + 1 \cdots = 0 \qquad (1.5)$$

Figure 1.9 is a clumsy way of representing the fluorite (CaF_2) structure. It is more elegantly shown by basing the representation of the lattice on the

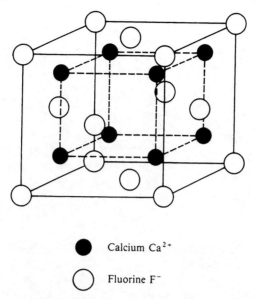

- ● Calcium Ca^{2+}
- ○ Fluorine F^-

FIGURE 1.10 Calcium cubic structure (fcc).

Ca^{2+} ions which form a face-centered cubic (fcc) cell. The F^- ions are located in tetrahedral interstices on the fcc lattice. Thus the cubic coordination of the F^- anions around the Ca^{2+} cations is retained, but only tetrahedral coordination of cations around anions is possible because of electrical neutrality requirements. A representation of this lattice is shown in Figure 1.10. The calculation of the radius ratio and charge balance for this CaF_2 structure is presented in Problem 1.1 along with other ceramic compounds.

1.4.4 Ionic Bonding

As mentioned earlier, the bonding in ceramics is a combination of ionic and covalent bonding. The ionic contribution is large in many crystal compounds of interest so let us examine the physical nature of this bond and the role it plays in establishing the physical properties of ceramics.

Consider two oppositely charged ions of charge $z_1 e$ and $z_2 e$, where z_1, and z_2 are the valence of the ions and e is the electronic charge per electron. When the distance of separation is large ($r = \infty$), there is little attraction between the ions. We adopt this state as one of zero energy. As the ions are brought closer together, *Gauss's law of electro-statics* requires that the electro-static or Coulomb attraction increases rapidly, as

$$F = -\frac{z_1 e z_2 e}{4\pi\epsilon_0 r^2} \tag{1.6}$$

where ϵ_0 is the permittivity of free space (see Chapter 5). Thus, the potential energy of the system, in electron volts (eV), will be lowered as the ions are brought closer together,

$$V = \int_\infty^r F\, dr = -A \frac{z_1 z_2 e^2}{4\pi\epsilon_0 r} \tag{1.7}$$

where the Madelung constant A depends on the packing arrangement of the ions. For a close-packed NaCl-type lattice, A is the range of 1.7, whereas for the fluorite-type lattice it is around five, and for Al_2O_3 it is 25.

The dependence of the potential energy of attraction is shown as a function of distance of separation of the ions in Figure 1.11. When the ions

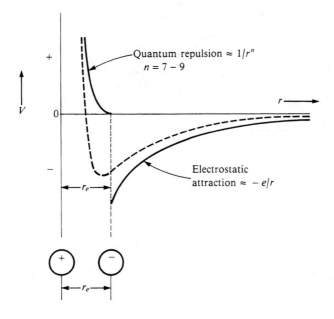

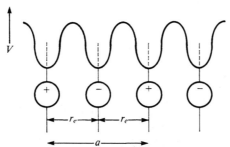

FIGURE 1.11 Lattice potential energy as a function of distance of separation of anions and cations.

come sufficiently close together, the outer electron shells begin to overlap. Appreciable overlapping of filled shells is highly unfavored, because there are no empty energy states to accommodate the electrons. Consequently, the only way in which the ions can come closer is by electrons occupying the same energy state. This is a violation of the *Pauli exclusion principle* which states that no two particles can possess the same four quantum numbers (see Chapter 2). This effect shows up as a strong repulsive force and contributes a positive repulsion term to the potential energy of the system, $+B(1/r^n)$. The total energy is a sum of the electro-static attraction and the quantum-mechanical repulsion:

$$V = -\frac{Az_1z_2e^2}{4\pi\epsilon_0 r} + B\frac{1}{r^n} \tag{1.8}$$

The values of B and n can be obtained from compressibility and cohesive energy experiments. The value of n is typically in the range of seven to nine, because the quantum mechanical repulsion is very strong.

The potential energy is a minimum at the equilibrium distance of separation, r_e. Consequently, the first derivative of equation (1.8) with respect to r must vanish (e.g.,)

$$\left(\frac{dV}{dr}\right)_{r=r_e} = 0 = +\frac{Az_1z_2e^2}{4\pi\epsilon_0 r_e^2} - \frac{nB}{r_e^{n+1}} \tag{1.9}$$

Solving for B and substituting in equation (1.9) and simplifying yields for the potential energy minimum,

$$V_{r=r_e} = -\frac{Az_1z_2e^2}{4\pi\epsilon_0 r_e}\left(1 - \frac{1}{n}\right) \tag{1.10}$$

which is shown graphically in Figure 1.11 as the bold dashed line. This simple ionic bonding model, known as the *Born and Lande theory*, can be extended to include a van der Waals term, $-C/r^6$, and a small contribution, δ, expressing the energy of the structure at a temperature of absolute zero (see Chapter 2). These modifications by Born and Mayer combined with a modification of the repulsive energy term finally yield equation (1.11) for the potential energy of an ionic bond (see Kittel for additional details).

$$V = -\frac{Az_1z_2e^2}{4\pi\epsilon_0 r} + Be^{-r/n} - \frac{C}{r^6} + \delta \tag{1.11}$$

Consequently, crystal compounds have an array of potential energy V wells with the ions residing at (r_e) with

$$2(r_e) = a \tag{1.12}$$

where a is defined as the lattice parameter of the crystal. The periodic potential wells for a crystalline structure are illustrated in Figure 1.11. The depth of the potential wells is a function of the product $(z_1 z_2 e^2)$. Therefore, the potential $V_{(r_e)}$ for MgO is greater than the potential $V_{(r_e)}$ for NaCl because

$$(\text{NaCl})z_1 z_2 = 1 < (\text{MgO})z_1 z_2 = 4 \tag{1.13}$$

Therefore, the potential wells are less deep for Na^+ and Cl^- ions than the Mg^{2+} and O^{2-} ions of MgO (Figure 1.12). The physical properties of MgO are consequently different than NaCl, even though the crystal structure is the same.

1.5 THERMAL EXPANSION

The classical model for the thermal expansion of a material is directly related to the bonding potentials of the atoms of the material. The expansion that occurs with increased temperature is found from the average separation, a_0, as seen in Figure 1.12. At 0°K there is still a finite energy of the molecule and therefore a range of separations that are allowed.

As the temperature is increased from 0°K to kT_2, the range of allowed separations also increases. The mean is shown as a_2 and is shown shifted toward increasing separation. Molecular dynamical calculations based on this model are able to predict the coefficient of thermal expansion (CTE), glass transition temperature (T_g), melting points, and boiling points of various materials. Strongly bonded materials show a reduced CTE as indicated in Figure 1.12*b* when compared with more weakly bonded materials. Likewise melting and boiling points tend to increase with increased bond strength.

The deeper potential wells of strongly bonded ions also increase the energy barrier to ionic migration and result in a higher electrical resistivity, as discussed in Chapter 4, and lower dielectric losses as described in Chapter 5.

1.6 GLASS

Glass is a solid which does not possess a long-range ordered structure. Inorganic glasses consist of ions which are bonded together with ionic-

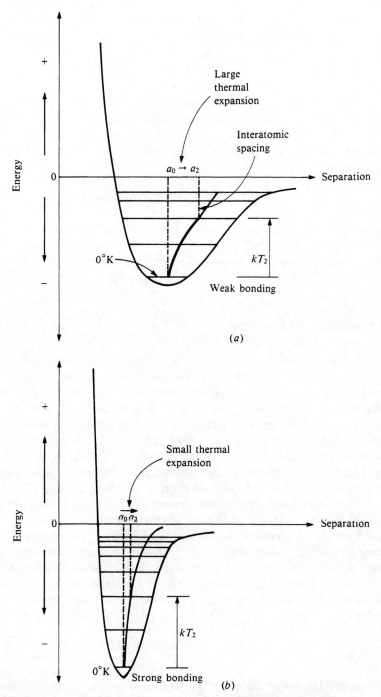

FIGURE 1.12 Thermal expansion depends on the depth of the potential energy wells (*a*) NaCl (*b*) MgO.

covalent bonds similar to crystalline ceramics. Because long-range order is not required of a glass structure, a glass composition does not necessarily correspond to a particular chemical compound formula. For example, it is possible to make a glass of the composition 60% SiO_2, 20% CaO, and 20% Al_2O_3, but there is no crystal structure containing exactly those percentages of oxide compounds. A glass of this composition will eventually change into a polycrystalline structure, because it lowers its free energy by doing so. Equilibrium crystal structures are more stable than glass structures. However, for many compositions, the change from glass to crystal is very slow. Consequently, a glass can exist as a useful material at room temperature for hundreds of years with only infinitesimal changes taking place in its properties.

SiO_2 (silica) is the fundamental component of most commercial glasses. It is inexpensive, can be purified easily, and most important for glass manufacture, it is a strong *network former*. A network former is a cation that has sufficient anion bonds that a continuous connecting structure can be formed. The Si^{4+} cation bonds with four O^{2-} ions forming a SiO_4^{4-} tetrahedron which is the basic building unit of glass technology as well as silicate chemistry. Crystalline and glassy silica structures are schematically shown in Figure 1.13.

Oxygen ions in the SiO_4^{4-} tetrahedra may be shared with another Si^{4+} ion or other network formers. Oxygen ions that are shared between two network formers provide a connecting link in a structure and are called *bridging oxygens*. Network formers are normally cations with a M^{4+} or M^{3+} valence. Cations with a valence of M^{1+} or M^{2+} cannot form a stable continuous random network structure. When oxygen ions in a SiO_4^{4-} tetrahedron are

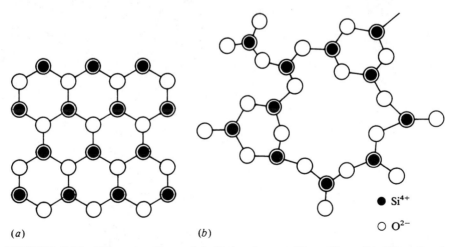

FIGURE 1.13 Silica structures. (*a*) Ordered crystalline silica. (*b*) Disordered glassy silica. (Note the fourth oxygen for each tetrahedron which would be either above or below the plane of the page.)

bonded with a low valence cation, direct links between tetrahedra are not possible. These oxygen ions are consequently termed *nonbridging oxygens*. The presence of broken structural links and nonbridging oxygens greatly change the properties of a glass. Consequently, the low valence cations are called *network modifiers*.

The properties of glasses are changed by varying the percentage and type of network formers and network modifiers in the glass composition. Monovalent (M^{+1}) modifiers lower the melting temperature required to produce glass and the viscosity of the molten glass, and thereby generally decrease the cost of glass manufacture. Compounds containing Na^+ are widely used as a glass flux. However, a glass containing only monovalent modifiers is very susceptible to corrosive attack by water and other chemicals. CaO or MgO is often added in glass compositions to increase the chemical stability. An even greater increase in chemical stability needed for special requirements is achieved with the addition of the trivalent cations Al^{3+} or B^{3+}. Most commercial glasses are primarily composed of SiO_2, Na_2O, CaO, MgO, Al_2O_3, and B_2O_3. The average compositions of two typical glasses are:

Glass type	SiO_2 (%)	Al_2O_3 (%)	CaO (%)	MgO (%)	Na_2O (%)	B_2O_3 (%)
Bottle glass	73	1.5	5	4	1.5	—
Soda borosilicate chemical ware	80.5	2.2	—	—	4	13

The relative influence of network modifiers on the properties of glass is illustrated in Figure 1.14 and shown schematically in Figure 1.15.

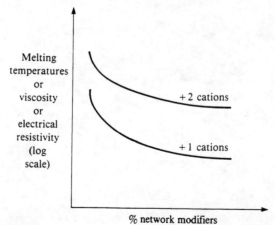

FIGURE 1.14 Effect of network modifiers on the physical properties of silicate glasses.

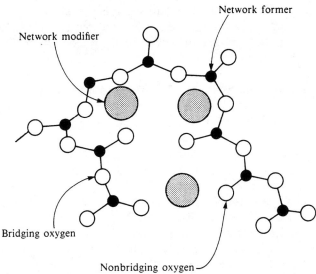

FIGURE 1.15 Schematic structure of a soda-silicate glass. Na$^+$ ions are network modifiers which create nonbridging oxygen ions.

1.7 PROCESSING AND MICROSTRUCTURES

Although there is an enormous variety of ceramic products, their microstructures can be classified into five categories:

1. Glass,
2. Cast polycrystalline ceramic,
3. Liquid-phase sintered (vitrified) ceramic,
4. Solid-state sintered ceramic, and
5. Polycrystalline glass-ceramic.

Most electronic ceramics are either solid-state-sintered or glass, although liquid-phase-sintering can be used if very careful control is exercised.

The difference in the microstructures of the five categories is primarily a result of the different thermal processing steps required to produce them. The top half of Figure 1.16 summarizes the relationships between the various processing steps involved in making electronic ceramics. Most electronic ceramics are made by fabricating the product from fine-grained particulate solids. For example, a desired shape may be obtained by mixing the particulates with water and an organic binder, then pressing in a mold. This is termed *forming*. The formed piece is called *green ware*. Subsequently, the temperature is raised to evaporate the water (i.e.,

22 CHAPTER 1: INTRODUCTION

drying), and the binder is burned out, resulting in *bisque ware*. At a very much higher temperature, the ware is densified during firing. After cooling to ambient temperature, one or more finishing steps may be applied, as indicated in Figure 1.16, resulting in the finished product with desired properties.

Control over the final properties of an electronic ceramic involves control of each processing step, in sequence. In order to ensure that the desired final properties are achieved, it is essential to characterize the electronic ceramic.

Characterization has been defined by the Materials Advisory Board of the National Research Council as: "Characterization describes those features of the composition and structure (including defects) of a materials that are significant for a particular preparation, study of properties, or use, and suffice for the reproduction of the material."

In order to characterize a ceramic material, therefore, it is necessary to evaluate its composition and structure sufficiently so that the material and its properties can be reproduced.

The bottom half of Figure 1.16 illustrates five major classes of characterization. The arrows relate the characterization features to the appropriate output in the ceramic processing sequence. It is important to emphasize two points here. First, each characterization feature indicated can be quantitatively evaluated. Second, characterization is directed towards the output of the processing steps and not to the processing method itself. Of course, the same analytical techniques may often be used to invesigate processing parameters, but it is not materials characterization. This is an important distinction, since different processing steps can lead to similar properties. Therefore, in order to establish the physical origin of properties, it is necessary to determine the composition and structure of the product and not just the specifics of the process used in the product's manufacture.

The physical properties important for electronic, insulation, optical, or magnetic applications of ceramics depend upon all five characterization features shown in Figure 1.16. The primary concern of this book is the dependence of electronic properties on chemical composition, phase state, and structure of the material. We will discuss microstructural and surface dependence of properties to a minor extent. However, it is essential to have a rudimentary knowledge of the interrelation between microstructure and thermal processing, as summarized in Figure 1.17. Figure 17a is a binary phase diagram consisting of SiO_2 (silica) and some arbitrary network modifier oxide MO. When a powdered mixture of MO and SiO_2 is heated to the temperature T_M, the entire mass will have melted and will be liquid (L). The liquid will become homogeneous when held at this temperature for a sufficient length of time. When the liquid is cast (paths 1, 2, 5), forming the shape of the object during the casting, either a glass or a polycrystalline microstructure will result.

If the starting composition contained a sufficient quantity of network

1.7 PROCESSING AND MICROSTRUCTURES

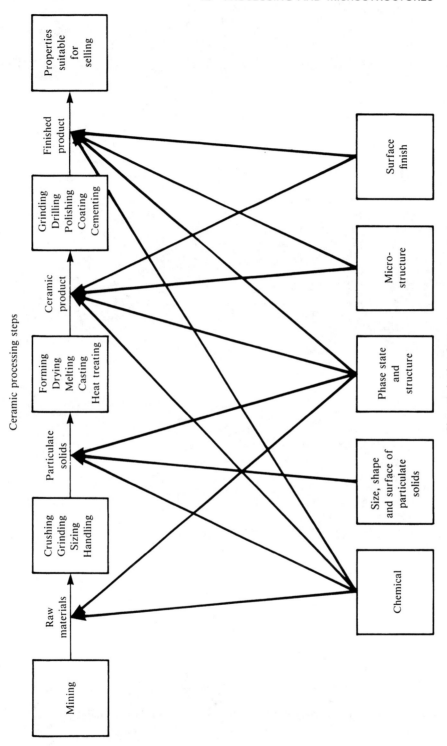

FIGURE 1.16 The relationship between characterization features and products of the ceramics processing sequence.

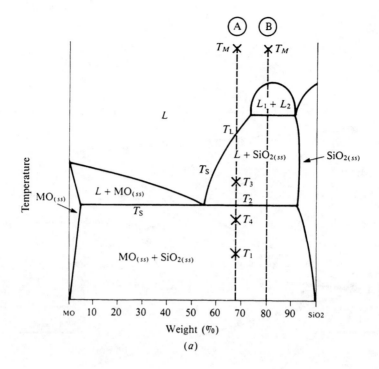

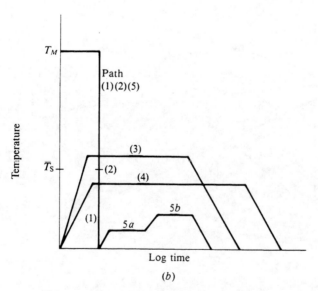

FIGURE 1.17 (*a*) Composition A: Microstructure: (1) glass; (2) cast polycrystalline (large-grained); (3) liquid-phase-sintered (vitrified); (4) solid-state sintered; (5) polycrystalline glass-ceramic. (*b*) Composition B: (1) Phase-separated glass. (2)–(5) Same as (*a*). (ss) = solid solution, T_S = solidus line.

former (SiO$_2$), and the casting rate was sufficiently slow, a glass will result (path 1). The viscosity of the melt increases greatly as it is cooled until at approximately T_1, the glass transition point, the material is transformed to a solid.

If either of these conditions is not met, a polycrystalline microstructure will be obtained. The crystals begin growing from T_L and below. The final material consists of the equilibrium crystalline phases predicted by the phase diagram (path 2). This type of cast object is not often used commercially because of the large shrinkage cavity produced during cooling. Since cold working of ceramics is not possible because of their brittleness, a large shrinkage cavity in the final product cannot usually be tolerated. Control over microstructure during the casting process is also difficult. The grains usually grow large and are highly oriented. Both of those features are undesirable from the standpoint of both mechanical and electrical properties.

If the MO and SiO$_2$ powders are first formed into the shape of the desired object and fired at a temperature T_3, a liquid-phase-sintered structure will result (path 3). Before firing, the composition will contain approximately 10–40% porosity depending upon the forming process used. A liquid will first be formed at grain boundaries at the eutectic temperature, T_2. The liquid will penetrate between the grains, filling the pores and drawing the grains together by capillary attraction. These effects decrease the volume of the powdered compact. Since the mass remains unchanged but only rearranged, the density increases. Should the compact be heated for a sufficient length of time, the liquid content can be predicted from the phase diagram. However, in most ceramic processes, liquid formation does not usually proceed to equilibrium, due to the slowness of the reaction and the expense of long-term heat treatments.

The microstructure resulting from liquid-phase sintering or vitrification, as it is commonly called, will consist of grains from the original powder compact surrounded by a liquid phase. As the compact is cooled from T_3 to T_2 the solidus temperature, T_S, the liquid phase will crystallize into a fine-grained matrix surrounding the original grains. If the liquid contains a sufficient concentration of network formers, the liquid will be quenched into a glassy matrix surrounding the original grains.

It is possible to densify a powder compact without the presence of a liquid phase by a process called solid-state sintering. This is the process usually used for manufacturing electronic ceramics. Under the driving force of surface energy gradients, solid material is moved to areas of contact between particles. The mechanisms by which the material is transported may be either grain boundary diffusion, volume diffusion, creep, or any combination of these depending upon the temperature of material involved. Because long-range migration of atoms is necessary, sintering temperatures are usually in excess of one-half of the melting point of the material: $T > T_L/2$ (path 4).

26 CHAPTER 1: INTRODUCTION

The material moves so as to fill up the pores and open channels existing between the grains of the powder (Figure 1.18). As the pores and open channels are closed during the heat treatment, the crystals become tightly bonded together and the density, strength, and electronic properties of the object improve greatly. The microstructure of a material that is prepared by sintering consists of crystals bonded together and a small amount of remaining porosity.

The relative rate of densification during solid-state sintering will be slower than that of liquid-phase sintering because material transport is slower in a solid than in a liquid. However, it is possible to solid-state sinter

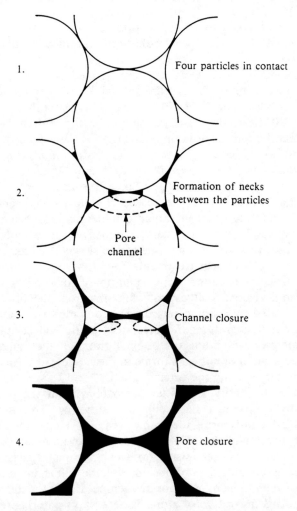

FIGURE 1.18 Steps in densification of a ceramic.

single component materials such as pure oxides, since liquid development is not necessary. Consequently, when high purity and uniform microstructures are required, such as for the electronics industry, solid-state sintering becomes essential.

The fifth class of microstructures is termed *glass-ceramics* because the object starts as a glass and ends up as a polycrystalline ceramic. This is accomplished by quenching the glass, as shown in path 1 of Figure 1.17. The transformation of the glass object into a ceramic occurs in two steps. First, the glass is heat treated at a temperature in the range of 500–700°C (path 5*a*) to produce a large concentration of nuclei from which crystals can grow. When sufficient nuclei are present to ensure that a fine-grained structure will be obtained the temperature of the object is raised to a range of 600–900°C which promotes crystal growth (path 5*b*). Crystals grow from the nuclei until they impinge and 100% crystallization is achieved. The resulting microstructure is non-porous and fine-grained randomly oriented crystals which may or may not correspond to the equilibrium crystal phases predicted by the phase diagram. When phase separation occurs, composition B in Figure 1.17, a non-porous glass-in-glass microstructure can be produced.

1.8 RANGE OF ELECTRONIC PROPERTIES

Understanding the full range of electronic properties of ceramics is one of the great challenges of physics and materials science. Figure 1.19 shows that electrical conductivity of oxides varies by 24 orders of magnitude. This variation is as large as the distance from Earth to the calculated edge of the universe (see Problem 1.5). At the upper limit of electrical conductivity are the high-T_c ceramic superconductors of $BaO-Y_2O_3-Cu_3O_{7-x}$, known as 1–2–3 superconductors due to the ratio of rare earth to alkaline earth to copper oxides in the chemical formula of the material. Below 90°K, the conductivity of these materials becomes infinitely high; (i.e.,) zero resistivity. The physical principles of ceramic superconductors are discussed in Chapter 12.

At the other extreme of the conductivity spectrum lies high-purity Al_2O_3 (alumina) with the value of 10^{-15} (ohm cm)$^{-1}$ even at 500°K. The very small amount of charge transport in alumina is due to migration of ions and thermally induced defects, often in the grain boundaries, as discussed in Chapter 4.

Addition of lower valence cations to a crystal or glass usually increases the ease of ionic transport, as shown in Figure 1.19 by comparing the conductivity of SiO_2 and soda-silicate glasses.

When stronger chemical bonds are formed, such as occurs when Si is reacted with O_2 to form SiO_2, the electrical conductivity is greatly

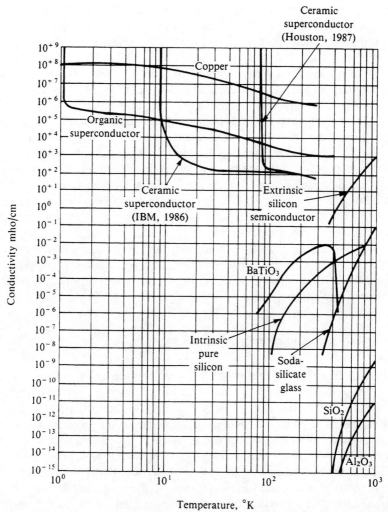

FIGURE 1.19 Electrical conductivity vs. temperature for various materials.

decreased, as indicated in Figure 1.19. In fact, the electronic conductivity of Si semiconductors is nearly eliminated when its oxide is formed, i.e., the semiconductor becomes an insulator. The physical principles of elemental and compound semiconductors are discussed in Chapter 3.

In addition to a broad range of electrical conductivity, electronic ceramics also can possess dielectric constants ranging from 5 to >5000 (Chapters 5 and 6), may be diamagnetic, antiferromagnetic or ferromagnetic (Chapter 7), and have optical transmission over varying portions of the electromagnetic spectrum ranging from 150 nm to 8000 nm (Chapters 8 and 9). One of the most important aspects of electronic ceramics is the cooperative interaction of electrical, mechanical, optical, and magnetic properties. These interactions are a consequence of complex cationic–anionic crystal compound structures which have anisotropic characteristics. Devices and behavior of such materials are described in Chapters 6, 7, 10, and 11.

In order to understand the principles of the properties of electronic ceramics in response to various applied fields, it is necessary to understand the nature of the structural bonds between metal cations and nonmetal anions and the electronic states resulting from the bonding. Chapter 2 reviews the quantum mechanics background of chemical bonding and energy bands. Atomic theory is used to interpret the electronic, magnetic, and optical behavior of ceramics. Application of these concepts occurs throughout the book, including the theory of semiconductors (Chapter 3), lasers (Chapter 11), and ceramic superconductors (Chapter 12).

1.9 ELECTRONIC CERAMICS MARKET

Understanding the principles of electronic ceramics is not only an exciting scientific challenge but is also an important commercial objective. Figure 1.20 shows the enormous size of the market for these important high

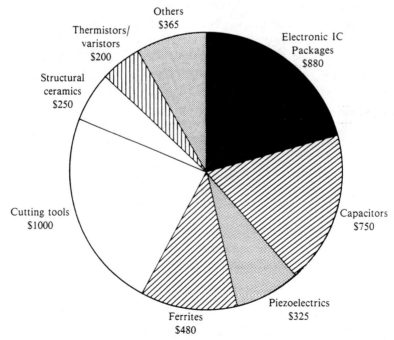

FIGURE 1.20 High technology ceramics market. Products and world sales (millions $). Total market in: 1980, $4250 million; 1985, $5500 million; 2000, $15,000 million.

TABLE 1.3 Photonic Materials Market (millions)

	1985	2000
Lenses, mirrors, filters	$ 850	$ 3,500
Laser systems		
(Overall)	$2,000	$ 8,000
(Add-on components)	$1,700	$ 7,000
Fiber optics and components	$ 750	$ 6,000
Military		
(Not included above)	$ 100	$ 1,000
(SDI components)	$ 100	$ 2,000
Optical computers and		
electro-optic components	$ 20	$ 2,000
Total	$4,570	$22,500

technology materials in 1980, 1985, and the 300% projected growth by the year 2000. The projected growth rate is one of the largest of any field of technology.

The projected growth of the photonic materials market, which is largely ceramics and glasses, is even larger, at about 400% (see Table 1.3). Thus, there is considerable incentive to investigate new electronic and photonic materials to take advantage of this growth potential.

PROBLEMS

1.1 *(a)* Calculate the radius ratio and charge balance for the following:

$$\begin{array}{llll} BaO & CaF_2 & Cr_2O_3 & SiO_2 \\ TiO & Al_2O_3 & MgO & SiC \\ CaO & HfO_2 & NaCl & GaAs \end{array}$$

(b) What is the theoretical coordination number of the above?

(c) Draw a schematic crystal lattice of $BaTiO_3$ based on *(a)* and *(b)*. (See Chapter 6 if necessary.)

1.2 *(a)* Calculate the radius ratio of U and O.

(b) Draw a schematic crystal structure of UO_2.

1.3 *(a)* Draw a schematic of a (100) crystal plane of MgO.

(b) Draw a schematic of a (111) crystal plane of MgO.

(c) What are the Miller indices (i, j, k) of the crystal plane of alumina shown in Figure 1.8?

1.4 (a) Solve equation (1.10) for the equilibrium position, $r(e)$.
(b) From the equation for the equilibrium position, $r(e)$, calculate the atomic separation of NaCl, where

$$n = 9$$
$$A = 1.7$$
$$V_{min} = 4.24 \text{ eV}$$

(c) What is the atomic separation for NaCl based on the ionic radius model? Why is it different from the separation calculated in (b)? What is the actual value of the separation?

1.5 Prepare a graph or table that compares the range of magnitudes of electrical conductivity of ceramic materials to that of the magnitudes of physical objects in the known universe.

1.6 List the materials in problem 1.1 in order of their bond strengths.

1.7 (a) Considering Figure 1.17a and composition A, what happens when a liquid is cooled from a temperature T_M to T_2?
(b) What happens when composition B is cooled from T_M to T_2?

1.8 (a) Considering a SiO_2–Al_2O_3 composition, (see Bergeron and Risbud, p. 74), discuss the effect of 40 mole% silica on the solid-state-sintering temperature. What happens if the temperature is raised 600 degrees? 800 degrees?
(b) What happens to the sintering behavior for 60 mole% silica?
(c) What happens at 90 mole% silica?

1.9 (a) What is the annually compounded rate of growth required for electronic ceramics and photonic ceramics to reach their values by the year 2000?
(b) At this rate what is the projected market for both at the years 2010 and 2020?

READING LIST

Bergeron, C. G. and Risbud, S. H., (1984). *Introduction to Phase Equilibria in Ceramics*, The American Ceramic Society, Columbus, OH, Chapters: 1–5.
Clark, J. P. and Flemings, M. C., (1986). *Sci. Amer.* **225**(4), 50.
Compton, W. D. and Gjostein, N. A., (1986). *Sci. Amer.*, **225**(4), 92.
Elliott, S. R., (1983). *Physics of Amorphous Materials*, Longman, London, Chapers: 1, 2, 3.

Fricke, J., (1988). *Sci. Amer.*, **258**(5), 92.

Guy, A. G., (1972). *Introduction to Materials Sciences*, McGraw-Hill, New York, Chapters: 1, 2, 3.

Harris, W. F., (1977). *Sci. Amer.*, **237**(6), 130.

Hummel, R. E., (1985). *Electronic Properties of Materials*, Springer-Verlag, Berlin, Chapters: 1, 2.

Kingery, W. D., Bowen, H. K., Uhlmann, D. R., (1976). *Introduction to Ceramics*, 2nd edition, Wiley, New York, Chapters: 1, 2, 3, 7, 8, 12.

Mayo, J. P., (1986). *Sci. Amer.*, **255**(4), 58.

Steinberg, M. A., (1986). *Sci. Amer.*, **255**(4), 66.

Tanaka, T., (1981). *Sci. Amer.*, **244**(1), 124.

Van Vlack, L., (1985). *Elements of Materials Science and Engineering*, 5th edition, Addison Wesley, Reading, MA, Chapters: 1, 2, 3, 7, 10, 11.

Wolsky, A. M., Giese, R. F., and Daniels, E. J., (1989). *Sci. Amer.*, **260**(2), 60.

2

QUANTUM MECHANICS AND THE BAND THEORY OF SOLIDS

2.1 GENERAL APPLICATIONS OF QUANTUM MECHANICS

The early successes of quantum mechanics were remarkable. The ideas which explained the photoelectric effect, the ultraviolet catastrophe, the Compton effect, Bragg diffraction and the spectrum of the hydrogen atom led to the understanding that matter has both a particle nature and a wave nature. The theory that formalized these ideas is known as *quantum mechanics*. Schrödinger, Heisenberg, Dirac, and others formalized the theory as a set of laws, more complex than Newton's laws of motion but no more difficult to express and apply than Maxwell's equations of electrodynamics.

Following the early success of quantum mechanics, theorists tackled problems of more complex systems. However, things became difficult for systems of more than two or three electrons. The basic approach was to calculate the potential and wavefunctions for a single electron in a potential for the total atomic nucleus. A second electron was added and the perturbations to the wavefunctions were calculated, then the third electron was added and the calculations continued until all the electrons had been added to the atomic nucleus. These calculations are similar to those made by astronomers to predict the orbits of minor bodies in the solar system. This approach is successful as long as there is a clearly defined dominant potential for the entire system, (i.e., the sun). For example, planets even as large as Jupiter have a very small effect on the overall dominant potential created by the sun.

When this approach was applied to a system of atoms to calculate the electronic characteristics of a solid material, numerical errors in the computer calculations yielded totally unreasonable predictions. The problem encountered was the misapplication of perturbation theory and computer roundoff errors. These difficulties nearly destroyed the hopes of the early theorists to achieve broad application of quantum mechanics for materials.

In order to understand the difficulty of calculating the wavefunctions for large atoms, a simple numerical study can help. The element cobalt has 27 electrons to balance the electric potential of the nucleus. Each wavefunction perturbs the next as the 27 electrons are added to the system. Any errors or approximations early in the calculations compound rapidly.

A simple exercise of squaring a number slightly larger than one illustrates the problem encountered by quantum theorists. If the number

$$n = 1.0000001$$

TABLE 2.1 Result of Squaring 1.0000001 Repeatedly with Different Precision

Number of squarings	Power	Precision	15-Digit precision
0	1	1.0000001	1.00000010000000
1	2	1.0000002	1.00000020000001
2	4	1.0000004	1.00000040000006
3	8	1.0000008	1.00000080000028
4	16	1.0000016	1.00000160000120
5	32	1.0000032	1.00000320000496
6	64	1.0000064	1.00000640002016
7	128	1.0000128	1.00001280008128
8	256	1.0000256	1.00002560032640
9	512	1.0000512	1.00005120130818
10	1,024	1.0001024	1.00010240523794
11	2,048	1.0002048	1.00020482096271
12	4,096	1.0004096	1.00040968387705
13	8,192	1.0008192	1.00081953559497
14	16,384	1.0016391	1.00163974282853
15	32,768	1.0032809	1.00328217441361
16	65,536	1.0065726	1.00657512149610
17	131,072	1.0131884	1.01319347521490
18	262,144	1.0265507	1.02656101821804
19	524,288	1.0538063	1.05382752412486
20	1,048,576	1.1105077	1.11055245060312
21	2,097,152	1.2332274	1.23332674554061
22	4,194,304	1.5208498	1.52109486126578
23	8,388,608	2.3129841	2.31372957696917
24	16,777,216	5.3498954	5.35334455534193
25	33,554,432	28.621381	28.6582979282091
26	67,108,864	819.18345	821.298040141993
27	134,217,728	671062.52	674530.470741078

TABLE 2.2

	Result	Percent error
Calculator		
Texas Instruments SR-52	674520.6053	0.00146
Hewlett–Packard 33, 67, 41C	674494.0561	0.00540
Sharp electronics EL506	674492.75	0.00559
Monroe Calculator 1930	674383.1672	0.02183
Texas Instruments 30	674363.69	0.02473
Computer and language		
Double-precision Fortran (CDC Cyber)	674530.5363	0.00000973
Eight-digit Fortran (CDC Cyber)	674530.5765	0.00001568
Apple II BASIC	22723.9709	96.63114
IBM Personal Computer BASIC	8850273	1212.06423
Ontel BASIC	8886690	1217.46401
Lotus 123 (IBM System II)	674530.4755	7.116 E-7
TRS 80 PC 1 BASIC	674494.056	5.4 E-3

is squared 27 times, then errors compound depending upon the number of digits used in the calculations.

Table 2.1 shows that a 0.5% error occurs between eight-digit calculations and 15-digit calculations. This does not appear to be significant; however, the errors compound rapidly beyond this point. In fact, doing the same calculation using IBM PC BASIC, the errors compound to over 1000%. Table 2.2 shows these calculations for various calculators, computers and languages.

What quantum theorists found was that calculations involving 100 atoms yielded results several thousands percent off. Any more than a few dozen sequential squarings was basically impossible to accomplish.

Computers, by design, are limited to a fixed number of digits of accuracy. This limit has no foreseeable solution and thus the limit of simple atoms with only a few electrons was imposed on the early theorists.

The solution to this problem came in the form of a novel idea. The inner electrons are so tightly bound to the nucleus that they can be considered part of a "frozen" atomic core. The negatively charged electrons in the atomic core neutralize or screen part of the positive charge of the nucleus. Consequently, the valence electrons see only a partial nucleus or core potential. This approach is called the *molecular orbital* theory.

In 1939, a researcher from Princeton University showed that the effects of the "frozen" atomic core on valence electrons are minimal. W. Conyers Herring observed that the wavefunction of a valence electron must have one more eigenvalue (zero value) than the wavefunction of the outermost core electron. The extra value approximates the value of the outermost electron of the core. This yields an amplitude of the valence-electron wavefunction near zero for this inner boundary condition. Therefore, the square of the amplitude, or probability of finding an electron, is also nearly zero. Thus, valence electrons tend not to be found near the surface of this frozen core.

Expanding on this idea, later researchers developed the idea of discarding the atomic cores while solving for the wavefunctions of the valence electrons. Their computational strategy was to combine and modify the wavefunctions directly, then systematically and sequentially eliminate the atomic cores. The new strategy considers a valence electron to be moving under the influence of a potential called a *pseudoatom* which carries a positive charge smaller than the charge of the nucleus. The pseudoatom charge is approximately that of the nuclear charge minus that of the core electrons.

This methodology opened up new opportunities for quantum mechanics. The pseudopotential approach is appropriate for evaluating the structure of single crystalline solids. Pseudopotentials have been constructed for most of the elements of the periodic table and also account for the nature of the electronic bonding in alloys and binary solid compounds. Pseudopotential calculations in materials that are partially ionic and partially covalent in nature indicate a high probability that the valence electrons would be found between the two atoms. This is typically called the *bonding region* and is calculated by the superposition of the wavefunctions for these bonding electrons. The structures of more than 60 binary compounds have been successfully predicted by these methods.

2.2 POSTULATE THE FIVE RULES OF QUANTUM MECHANICS

There are several ways to establish a theory. Einstein, when publishing his special theory of relativity, simply postulated that the speed of light is constant. He gave no explanation but made detailed predictions from this theory by deduction. Likewise, one of the best ways to establish the concepts of quantum mechanics is simply to postulate the rules of analysis. In stating these rules, we will consider for the sake of simplicity only one dimension x and only time t (for a three-dimensional and relativistic treatment see Slater).

Rule 1: For each system with one degree of freedom there exists a corresponding wavefunction $\psi(x, t)$.

Rule 2: The total energy E of a system is classically expressed as the sum of the kinetic and potential energy.

$$E = \tfrac{1}{2}mv_x^2 + V(x) \tag{2.1}$$

$$E = \frac{p_x^2}{2m} + V(x) \tag{2.2}$$

2.2 POSTULATE THE FIVE RULES OF QUANTUM MECHANICS

where m is the mass, v the velocity and therefore p_x is the momentum mv_x and $V(x)$ is the potential energy. This equation can be converted into a wave equation by substituting operators for the dynamic variables.

Thus,

$$x \text{ remains } x \text{ in equations (2.1) and (2.2), and} \qquad (2.3)$$

$$f(x) \text{ remains } f(x) \qquad (2.4)$$

but

$$p_x \text{ becomes } \frac{\hbar}{i}\frac{\partial}{\partial x} \text{ where } \hbar = \frac{h}{2\pi} \qquad (2.5)$$

and

$$E \text{ becomes } -\frac{\hbar}{i}\frac{\partial}{\partial t} \qquad (2.6)$$

Next, insert the wavefunction, $\psi(x, t)$ into both sides of equation (2.2),

$$\left[\frac{p_x^2}{2m} + V(x)\right] \cdot \psi(x, t) = E \cdot \psi(x, t) \qquad (2.7)$$

Then, substitute the operators for the dynamic variables, in equation (2.7) which yields

$$+\frac{1}{2m}\left(\frac{\hbar}{i}\frac{\partial}{\partial x}\right)^2 \cdot \psi(x, t) + V(x)\psi(x, t) = -\frac{\hbar}{i}\frac{\partial}{\partial t}\psi(x, t) \qquad (2.8)$$

Finally, combine terms to obtain

$$-\frac{\hbar^2}{2m}\frac{\partial^2}{\partial x^2}\psi(x, t) + V(x)\psi(x, t) = -\frac{\hbar}{i}\frac{\partial}{\partial t}\psi(x, t) \qquad (2.9)$$

This is the Schrödinger wave equation expressed as a function of time and the x-dimension. The potential energy is a function of x only.

Rule 3: The functions $\psi(x, t)$ and $\partial/\partial x\, \psi(x, t)$ must be continuous, finite, and single-valued throughout the region of space being considered.

Rule 4: The integral of the conjugate of the wavefunction $\psi^*(x, t)$ times the wavefunction $\psi(x, t)$ is normalized to the value one and represents the probability of location of a particle within x:

$$\int_{-\infty}^{\infty} \psi^*\psi\, dx = 1 \qquad (2.10)$$

Rule 5: The expectation value, $\langle \alpha \rangle$, of any dynamic variable α (and its corresponding operator, O_α) is determined by the following formula:

$$\langle \alpha \rangle = \int_{-\infty}^{\infty} \psi^* O_\alpha \psi \, dx \qquad (2.11)$$

These five rules represent the essentials of quantum mechanics.

2.3 DE BROGLIE'S WAVELENGTH

The analysis of electrons under the influence of a potential is of primary importance in understanding the electronic behavior of solids. The wave nature of electrons had its beginnings early in this century. In 1926, Louis de Broglie summarized his ideas concerning the wave nature of matter.

I shall take it that there is reason to suppose the existence, in a wave, of points where energy is concentrated, of very small corpuscles whose motion is so intimately connected with the displacement of the wave that a knowledge of the laws regulating one of these motions is equivalent to a knowledge of the laws governing the other. Conversely, I shall suppose that there is reason to associate wave propagation with the motion of all the kinds of corpuscles whose existence has been revealed to us by experiment. ... I shall take the laws of wave propagation as fundamental, and seek to deduce from them, as consequences which are valid in certain cases only, the laws of dynamics of a particle.

De Broglie assumed that matter waves existed that could be represented by a characteristic frequency, v. This frequency was related to the total energy of a photon described by Einstein:

$$E = hv \qquad (2.12)$$

The momentum p of such a particle is also given by:

$$p = \frac{E}{c} \qquad (2.13)$$

where c is the velocity of light.
Combining these relationships yields

$$p = \frac{hv}{c} \quad \text{and} \quad v = \frac{cp}{h} \qquad (2.14)$$

and since the wavelength λ is given by

$$\lambda = \frac{c}{v}, \qquad (2.15)$$

this leads to the de Broglie wavelength λ:

$$\lambda = \frac{h}{p} = \frac{h}{mv} \qquad (2.16)$$

2.4 ELECTRON SCATTERING

Classical wave diffraction yields a pattern of maxima and minima. This pattern can be represented by the equation:

$$n\lambda = 2d \sin \theta \qquad (2.17)$$

where n = maxima number (integer 1, 2, ...)
d = separation between slits
θ = angle of observation
λ = wavelength

When monoenergetic electrons of 54 eV are beamed onto the face of a crystal a similar pattern is generated. If electrons are considered as classical particles this result cannot be explained. However, by employing the de Broglie wavelength in equation (2.16) we can determine how an electron will diffract like a wave.

Using the classical kinetic energy of an electron:

$$E = \tfrac{1}{2}mv^2 \qquad (2.18)$$

Also, using the energy of an electron accelerated over a potential V with charge e:

$$E = Ve \qquad (2.19)$$

Then by setting these energies equal to one another we obtain

$$\tfrac{1}{2}mv^2 = Ve \qquad (2.20)$$

or

$$mv^2 = 2Ve \qquad (2.21)$$

Multiplying both sides by m yields

$$(mv)^2 = 2mVe \qquad (2.22)$$

or

$$mv = (2mVe)^{1/2} \qquad (2.23)$$

Now we can use equation (2.16):

$$\lambda = \frac{h}{mv} \qquad (2.24)$$

Substituting equation 2.23 yields

$$\lambda = \frac{h}{(2mVe)^{1/2}} \qquad (2.25)$$

where $h = 6.625 \times 10^{-27}$ erg sec
$m = 9.11 \times 10^{-28}$ g
$e = 4.8 \times 10^{-10}$ (esu) or stat coulombs
V = stat volts (esu) = 300 V

Substituting these values into equation (2.25) we obtain:

$$\lambda = \frac{(6.625 \times 10^{-27})}{\left[(2)(9.11 \times 10^{-28})(4.8 \times 10^{-10})\left(\frac{V}{300}\right)\right]^{1/2}} \qquad (2.26)$$

This reduces to

$$\lambda = \frac{12.27}{\sqrt{V}} \times 10^{-8} \text{ cm} \qquad (2.27)$$

which yields: $\lambda = 1.67 \times 10^{-8}$ cm for a potential of 54 V.

Substitute this value of λ into the diffraction equation (2.17) with $n = 2$ for the secondary maxima. The experimental separation between the scattering points is $d = 2.15 \times 10^{-8}$ cm. Thus, the second electron diffraction peak occurs at the following angle:

$$\theta_{n=2} = 50° \qquad (2.28)$$

The diffraction of particles is known as *Bragg diffraction*. The theory predicts values within 1% of the experimental values. There are many other examples to indicate the wave nature of matter. In order to understand the details of this dual wave nature of particles, it is instructive to analyze a few simple systems based upon a single particle.

2.5 SIMPLE HARMONIC OSCILLATOR (SHO)

Consider a harmonic motion of the form

$$y = A \cos(2\pi f t) \qquad (2.29)$$

2.5 SIMPLE HARMONIC OSCILLATOR (SHO)

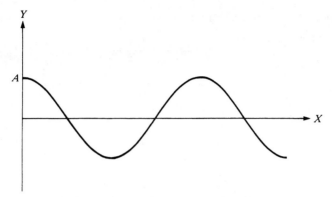

FIGURE 2.1 Simple harmonic oscillator.

shown in Figure 2.1, where there is an equivalent way of expressing the angle $2\pi ft$. The units of the angle must be radians, therefore

$$\left(\frac{2\pi \text{ radians}}{\text{cycle}}\right)\left(\frac{\text{cycle}}{\text{sec}}\right)(\text{sec}) \equiv (\text{radians}) \tag{2.30}$$

can be equivalently expressed in terms of the wavelength and distance.

$$\left(\frac{2\pi \text{ radians}}{\text{cycle}}\right)\left(\frac{\text{cycle}}{\text{cm}}\right)(\text{cm}) \equiv (\text{radians}) \tag{2.31}$$

Using wavelength λ and distance x for the angle in equation (2.29) and differentiating it twice, we obtain

$$y' = -\frac{2\pi A}{\lambda}\sin\left(\frac{2\pi}{\lambda}x\right) \tag{2.32}$$

$$y'' = -(2\pi)^2 \frac{1}{\lambda^2} A \cos\left(\frac{2\pi}{\lambda}x\right) \tag{2.33}$$

therefore, by comparing equation (2.33) with equation (2.29), we obtain

$$y'' = -\left(\frac{4\pi^2}{\lambda^2}\right)y \tag{2.34}$$

Let y represent the wavefunction ψ and changing notation:

$$\frac{d^2}{dx^2}\psi(x) = -\frac{4\pi^2}{\lambda^2}\psi(x) \tag{2.35}$$

The energy of the system can be expressed as

$$E = \tfrac{1}{2}mv^2 + V(x) \qquad (2.36)$$

$$E = \frac{p^2}{2m} + V(x) \qquad (2.37)$$

or by rearranging

$$p^2 = 2m(E - V(x)) \qquad (2.38)$$

Imposing the de Broglie wavelength, $\lambda = h/p$, as

$$\frac{1}{\lambda^2} = \frac{p^2}{h^2} \qquad (2.39)$$

Then equation (2.35) becomes

$$\frac{d^2}{dx^2}\psi(x) = -\left(\frac{4\pi^2}{h^2}\right)p^2\psi(x) \qquad (2.40)$$

Rearranging equation (2.37) and substituting it into (2.40) yields

$$\frac{d^2}{dx^2}\psi(x) = -\frac{8\pi^2 m}{h^2}(E-V)\psi(x) \qquad (2.41)$$

and defining

$$\hbar \equiv \frac{h}{2\pi} \qquad (2.42)$$

yields the time-independent Schrödinger equation:

$$\frac{d^2}{dx^2}\psi(x) + \frac{2m}{\hbar^2}(E-V)\psi(x) = 0 \qquad (2.43)$$

Compare with equation (2.9) and see that the time dependence has been removed.

2.6 PROBABILITY

Solutions to the Schrödinger equation are probability amplitudes. Therefore, $dP = \psi^*\psi\, dx$ gives the probability of finding a particle within dx; (i.e., Rule 4). This relationship is similar to light intensity which is proportional to amplitude squared.

2.7 SUPERPOSITION

The reason that the Schrödinger equation solutions are probability amplitudes is because the superposition of waves provides the localization of the wavefunction needed to quantify a particle. This is seen in Figure 2.2.

In this example, we see how groups of waves can be formed from simple sine waves by superimposing wavetrains of different frequency (f_1 and f_2) but the same amplitude (A):

$$u_1(x, t) = A \cos\left[2\pi f_1\left(\frac{x}{c} - t\right)\right] = A e^{2\pi i f_1 x} e^{-2\pi i f_1 t} \qquad (2.44)$$

$$u_2(x, t) = A \cos\left[2\pi f_2\left(\frac{x}{c} - t\right)\right] = A e^{2\pi i f_2 x} e^{-2\pi i f_2 t} \qquad (2.45)$$

$$u_1(x, t) + u_2(x, t) = A \cos\left[2\pi f_1\left(\frac{x}{c} - t\right)\right] + A \cos\left[2\pi f_2\left(\frac{x}{c} - t\right)\right] \qquad (2.46)$$

Recalling the trigonometric identities of

$$\cos(x + y) = \cos x \cos y - \sin x \sin y \qquad (2.47)$$

and

$$\cos(x - y) = \cos x \cos y + \sin x \sin y \qquad (2.48)$$

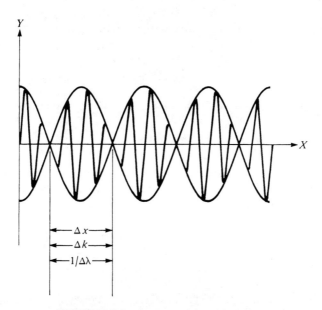

FIGURE 2.2 Superposition of waves.

44 CHAPTER 2: QUANTUM MECHANICS AND THE BAND THEORY OF SOLIDS

By adding these identities we create a new one:

$$\cos(x+y) + \cos(x-y) = 2\cos x \cos y \qquad (2.49)$$

By comparing equation (2.46) with (2.49), we see that (2.46) may be rewritten in the form:

$$u_1(x,t) + u_2(x,t) = 2A\left\{\cos\left[\frac{2\pi f_1\left(\frac{x}{c}-t\right) - 2\pi f_2\left(\frac{x}{c}-t\right)}{2}\right]\right.$$

$$\left.\times \cos\left[\frac{2\pi f_1\left(\frac{x}{c}-t\right) + 2\pi f_2\left(\frac{x}{c}-t\right)}{2}\right]\right\} \qquad (2.50)$$

$$u_1(x,t) + u_2(x,t) = 2A\left[\cos(f_1 - f_2)\left(\frac{x}{c}-t\right)\cos(f_1 + f_2)\left(\frac{x}{c}-t\right)\right] \qquad (2.51)$$
$$\qquad\qquad\qquad\qquad\qquad (1) \qquad\qquad\qquad (2)$$

The above equation is shown graphically in Figure 2.2. The superposition results in the formation of "wavepackets".

2.8 HEISENBERG UNCERTAINTY

The wavenumber k is defined as $k \equiv 1/\lambda$, as illustrated in Figure 2.2. The uncertainty in wavenumber Δk is such that, over the region Δx, it leads to a difference in the number of peaks of at least one:

$$\Delta k \, \Delta x \geq 1 \qquad (2.52)$$

and since from equation (2.16), $p \neq h/\lambda = hk$, this leads to:

$$\Delta p \, \Delta x \geq h \qquad (2.53)$$

This is one part of the *Heisenberg uncertainty principle*. The significance of this relationship is that one cannot simultaneously define the momentum and location of a particle. Also, Figure 2.2 shows that

$$\Delta v \, \Delta t \geq 1 \text{ and since } E = hv \text{ then } \Delta E \, \Delta t \geq h \qquad (2.54)$$

The importance of this is that one cannot simultaneously establish the energy and the time of a particle.

The consequences of these uncertainty relationships are important. If a particle moves slowly, its total momentum is small. Therefore its wavefunction ψ extends over a considerable space, so the particle can interact with anything in the region. That is, a slow-moving neutron acts as if it were larger than a fast neutron. As we shall see later, differences in large and small polarons are related to this uncertainty relationship and electron–lattice interactions.

If a molecule or activated chemical species exists for a short time, the uncertainty in time is necessarily small, so its uncertainty in energy is correspondingly large. This uncertainty is noticed at about 0.03 cal/mole or 3.3×10^{-9} sec.

2.9 QUANTIZATION OF WAVEFUNCTIONS AND ENERGIES

Imposing boundary conditions upon the wave equation results in quantization of both the wavefunctions, termed *eigenfunctions*, and the energies associated with the waves, termed *eigenvalues*. One of the simple examples that illustrates this concept is that of a particle in a one-dimensional square well (see Figure 2.3).

In regions 2 and 3:

$$V \to \infty \quad \text{therefore} \quad \psi = 0 \quad (2.55)$$

because applying Rule 5 for the boundary conditions:

$$\psi^*\psi = |\psi|^2 = 0 \quad (2.56)$$

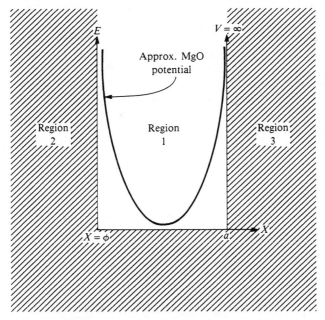

FIGURE 2.3 Infinite-square-well potential.

In region 1:

$$\frac{d^2\psi}{dx^2} = \left(\frac{8\pi^2 m}{h^2}E - V\right)\psi \quad \text{with} \quad V = 0 \tag{2.57}$$

Therefore

$$\frac{d^2\psi}{dx^2} = \alpha^2\psi, \quad \alpha^2 = \frac{8\pi^2 mE}{h^2} \tag{2.58}$$

The two solutions to this equation are:

$$\psi_{(x)} = Ae^{+i\alpha x} \tag{2.59}$$

and

$$\psi_{(x)} = Be^{-i\alpha x} \tag{2.60}$$

Therefore

$$\psi_{(x)} = Ae^{i\alpha x} + Be^{-i\alpha x} \tag{2.61}$$

Applying the boundary conditions results in a quantization of the energy of the eigenfunctions. The boundary conditions are: ψ must be single-valued and continuous (Rule 3), so

$$\psi_{(x)} = 0; \quad \psi_{(a)} = 0 \tag{2.62}$$

therefore

$$\psi_{(0)} = A + B = 0; \quad B = -A \tag{2.63}$$

so

$$\psi_{(x)} = (e^{i\alpha x} - e^{-i\alpha x}) = N \sin \alpha x \tag{2.64}$$

Apply the second boundary condition at $x = a$:

$$\psi_{(a)} = N \sin(\alpha a) = 0 \tag{2.65}$$

or

$$\alpha a = n\pi, \quad n = 1, 2, 3, \ldots \tag{2.66}$$

$$\alpha = \frac{n\pi}{a} \tag{2.67}$$

and

$$\alpha^2 = \frac{8\pi^2 mE}{h^2} \tag{2.68}$$

Thus

$$E = \alpha^2\left(\frac{h^2}{8\pi^2 m}\right) \tag{2.69}$$

Substituting $\alpha = n\pi/a$ yields

$$E = n^2\left(\frac{h^2}{8ma^2}\right) \tag{2.70}$$

which are eigenvalues of energy at $n = 1, 2, 3, \ldots$. Also,

$$\psi_{(x)} = N \sin(\alpha x) = N_n \sin\left(\frac{n\pi x}{a}\right) \tag{2.71}$$

which are eigenfunctions of $n = 1, 2, 3, \ldots$. Consequently, if

$$n = 1, \quad E_m = 1\left(\frac{h^2}{8ma^2}\right) \tag{2.72}$$

$$n = 2, \quad E_m = 4\left(\frac{h^2}{8ma^2}\right) \tag{2.73}$$

$$n = 3, \quad E_m = 9\left(\frac{h^2}{8ma^2}\right) \tag{2.74}$$

$$n = 4, \quad E_m = 16\left(\frac{h^2}{8ma^2}\right) \tag{2.75}$$

These energy levels are shown in Figure 2.4.

2.10 PROBABILITY OF LOCATING A PARTICLE

The probability of locating a particle within the box of width a in Figures 2.3 and 2.4 is determined by integrating over the dimension of the box (i.e.,)

$$\psi_{m(x)} = N_{(m)} \sin\left(\frac{n\pi x}{a}\right) \tag{2.76}$$

$$\int_0^a \psi_{n(x)} \psi_{n(x)} \, dx = 1 \tag{2.77}$$

so

$$1 = N_m^* N_m \int_0^a \sin^2\left(\frac{n\pi x}{a}\right) dx = \tfrac{1}{2}|N_m|^2 \int_0^a \left[1 - \cos\left(\frac{2\pi x}{a}\right)\right] dx \tag{2.78}$$

$$1 = \tfrac{1}{2}|N_m|^2 \left[x - \left(\frac{a}{2n\pi}\right) \sin\left(\frac{2n\pi x}{a}\right)\right]_0^a \tag{2.79}$$

$$1 = \tfrac{1}{2}|N_m|^2 a \tag{2.80}$$

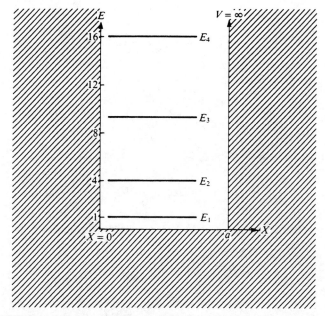

FIGURE 2.4 Energy eigenvalues for a particle in an infinite square well.

Therefore

$$|N_m| = \left(\frac{2}{a}\right)^{1/2} \tag{2.81}$$

and

$$\psi_{m(x)} = \left(\frac{2}{a}\right)^{1/2} \sin\left(\frac{n\pi x}{a}\right) \tag{2.82}$$

Equation (2.82) results in the following eigenfunctions:

$$\psi_{0(x)} = \left(\frac{2}{a}\right)^{1/2} \sin 0 = 0 \tag{2.83}$$

$$\psi_{1(x)} = \left(\frac{2}{a}\right)^{1/2} \sin\left(\frac{\pi x}{a}\right) \tag{2.84}$$

$$\psi_{2(x)} = \left(\frac{2}{a}\right)^{1/2} \sin\left(\frac{2\pi x}{a}\right) \tag{2.85}$$

$$\psi_{3(x)} = \left(\frac{2}{a}\right)^{1/2} \sin\left(\frac{3\pi x}{a}\right) \ldots \tag{2.86}$$

Graphically, the eigenfunctions and the resulting probabilities of locating

2.10 PROBABILITY OF LOCATING A PARTICLE

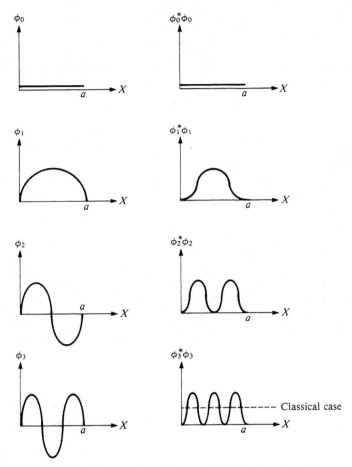

FIGURE 2.5 Eigenfunctions and resulting probability functions for a particle in infinite square-well potential.

the particle are shown in Figure 2.5, compared with the classical case. In classical mechanics there is equal probability that the particle can be anywhere between zero and a. As n becomes very large, the probability of locating the particle approaches that of the classical case. There are a number of important points illustrated by this simple example. These include:

1. Zero point energy; the lowest value of energy is $E = h^2/8ma^2$,
2. The existence of a zero point energy is understood in terms of the uncertainty principle; (e.g.,) since we know that the particle is in the box, the uncertainty of x is $|a|$, a finite quantity. Therefore, there must be uncertainty in momentum, so p cannot be equal to zero and the particle must have some finite value of kinetic energy, and

3. The energy of a particle confined to a finite space is quantized. The energy levels increase proportional to n^2 (see equation 2.70).

Comparisons of bound and free states of a quantized particle yield important differences that are observed in the electronic behavior of materials, especially the interfaces between phases (see Figure 2.6).

In current studies of nonlinear optics (see Chapter 10), efforts are directed toward materials that exhibit changes in index of refraction. One technique being evaluated is that of *quantum confinement*. This effect is used to spread out the energy eigenvalues listed in equations (2.57) through (2.60). Here the energy levels are universally proportional to the square-well dimension, a. As this dimension is reduced, the differences between the

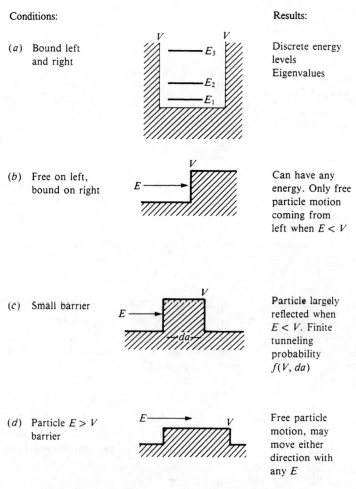

FIGURE 2.6 Bound and free states of electrons.

energy levels increase. This quantum effect has been demonstrated in GaAs heterojunctions and in small crystals of CdS in silicate glass, for example. For very small crystals, the square-well walls literally become the boundary of the crystal. This effect is caused by the particle being confined in the equivalent of square-well potential, where the dimensions of the particle (an exciton) and the dimensions of the crystal (CdS) are approximately the same size.

2.11 BOHR MODEL OF THE HYDROGEN ATOM

At this point, it is useful to examine the original model which first quantized the energy levels of the atom. It was learned during the time of Niels Bohr (1922) that the emission spectra of various materials showed characteristic lines which could be used to identify various elements. The Balmer series of lines, for example, could be identified for excited elements such as hydrogen. The spacing and wavelengths of these lines were quite specific and spectroscopic "rules" were developed to characterize different elements by their emission lines.

Bohr suggested that these spectra were the result of a much more basic phenomenon. He postulated that electrons were in "orbit" about a central nucleus and that the angular momentum was quantized as follows:

$$mvr \equiv \frac{nh}{2\pi} = n\hbar \qquad (2.87)$$

where m = mass of the electron of charge (e^-)
v = orbital velocity
r = orbital radius
h = Planck's constant = 6.625×10^{-34} joule · sec
n = integer 1, 2, 3, ...

The inward and outward forces on the electrons must be balanced; (e.g.,) the outward force

$$F = ma \qquad (2.88)$$

or

$$F = \frac{mv^2}{r} \text{ (outward force)} \qquad (2.89)$$

must equal the attractive (Coulomb) force on the electron (see Chapter 1)

$$F = \frac{e^2}{r^2} \text{ (inward force)} \qquad (2.90)$$

for the orbit to be stable. Thus

$$\frac{mv^2}{r} = \frac{e^2}{r^2} \tag{2.91}$$

and solving for mv^2

$$mv^2 = \frac{e^2}{r} \tag{2.92}$$

Secondly, the kinetic energy of the electron is given by

$$\text{K.E.} = \frac{mv^2}{2} \tag{2.93}$$

which becomes

$$\text{K.E.} = \frac{e^2}{2r} \tag{2.94}$$

Since the potential energy for the electron is given by

$$V = -\frac{e^2}{r} \tag{2.95}$$

Then the total energy T becomes

$$T = \text{K.E.} + V \tag{2.96}$$

$$T = \frac{e^2}{2r} - \frac{e^2}{r} \tag{2.97}$$

$$T = -\frac{e^2}{2r} \text{ (binding energy)} \tag{2.98}$$

From equation (2.92) the expression for kinetic energy can be rearranged as

$$mv^2 r = e^2 \tag{2.99}$$

Now we use the Bohr postulate that the orbital angular momentum is quantized as (2.87):

$$mvr = n\hbar$$

Squaring both sides yields

$$(mvr)^2 = n^2\hbar^2 \qquad (2.100)$$

By dividing equation (2.99) by (2.100) we have

$$\frac{mv^2r}{(mvr)^2} = \frac{e^2}{n^2\hbar^2} \qquad (2.101)$$

or

$$\frac{1}{mr} = \frac{4\pi^2 e^2}{h^2} \frac{1}{n^2} \qquad (2.102)$$

then

$$\frac{1}{r} = \frac{4\pi^2 e^2}{h^2} \frac{m}{n^2} \qquad (2.103)$$

By redefining the total energy of the electron (equation 2.98) as $T \equiv E_n$, then

$$E_n = -\frac{e^2}{2r} \qquad (2.104)$$

and substituting for $1/r$ from equation (2.103), the electron energy becomes

$$E_n = -\frac{2\pi^2 e^4}{h^2} \frac{m}{n^2} \qquad (2.105)$$

The result shows that the electron energy is also quantized and is a function of n^2, similar to the particle in a box (equation 2.70).

Remember that

$$E = h\nu \qquad (2.106)$$

where ν is the frequency, so the frequency of the light emitted from the Bohr atom corresponds to changes in the quantized orbital energies, E_n, of the electron. Therefore a change in energy of E_n to $E_{n'}$ would correspond to the frequency $\nu_{nn'}$:

$$\nu_{nn'} = \frac{E_n - E_{n'}}{h} \qquad (2.107)$$

thus, by substituting equation (2.105) we obtain

$$\nu_{nn'} = \frac{2\pi^2 e^4 m}{h^3} \left(\frac{1}{n'^2} - \frac{1}{n^2} \right) \qquad (2.108)$$

This relationship for the emission spectra of atomic hydrogen fits the experimental results very well. Trouble developed, however, when this model was applied to more complex atoms. The more formalized approach of quantum mechanics produced better results.

2.12 QUANTUM NUMBERS

As we have just seen, the Bohr atomic energy and the energy of a particle in an infinite square well results in functions of n^2. The number n has been defined as the primary quantum number and defines the energy of the particle (or electron).

The quantum number n is only one of *four* quantum numbers used to describe the state of an electron in an atom. The electron is basically held in a finite potential well created by the attractive electro-static forces of the nucleus.

The second quantum number is the orbital angular momentum (l). The quantum number l can have any integer value between $(n-1)$ and zero.

$$0 \leq l \leq (n-1) \qquad (2.109)$$

The third quantum number m is the magnetic moment of the electrons. This quantum number is also associated with the orbital angular momentum. The orbits will orient themselves with respect to an external magnetic field depending upon their momentum. The magnetic quantum number m can take on integral values between $\pm l$.

$$-l \leq m \leq l \qquad (2.110)$$

The fourth quantum number s defines the spin of the electron itself. This quantum number can take on only two values:

$$s = \pm \tfrac{1}{2} \qquad (2.111)$$

The electron is spinning either one way or the other.

The electrons will tend to seek the lowest possible energy state. At the same time, the Pauli exclusion principle allows only one electron to have any given state characterized by the four quantum numbers n, l, m, and s. Since s can have only two values, $\pm\tfrac{1}{2}$, then only two electrons may have the same value of n, l, and m. This pairing of electrons with opposite spin results in a zero spin dipole moment.

Likewise, if all the states for a given n and l are occupied, the resulting orbital dipole moment will be zero. There will be a $+m$ electron for every $-m$ electron in the orbital shell. This configuration of electrons in an atomic structure is known as a *closed shell*.

Table 2.3 displays the relationships between the quantum numbers,

TABLE 2.3 Quantum Numbers

Energy n	Orbital momentum l	Orbital state	Maximum number of electrons	Closed shell element
1	0	$1s$	2	He
2	0	$2s$ (or 2σ)	2	Ne
2	1	$2p$ (or 2π)	6	
3	0	$3s$ (or 3σ)	2	
3	1	$3p$ (or 3π)	6	Ar
3	2	$3d$ (or 3Δ)	10	
4	0	$4s$ (or 2σ)	2	
4	1	$4p$ (or 2π)	6	Kr
4	2	$4d$ (or 2Δ)	10	
4	3	$4f$	14	

orbital states, and closed shell elements. Elements with similar outer shell configurations tend to have similar chemical and physical properties.

2.13 MOLECULAR ORBITALS

Up to now our discussions have focused on the quantum mechanics of a single electron. Real materials, however, are made up of large numbers of atoms with a corresponding large number of electrons. These atoms are bound together by the interaction of electrostatic forces. The type of bond, its strength, and physical orientation depends on the interaction of the electrons of the bound atoms.

The study of molecular orbitals gives us insight into the nature of chemical bonds. For example, consider the overlap of the $1s$ electrons of two hydrogen atoms as they form the bonding of the diatomic hydrogen molecule. The probability of finding a $1s$ electron for each of the atoms is a distribution and is calculated from the wavefunction of the $1s$ electron. There is an overlap region that is slightly more negative because both electrons of H_2 may occupy this region. The positive nuclei are therefore attracted by this negatively charged area. The hydrogen atoms are thereby bound together to form what is termed a *sigma bond* (Figure 2.7a).

Hydrogen is the simplest atom but carbon is the most abundant. Consequently, the field of chemistry is often divided into organic (carbon-based) and inorganic (all other elements). Figure 2.7b shows the overlap region for a carbon–carbon pi bond. These bonds are characteristic of

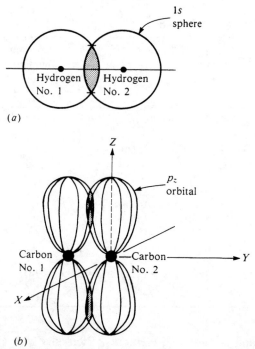

FIGURE 2.7 Electronic bonding in (*a*) H_2 (overlap of the two $1s$ orbitals) and (*b*) C_2 (overlap of the two $2p_z$ orbitals).

most unsaturated organic compounds. Sigma bonds are graphically represented as a sphere, see Figure 2.7a. The *pi bonds* are formed by the overlap of the p_z orbitals. The p_z orbitals are figure-eight-shaped probability distributions for the two *p* electrons of carbon.

The $2s$ electrons of carbon form spherical orbitals. These are 120° apart and are available to form sigma type bonds with hydrogen. An example is shown in Figure 2.8a. The thin lines indicate the sigma bonds, while the bold line indicates the pi bond.

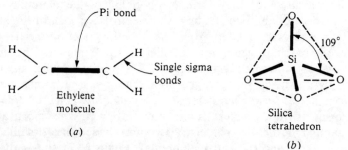

FIGURE 2.8 Bonds in ethylene (*a*) and silica tetrahedron (*b*).

The outer four electrons of silicon can form four sp^3 hybrid bonds with oxygen as shown in Figure 2.8b. These orbitals have approximately one-quarter sigma character and three-quarters pi character. The orbitals of oxygen and those of silicon overlap to form the familiar silica tetrahedron. The bond angles average near 109° in three-dimensional space.

Returning to the carbon–carbon bond, imagine a large number of p_z orbitals stacked together on line. Each $+z$ and $-z$ overlap area allows electrons to move from one carbon atom to the next. The only limitation for this movement is that each p_z orbital must have exactly the same set of quantum numbers in the same physical space. Thus in the p_z orbital for each atom only two electrons can coexist together, one with spin $= +\frac{1}{2}$ and the other with spin $= -\frac{1}{2}$. This pair of electrons forms a *pi molecular orbital* (πMO).

2.14 AB-INITIO CALCULATIONS

The extreme complexity of large molecules makes it difficult to calculate the MOs exactly. Nearly all computer programs make some simplifying assumptions in order to make computations easier.

The basic problem is the solution of the Hamiltonian eigenvalue problem: (Schrödinger equation 2.9)

$$H_n \psi = E_n \psi \qquad (2.112)$$

where H_n = Hamiltonian or energy matrix = kinetic + potential energy operator
ψ = wavefunction matrix or eigenvector matrix
E_n = energy eigenvalue

The limitations to the solution of finding the energy eigenvalues and eigenvectors seems to be limited by computer power and understanding of the basic physics involved in bulk materials.

There are basically two quantum chemistry approaches to this problem. A dilemma arises from the fact that the two methods generally do not agree.

The basic physics approach is known as an *ab-initio* calculation. The Hamiltonian and eigenvectors are built up from individual wavefunctions of each particle in the system under study. The problem here is the large number of parameters and the few experimental data points. In this case, the researcher must make choices concerning the importance of the various parameters. Most agree that there is significant disagreement on these choices. This leads to the discrepancies mentioned earlier.

2.15 SEMI-EMPIRICAL CALCULATIONS

Semi-empirical methods usually result in very good agreement between experiment and theory when they are used for the specific system for which they were developed. In 1953, Pople developed a semi-empirical method based on the "neglect of differential overlap" (NDO) or the Pariser–Parr–Pople method. In this process, electrons that create a primary overlap region between the molecular orbitals of bonding atoms are the only ones considered. All second- and higher-order overlap regions are neglected. Referring back to the discussion on πMO bonding, only nearest bonding MOs are considered. In Figure 2.9, the structure of C_4H_6 (butadiene) is shown schematically. The pi bonds result from the small overlap regions in the p_z molecular orbitals between carbon atoms C_1 and C_2, C_2 and C_3, and C_3 and C_4. There are also interactions or differential overlap of the molecular orbitals for atoms C_1 and C_3 and for atoms C_1 and C_4. These overlap regions are real and finite but are also very small. For an *ab-initio* calculation these differential overlaps are not neglected, but for the NDO semi-empirical approach they are simply defined as zero which greatly simplifies calculations. These studies have led to several different semi-empirical models:

CNDO $\equiv$ Complete Neglect of Differential Overlaps,
INDO $\equiv$ Intermediate Neglect of Differential Overlap,
MINDO $\equiv$ Modified Intermediate Neglect of Differential Overlap,
NDDO $\equiv$ Neglect of Diatomic Differential Overlap,
and others.

These models have been successful in yielding conceptual ideas of molecular structures and predict experimental results quite accurately. However, they

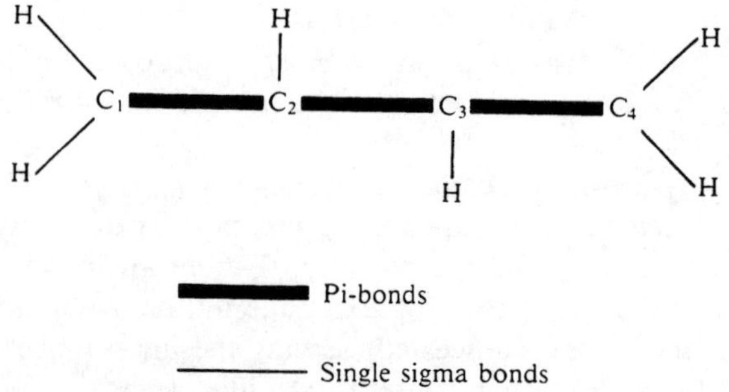

FIGURE 2.9 Structure of butadiene.

are less effective when extended beyond the system for which they were originally constructed.

2.16 HUCKEL MOLECULAR ORBITAL THEORY

In all quantum chemical calculations, either *ab-initio* or semi-empirical, there are two primary steps. First, the Hamiltonian or energy matrix must be generated. Nothing can be done until this is defined. The matrix is an energy set for each of the n electrons in the molecule.

$$H_n = \text{Kinetic energy} + \text{Potential energy} + \text{Interactions} \quad (2.113)$$

The Hamiltonian operator will yield the energy eigenvalues according to the Schrödinger equation (2.97)

$$H_n \psi = E_n \psi \quad (2.114)$$

Once the Hamiltonian is formed, then the second phase of the problem is to diagonalize it by an appropriate transformation matrix. The diagonalization operator or transformation matrix makes all the nondiagonal elements of H_n equal to zero. The resulting diagonalized matrix represents the energy eigenvalue matrix, E_n.

The transformation matrix T represents the eigenvector matrix ψ of the molecule. T is also the square root of the probability of finding an electron in a specific molecular orbital for a specific atom and is related to the eigenvector matrix through a delta (Δ) operator:

$$T = \Delta \psi \quad (2.115)$$

where Δ represents the nonorthogonality of the molecular orbitals under study.

Finally,

$$T = |\psi^* \psi|^{1/2} \quad (2.116)$$

which is the square root of the probability distribution of the electrons. The following simplifying assumptions are made for the Huckel molecular orbital theory (HMO):

1. The πMOs are the only orbitals considered in the calculations. All other electrons in the molecule are neglected.
2. It is assumed that -11.26 eV of energy is required to remove an electron from a p_z orbital of carbon. This is called α_i in the standard HMO theory. This energy eigenfunction becomes the diagonal of the Hamiltonian.

3. The bond energy of the pi bond is -2.5 eV. This HMO model neglects all other bond energies (differential overlaps) between nonneighboring orbitals. This energy is defined as β_{ij} in the standard HMO theory. These energy values form the nondiagonal elements. As shown in equation (2.119), all other nondiagonals are zero. They represent higher order overlaps that are neglected.
4. Atoms other than carbon (such as oxygen and silicon) pi-bonded to carbon are called heteroatoms. Energy required to remove the p_z heteroatom electrons is assumed to be a linear function of that of carbon.

$$\alpha_i = h_i(-11.26) \text{ eV} \qquad (2.117)$$

and the bond energy

$$\beta_{ij} = k_{ij}(-2.5) \text{ eV} \qquad (2.118)$$

where h and k are called the heteroatom parameters. Typical values for the heteroatom parameters in Table 2.4 are determined from experimental bond strengths. The values are arrived at empirically and only conceptually represent the bond energy or binding energy of the electrons.

Referring again to the schematic of butadiene (Figure 2.9), an energy matrix or Hamiltonian can be constructed for this molecule using the HMO assumption:

$$H = \begin{bmatrix} -11.26 & -2.5 & 0 & 0 \\ -2.5 & -11.26 & -2.5 & 0 \\ 0 & -2.5 & -11.26 & -2.5 \\ 0 & 0 & -2.5 & -11.26 \end{bmatrix} \qquad (2.119)$$

The electron binding energies fall along the diagonal of the Hamiltonian.

TABLE 2.4 Heteroatom Parameters

Bond (ij)	k-Parameter	Atom (i)	h-Parameter
C—C	1.00	C	1.00
C—O	0.80	O	1.00
C—N	0.90	N	1.50
C=C	1.00	Si	0.80
C=O	1.414		
C=N	1.0		
C=Si	0.72		
O=Si	1.4		

2.16 HUCKEL MOLECULAR ORBITAL THEORY

For different atoms, the heteroatom parameter, h, would scale the -11.26 eV value.

For the bond strength, the -2.5 eV fall off the diagonal. The bond strength between atoms 1 and 2 becomes element 1-2. Likewise, the bond between atom 2 and 1 falls into element 2-1. No interaction is assumed between nonadjacent atoms of a molecule.

If butadiene forms a ring it is called cyclobutadiene. In this case there is an interaction between atoms 1 and 4. This would modify the Hamiltonian to have a -2.5 eV value for elements 1-4 and 4-1, i.e., the corner elements.

We can now calculate the eigenvector for butadiene from the transformation matrix through Δ:

$$\psi_n = \begin{bmatrix} -0.371748 & 0.601501 & -0.601501 & 0.371748 \\ 0.601501 & -0.371748 & -0.371748 & 0.601501 \\ -0.601501 & -0.371748 & 0.371748 & 0.601501 \\ -0.371748 & -0.601501 & -0.601501 & -0.371748 \end{bmatrix} \quad (2.120)$$

and the corresponding energy eigenvalues E_n are a result of the diagonalized Hamiltonian H:

$$E_n = \begin{bmatrix} 4.04509 & 0 & 0 & 0 \\ 0 & 1.54508 & 0 & 0 \\ 0 & 0 & -1.54508 & 0 \\ 0 & 0 & 0 & -4.04509 \end{bmatrix} \quad (2.121)$$

The element of E_n represents the energy of the four states above and below the center energy of the molecule. Figure 2.10 shows these states from the lowest to the highest πMO. The lowest state is 4.04509 eV below the central energy for the molecule.

The central energy for the HMO calculation is simply:

$$E_{\text{central}} = \alpha N \quad (2.122)$$

where N = number of carbons

$$\alpha \equiv -11.26 \text{ eV (from HMO model)}$$

Therefore

$$E_{\text{central}} = -45.04 \text{ eV} \quad (2.123)$$

For $E(2)$ the energy is -1.545 eV. This means that each electron in MO2 only contributes 1.545 eV to the bond strength. For the remaining two MOs, any electron in these orbitals takes energy from the bond. Therefore,

62 CHAPTER 2: QUANTUM MECHANICS AND THE BAND THEORY OF SOLIDS

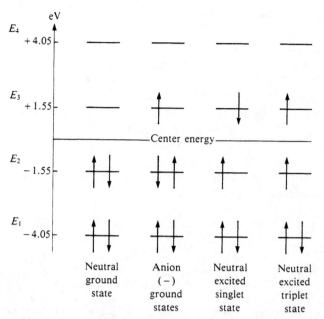

FIGURE 2.10 Occupation of molecular orbitals for butadiene.

these are called antibonding MOs. This means that the anion state and the two excited states shown in Figure 2.10 are 1.545 eV less bound than that of the neutral ground state.

The eigenvector ψ_n (equation 2.116) gives us the probability of finding electrons in the MOs of the different carbon atoms.

The element of the eigenvector matrix, ψ_n, can be used to calculate the probability of finding an electron associated with the corresponding atom:

$$
\begin{array}{lll}
\text{Atom} & & \text{Probability} \\
C_1 & (\psi_{11})^2 = (-0.371748)^2 & = 13.82\% \\
C_2 & (\psi_{12})^2 = (0.601501)^2 & = 36.18\% \\
C_3 & (\psi_{13})^2 = (-0.601501)^2 & = 36.18\% \\
C_4 & (\psi_{14})^2 = (0.371748)^2 & = 13.82\% \\
& & \overline{100\%}
\end{array}
\qquad (2.124)
$$

Therefore, the probability of finding the electron from C_1 over C_2, is 36.18%. Thus, the calculation shows quite clearly that the electrons are free to move through the overlap region in pi-bonded carbons.

The resonance energy is calculated by using the ground state Huckel energy of the molecule. The resonance energy is defined as the sum of the

2.16 HUCKEL MOLECULAR ORBITAL THEORY

energies of all occupied orbitals:

$$E_H = \sum_l N_l \cdot E(l) \qquad (2.125)$$

$$E_H = -11.26 \cdot \sum_n B_{nn} + (-2.5) \cdot \sum_{i \neq j} B_{ij} \qquad (2.126)$$

where the B_{ij} is the bond order matrix, which gives the relative percentage of a pi bond between each carbon atom.

$$B_{ij} = \sum_k N_k \cdot \psi_{ik} \cdot \psi_{jk} \qquad (2.127)$$

where N_k = number of e^- in the kth MO
ψ_{ik} = eigenvector matrix

For our example, butadiene, in the ground state, the electrons are occupying the first two MOs (see Figure 2.10).

	MO1	MO2	MO3	MO4
No. of e^-	2	2	0	0

Then the resulting bond order matrix is

$$B_{ij} = \begin{bmatrix} 1 & 0.894427 & 0 & -0.447214 \\ 0.894427 & 1 & 0.447214 & 0 \\ 0 & 0.447214 & 1 & 0.894477 \\ -0.447214 & 0 & 0.894427 & 1 \end{bmatrix} \qquad (2.128)$$

Note that B_{ij} is a symmetrical matrix and is usually shown in triangular form. In order to analyze the bonds, we define that if

$$B_{ij} = 1 \qquad (2.129)$$

there is a pure pi bond and if

$$B_{ij} = 0 \qquad (2.130)$$

then the pi bond does not exist at all. Thus we can determine the bond structure of butadiene. The pi bond between C_1 and C_2 is not as pure as one would suspect from the structural formula:

$$B_{12} = B_{34} = 0.894 \qquad (2.131)$$

Likewise, there is a weaker pi bond between C_2 and C_3

$$B_{23} = 0.447 \qquad (2.132)$$

By a simple electron count there should be no electrons in the bond between C_2 and C_3, i.e., B_{23} would be zero.

We are now in a position to calculate the ground state Huckel energy as given in equation (2.126) and from the bond order matrix:

$$E_H = -56.22 \text{ eV} \qquad (2.133)$$

This level is lower than one would expect if the electrons were distributed by the chemical formula for butadiene.

$$E' = 4 \cdot (-11.26) + 4 \cdot (-2.5) \qquad (2.134)$$

$$E' = -55.04 \text{ eV} \qquad (2.135)$$

The difference between E_H and E' is defined as the molecular resonance energy:

$$E_r = -1.180 \text{ eV} \qquad (2.136)$$

Therefore each electron makes the molecule more stable by -0.295 eV.

In the next section, we discuss the extension of electronic energy levels for crystalline materials. For this purpose, a crystal can be modeled as being a large molecule with a specific periodic structure. Using an HMO analysis similar to that described above for butadiene, we show the origin of the band gap in semiconductors such as SiC and its relationship to larger numbers of electrons.

In summary, the basic problem in calculating electronic structures is that of forming the Hamiltonian or energy matrix of the particular system under study. Various methods are used to do this. They include, but are not limited to, exact *ab-initio* formulations, and semi-empirical formulations of the energy of the electrons in a molecule or large cluster of atoms. Once the Hamiltonian has been established, it must be diagonalized. This can be done, for example, by the Jacobi method of diagonalization (see Slater 1968). The result is a diagonal matrix that represents the energy eigenvalues of the system and a transformation matrix that represents the eigenvectors of the system.

These results yield the energy states of the molecular orbitals, the probabilities of finding electrons in these orbitals, and the nature of the bond between atoms.

2.17 BAND THEORY OF SOLIDS

The valence electrons of metals and semiconductors can move freely between atoms. This was clearly seen even in the simple example of

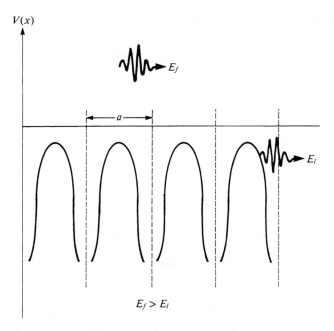

FIGURE 2.11 One-dimensional lattice.

butadiene. The wavefunction ψ of these electrons is subject to the potential of the core pseudoatoms of the crystal lattice. This potential must clearly have the same space symmetry and the same periodicity as the crystal itself. Therefore, the wavefunction ψ must reflect some of this symmetry.

Figure 2.11 shows a symmetrical one-dimensional lattice crystal (Kronig–Penney model). The lattice potential can be represented by a series of potential wells with a separation a, equal to the separation of the lattice atoms. This model is thus similar to having a periodic sequence of infinite square well potentials (Figure 2.3).

Free electrons have an energy E_f. Likewise, electrons can be associated with the lattice. They have an energy E_l, less than E_f. The lattice electrons interact with the potential wells and scatter like a diffracted wave.

The time-independent Schrödinger equation can be used to evaluate the wavefunction of the free electrons

$$-\frac{k^2}{2m}\frac{d^2}{dx^2}\psi_{(x)} = E_f\psi_{(x)} \tag{2.137}$$

with the solution

$$\psi_{(x)} = Ae^{ikx} \tag{2.138}$$

Remembering de Broglie's wavelength

$$p = \frac{h}{\lambda} \tag{2.139}$$

and

$$k = \frac{2\pi}{\lambda} \tag{2.140}$$

then

$$p = \frac{hk}{2\pi} \tag{2.141}$$

The kinetic energy of a free particle is given by

$$E = \tfrac{1}{2}mv^2 \tag{2.142}$$

$$E = \frac{p^2}{2m} \tag{2.143}$$

$$E = \frac{h^2 k^2}{2m(2\pi)^2} \tag{2.144}$$

$$E = \frac{\hbar k^2}{2m} \tag{2.145}$$

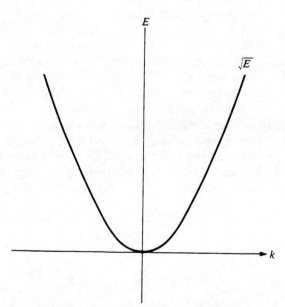

FIGURE 2.12 Free-electron energy.

2.17 BAND THEORY OF SOLIDS 67

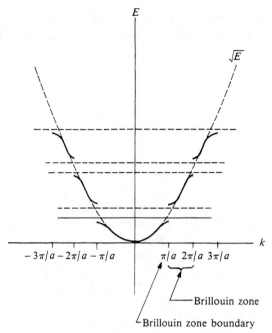

FIGURE 2.13 Energy of lattice electron.

Thus, the energy of the free electron is represented by the relation (see Figure 2.12):

$$k \approx E^{1/2} \tag{2.146}$$

The lattice electrons are diffracted or scattered according to Bragg diffraction discussed earlier, when

$$n\lambda = 2d \sin \theta \tag{2.147}$$

For the primary maxima (i.e., $\theta = \pi/2$ and $\sin \theta = 1$), the diffraction equation becomes

$$n\left(\frac{2\pi}{k}\right) = 2a \tag{2.148}$$

Therefore, when $k = n\pi/a$, Bragg diffraction occurs. Figure 2.13 shows the energy as a function of k. Solving the wave equation for each interval yields the energy of the lattice electrons:

$$E_1 \approx E_n + \tfrac{1}{2}\Delta_n \pm \tfrac{1}{2}A_n\left(k - \frac{n\pi}{a}\right)^2 \tag{2.149}$$

The lattice electrons have energies that nearly approximate those of the free electrons. Near the discontinuities, the lattice electron energy varies as

$$E_l \approx \left(k - \frac{n\pi}{a}\right)^2 \quad (2.150)$$

The reflection at $k = \pm\pi/a$ arises because the wave reflected from the $(p+1)$th atom interferes constructively with the original wave at the pth atom, the difference in phase being just ± 2 for the values of k. This region in k space is called the first Brillouin zone.

Thus, wavefunctions at values of k far removed from Brillouin zone boundaries, $\pm\pi/a$, may be represented as plane waves (e^{ikx}), but as the boundary at $\pm\pi/a$ is approached and reflection becomes imminent, (e^{ikx}) is gradually supplanted by an increasing admixture of the wave $\exp i[k - (2\pi/a)]x$ until, at the zone boundary $k = \pm n\pi/a$, the solution to the Schrödinger equation for this condition becomes:

$$\psi = e^{i\pi x/a} \pm e^{-i\pi x/a} \quad (2.151)$$

The above solution is really two standing-wave solutions:

$$\psi_1 \approx e^{i\pi x/a} - e^{-i\pi x/a} = \sin\frac{\pi x}{a} = \sin kx \quad (2.152)$$

and

$$\psi_2 \approx e^{i\pi x/a} + e^{-i\pi x/a} = \cos\frac{\pi x}{a} = \cos kx \quad (2.153)$$

The two standing-wave solutions result in two alternative energies at the Brillouin zone boundary, $\pm\pi/a$. This can be seen graphically by plotting $|\psi|^2$, the probability distribution of the electron charge, over the positive ion cores (see Figure 2.14). Therefore:

1. A plane wave distributes charge uniformly over the ion cores,
2. The standing wave $\psi_1 \cong \sin \pi x/a$ distributes charge preferentially midway between the ion cores. This gives rise to the overlap in molecular orbitals, and
3. The standing wave $\psi_2 \approx \cos \pi x/a$ distributes charge on the ion cores where V is a minimum.

Consequently, when we calculate the potential energy of the three configurations we find:

$$V_{\psi_2} < V_{\text{plane wave}} < V_{\psi_1}$$

The difference $\Delta V = V_{\psi_1} - V_{\psi_2}$ is the band gap.

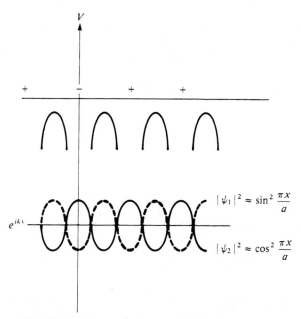

FIGURE 2.14 Eigenfunctions of a lattice electron.

An alternative approach is to bring an array of N free atoms together so that their individual wavefunctions overlap. Because of the exclusion principle, for N atoms there must be formed N levels for each level of the isolated atom. This splitting of levels leads to occupied and forbidden bands and is termed a *tight-binding approximation*. Figure 2.15 shows energy bands obtained by bringing six hydrogen atoms together in a tight-binding approximation. Thus, each energy level becomes highly degenerate and is split into a large number of distinct levels. As the atomic separation reduces, these new levels become very numerous, and their average energy difference is extremely small. It is really acceptable to consider these energy levels to be continuously distributed. Thus, each of the original electron energy levels become bands of permissible energies. These bands also become wide as atomic separation is decreased.

This effect is greatest with the least tightly bound electrons. Thus, the higher energy states are more likely to spread into bands than the lower energy states. Again, this reinforces the idea of pseudoatoms as a core producing a potential that influences the valence electrons.

2.18 SILICON CARBIDE HIGH-BAND-GAP SEMICONDUCTOR

As will be discussed in Chapter 3 on semiconductors, silicon carbide is a high-band-gap semiconductor. The objective of this section is to demon-

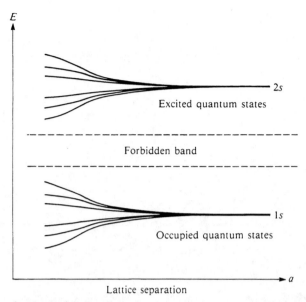

FIGURE 2.15 Energy levels spread into bands.

strate the formation of the band structure by increasing the number of atoms in a model lattice of SiC. The model lattice will be a linear Kronig–Penney model where the Hamiltonian, the eigenvalues, and the eigenvector will be calculated from the Huckel molecular orbital theory. Only pi bonds will be considered for both the silicon and the carbon atoms of the lattice.

The band gap is the difference between the highest occupied molecular orbital (HOMO) and the lowest unoccupied molecular orbital (LUMO). The heteroatom parameters $h = 0.80$ and $k = 0.72$ yield reasonable values near 3.0 eV for the SiC HOMO–LUMO gap. This indicates a slightly lower bond strength for SiC than for an unsaturated C—C molecule. It also correlates with differences in bond strengths published for SiC and C—C. The value of $h_{Si} < h_C$ indicates that the pi electrons of silicon are less tightly bound than those for carbon.

For diatomic SiC there exists only two energy levels for the two bonding pi electrons. For Si_2C_2 there are four energy levels. For Si_7C_7 there are 14 levels (see Figure 2.16). (See Problem 2.8 for an expanded development of the SiC calculation).

The energy between the ground state and the first excited state stays the same, while the number of levels available increases linearly with the number of atoms in the lattice. The energy differences between levels also decrease. These data represent relatively precise calculations of the eigenvalues for the molecular orbitals for this SiC lattice. It can be clearly seen that large numbers of energy levels are available even for this extremely

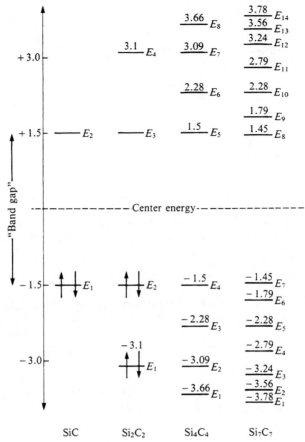

FIGURE 2.16 Kronig–Penney lattice, Huckel molecular orbital model of silicon carbide.

small linear lattice. For real materials with 10^{26} atoms/mole these molecular orbitals become a continuous band of available energy states. This is true for the ground states as well as the excited states.

PROBLEMS

2.1 Consider the numerical errors produced by repeated operation on the number n.

$$b = 1.0000001$$

Write a program or manually carry out 27 squaring operations on n. Do this with both six-digit accuracy and 16-digit accuracy and compare the results. What is the *actual value* of n squared five times. What causes the errors?

CHAPTER 2: QUANTUM MECHANICS AND THE BAND THEORY OF SOLIDS

2.2 Consider the diffraction of electrons from a crystal with an interatomic spacing of 1.5 Å.

   (a) If the energy of the electrons is 100 eV, at what angle does the first maxima occur?

   (b) If the energy of the electrons is 100,000 eV can a diffraction pattern be observed? Why or why not?

2.3 For an electron incident upon the crystal in Problem 2.2, what is the minimum uncertainty in the position of the electron if it is reflected directly back into the incident beam.

   (a) $E = 100$ eV

   (b) $E = 100,000$ eV

2.4 Einstein presented the solution to the photo-electric effect in 1905. He postulated that light has a particle nature defined by the following:

$$E = h\nu$$

Using this relationship he correctly modeled the emission of electrons from a metal when exposed to light.

Explain why extremely bright red light produces no electrons from a particular metal and extremely dim blue starlight could produce energetic electrons.

2.5 Explain the "ultraviolet catastrophe." Why was it important? What did it show about the nature of light? How did Max Planck solve the problem?

2.6 Generate the Hamiltonian for benzene using the assumptions of Huckel molecular orbital theory.

2.7 **(a)** The Balmer Series of Hydrogen represents electronic transitions between atomic orbitals: n to n' where $n' = 2$. Using equation (2.108) calculate the frequency and the wavelength of this series for values of $n = 3, 4, 5, 6, 7, 8$, and infinity.

   (b) The Lyman Series is for $n' = 1$. Calculate the frequency and wavelengths as in **(a)**.

   (c) Which series deals with the higher energies of electron transitions?

   (d) One of the two series of transitions above emits light in the visible and one in the ultraviolet. Which is which?

2.8 The formation of bands of allowed energies can be studied using the molecular orbitals calculated for SiC from HMO theory. Assume that the separation between MOs decreases as the log of the number of atoms in the cluster for the valence band. Do not use the band gap

(HOMO-LUMO gap), use the energies in filled MOs from Figure 2.16 for your analysis.

(a) How many atoms are required to reduce the MO separation to be less than kT for $T = 3$ K.
(b) For $T = 77$ K.
(c) For $T = 300$ K.
(d) For $T = 373$ K.
(e) What is the MO separation for Avagodro's number of atoms?

READING LIST

Azaroff, L. and Brophy, J. J., (1963). *Electronic Processes in Materials*, McGraw-Hill, New York, Chapters: 3, 5, 6, 7.

Bate, R. T., (1968). *Sci. Amer.*, **258**(3), 96.

Bohr, N., (1922). *The Theory of Spectra and Atomic Constitution*, Cambridge University Press, New York.

de Broglie, L., (1926). "Sur le parallelisme entre la dynamiquedupoint material et l'optique geometrique," *J. Phys. Radius*, **7**(1).

Cohen, M. L., Heine, V., and Phillips, J. C., (1982). "The quantum mechanics of materials," *Sci. Amer.*, **246**(6), 82–102.

French, A. P., (1966). *Principles of Modern Physics*, Wiley, New York, Chapters: 1, 5, 7, 8.

Gruenberger, F., (1984). "Computer recreations," *Sci. Amer.* **250**(4), 19.

Hench, L. L. and Dove, D. B., (1971). *Physics of Electronic Ceramics*, Part A, Dekker, New York, Chapters: 1.

Hummel, R. E., (1985). *Electronic Properties of Materials*, Springer-Verlag, Berlin, Chapters: 2, 3, 4, 5, 6.

Kittel, C., (1986). *Introduction to Solid State Physics*, 6th ed., Wiley, New York.

Leighton, R. B., (1959). *Principles of Modern Physics*, McGraw-Hill, New York, Chapters: 1–11.

Löwdin, P. O., (1986). "New directions in quantum chemical calculations—particularly as to the new materials," *Science of Ceramic Chemical Processing*, Wiley, New York, (L. L. Hench and D. R. Ulrich, eds.) pp. 577–582.

Pople, J. A., (1953). "Electron interactions in unsaturated hydrocarbons," *Trans. Faraday Soc.* **49**, 1375.

Pople, J. A. and Beveridge, D. L., (1970). *Approximate Molecular Orbital Theory*, McGraw-Hill, New York.

Rosenberg, H. M., (1978). *The Solid State*, 2nd ed. Oxford Phys. Ser., Oxford Univ. Press, New York and London, Chapters: 1, 2.

Sarnow, K., (1983). "Molecular matters," *80 Micro*, pp. 100–107.

Sherwin, C. W., (1966). *Introduction to Quantum Mechanics*, Holt, Rinehart, and Winston, New York, Chapters: 1, 2, 3, 4, 5.

Slater, J. C., (1968). *Quantum Theory of Matter*, 2nd ed., McGraw-Hill, New York.

Shimony, A., (1988). *Sci. Amer.*, **258**(1), 46.

3

SEMICONDUCTORS

3.1 INTRODUCTION

In the previous chapter, quantum mechanics has shown us that particles act like waves and waves act like particles. Also, we have seen that particles in matter have only discrete allowed energy levels which form into energy bands as the number of atoms increase and as atomic separations are reduced. We have analyzed the solution to the Schrödinger equation for a particle in a square potential. This problem represents reasonably accurately the motion and energy of an electron in a metal. The walls of the square well are analogous to the edges of the metal. Electrons within the metal are represented by standing waves of different electron energies within the square well.

The difference between a metal and an insulator is the separation between the conduction band and the valence band. In a metal they overlap. In an insulator, there is an energy gap between these bands that prevents the valence electrons from jumping to the conduction band. Halfway between these two bands is an energy level known as the *Fermi level*.

A semiconductor is just an insulator with a much narrower band gap. The band gap can be jumped by electrons if enough energy is available. The probability of electrons jumping the energy gap depends on the overall thermal energy of the semiconductor. Generally, as the temperature increases so does the conductivity, since more electrons are thermally excited into the conduction band.

The energy distribution of the electrons is also important. Electrons do not behave like a gas where all energies are allowed. The energy distribution of the electrons is not classical, but is modified by the Pauli exclusion principle.

3.2 THEORY OF SUPERPOSITION

We saw earlier that the superposition of two sine waves of equal amplitude propagating in the x-direction led to a quantized beat frequency. The beats propagate with a speed or group velocity, v_g:

$$v_g = \frac{2\pi f_1 - 2\pi f_2}{k_1 - k_2} \tag{3.1}$$

or

$$v_g = \frac{\Delta \omega}{\Delta \beta} \tag{3.2}$$

where $k_1 = \dfrac{2\pi}{\lambda_1}$

$k_2 = \dfrac{2\pi}{\lambda_2}$

$\omega = 2\pi f$

$\beta = \dfrac{1}{\lambda_1} - \dfrac{1}{\lambda_2}$

When ω_1 and ω_2 are close together, and taking the limit as $\Delta \beta \to 0$, then

$$v_g = \frac{d\omega}{dk} \tag{3.3}$$

Also, the frequency associated with a wavefunction of energy E is

$$E = \hbar \omega = \frac{h}{2\pi} \omega = \frac{h}{2\pi}(2\pi f) = hf = h\nu, \quad f = \nu \tag{3.4}$$

(from Hertz's photoelectric experiments in 1887). Thus,

$$\omega = \frac{E}{\hbar} \tag{3.5}$$

So the group velocity is

$$v_g = \frac{d\omega}{dk} = \left(\frac{dE}{dk}\right)\hbar^{-1} \tag{3.6}$$

or,

$$\frac{dE}{dk} = v_g \hbar \tag{3.7}$$

Now the work dW done on the electron, with v_g, by an electric field **E** in the time interval dt is:

$$dW = e\mathbf{E}v_g\, dt \tag{3.8}$$

However,

$$dW = \left(\frac{dE}{dk}\right) dk = \hbar v_g\, dk \tag{3.9}$$

So comparing the above:

$$e\mathbf{E}v_g\, dt = \hbar v_g\, dk$$

which yields

$$\hbar \frac{dk}{dt} = e\mathbf{E} = F \tag{3.10}$$

Thus, in a crystal where the electron interacts with the lattice and therefore has a standing wave character and an energy associated with its group velocity, then $\hbar\, dk/dt$ is equal to the external force applied to the electron; whereas in free space (or as a traveling wave) it is

$$F = e\mathbf{E} = m\frac{dv}{dt} \tag{3.11}$$

The significant point is that the electron in a crystal is subject to forces from the crystal lattice as well as forces from external sources. Thus,

$$v_g = \hbar^{-1}\frac{dE}{dk}; \tag{3.12}$$

$$\frac{dv_g}{dt} = \hbar^{-1}\frac{d^2E}{dk\, dt} = \hbar^{-1}\left(\frac{d^2E}{dk^2}\right)\frac{dk}{dt} \tag{3.13}$$

and substituting $dk/dt = e\mathbf{E}/\hbar$ from equation (3.10) we obtain

$$\frac{dv_g}{dt} = \frac{d^2E}{dk^2}\left(\frac{e\mathbf{E}}{\hbar^2}\right) \tag{3.14}$$

Thus on comparison with the classical or free-electron equation (3.11)

$$\frac{dv}{dt} = e\frac{\mathbf{E}}{m} \tag{3.15}$$

we see that the mass of the electron in a lattice becomes

$$m = \frac{\hbar^2}{\left(\dfrac{d^2E}{dk^2}\right)} \tag{3.16}$$

Since this factor plays the role of a mass for lattice electrons we call it the *effective mass*, m^*, where we define:

$$m^* \equiv \frac{\hbar^2}{\left(\dfrac{d^2E}{dk^2}\right)} \tag{3.17}$$

By solving for d^2E and integrating, the energy becomes

$$E = \left(\frac{\hbar^2}{2m^*}\right)k^2 \tag{3.18}$$

The physical basis for the difference between m and m^* is the interaction between the electron and the lattice. Thus, m^* is the mass a *free* electron would need in order for the velocity increment under the applied energy pulse to be equal to the actual velocity increment of a *lattice* electron under the same impulse.

Let us examine the consequence of this concept of effective mass, m^*. The total momentum change of the electron with the lattice is

$$\hbar \Delta k = \Delta p_{\text{electron}} + \Delta p_{\text{lattice}} \tag{3.19}$$

and since

$$\lambda = \frac{h}{p} \tag{3.20}$$

then

$$p = \frac{h}{\lambda} = h\frac{k}{2\pi} = \hbar k \tag{3.21}$$

so

$$\frac{\hbar \Delta k}{m^*} = \frac{\Delta p_{\text{electron}}}{m}$$

(conduction   (free electron)
electron)

Consider the E vs. k relation of the energy band shown in Figure 3.1. In region (a) the electron behaves as a plane wave e^{ikx} with $m^* \simeq m$. In region (b)–(c) the reflected component is large and continuously increases until it is equal in amplitude at (d). Thus, at (d)–(e) the amplitudes are equal at

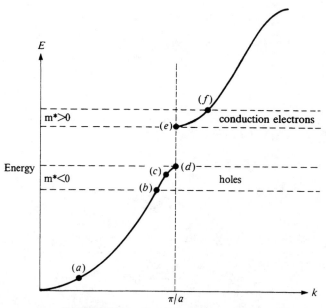

FIGURE 3.1 Lattice electron energy.

$k = \pm\pi/a$ and the eigenfunctions are standing waves and m^* becomes negative. A negative effective mass, $m^* < 0$, means only that going from state k to $k + \Delta k$, the momentum transfer to the lattice is opposite to and larger than the momentum transfer to the electron.

Although k is increased by Δk due to the electric field **E**, the consequent Bragg reflections result in an overall decrease in momentum of the electron, i.e., m^* is described as negative.

As the wavefunction proceeds from region (e)–(f) the amplitude of the reflected wave decreases and m^* assumes a small positive value. The increase in electron velocity from the applied field **E** is larger than a free electron would experience. The lattice makes up the difference through recoil.

Thus, at the bottom of a nearly empty conduction band, the electrons have an effective mass m^* which is nearly equal to m and a negative electric charge. They are therefore termed *conduction electrons,* or just electrons.

At the top of a nearly filled valence band, however, the vacant levels have negative effective masses, $m^* < 0$, and an effective positive charge. They are therefore termed *holes*.

Conduction electrons and holes have an inverse k dependence as shown in the following functional relationships (Figure 3.2).

The energy for electrons is

$$E = \left(\frac{\hbar^2}{2m^*}\right)k^2 \qquad (3.22)$$

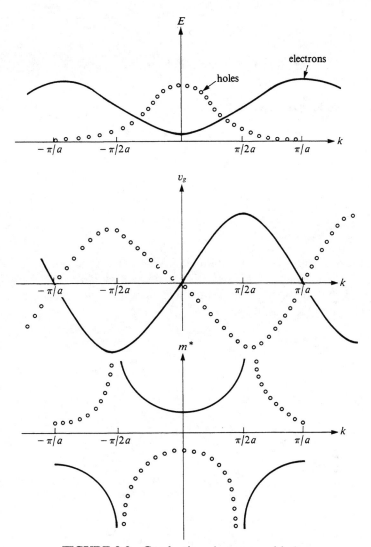

FIGURE 3.2 Conduction electrons and holes.

and the energy of the holes is

$$E = -\left(\frac{\hbar^2}{2m^*}\right)k^2 \tag{3.23}$$

The drift velocity for electrons is

$$v_g = \hbar^{-1}\frac{dE}{dk} \tag{3.24}$$

and for holes is

$$v_g = -\hbar^{-1}\frac{dE}{dk} \qquad (3.25)$$

Finally, the effective mass for electrons is

$$m^* = \hbar^2\left(\frac{d^2E}{dk^2}\right)^{-1} \qquad (3.26)$$

and for holes it is

$$m^* = -\hbar^2\left(\frac{d^2E}{dk^2}\right)^{-1} \qquad (3.27)$$

These relations yield the following schematic representation for the conduction and valence band of semiconductors with electrons at the bottom of the conduction band and holes at the top of the valence band.

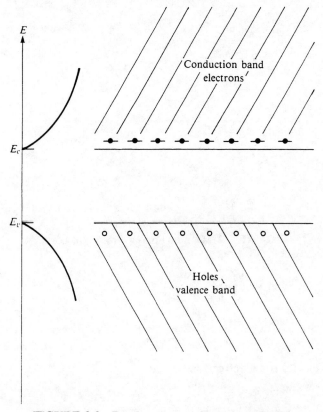

FIGURE 3.3 Band structure of a semiconductor.

The E vs. k relation that corresponds with the electron and hole states is also shown in a shorthand manner in Figure 3.3.

Electrons near the top of the valance band have

$$\frac{d^2E}{dk^2} < 0 \tag{3.28}$$

and therefore,

$$\frac{d\mathbf{v}}{dt} < 0; \quad \text{and consequently } F > 0 \tag{3.29}$$

That is, the applied force F retards these electrons. Consider the effect as one where the electron moves into a state of lower velocity and thereby frees a state of higher velocity. The freed state is termed a hole and can be assigned a positive charge. The field thus drives the hole towards an increase in energy. This is against the field which is the correct direction. The energy of a hole increases from the top of the band down to lower levels. The electric current which results is equivalent to an electron with a positive charge and a positive mass.

3.3 HALL EFFECT

The *Hall effect* makes it possible to measure experimentally the concentration of current carriers and their sign. A typical experimental configuration is as shown in Figure 3.4. In order to describe the Hall effect, we use the complete form of the vector equation known as the *Lorentz force* $\mathbf{F}_L$ on a charge:

$$\mathbf{F}_L = e\mathbf{E} + e\mathbf{v} \times \mathbf{B} \tag{3.30}$$

where e = electron charge
$\mathbf{E}$ = electric field vector
$\mathbf{v}$ = electron velocity vector
$\mathbf{B}$ = magnetic field strength vector

The vectors in the x-, y-, and z-directions are represented by the unit vectors $\mathbf{i}$, $\mathbf{j}$, and $\mathbf{k}$, respectively. [Note that in equation (3.10), only the first term in the vector equation was used in deriving the effective mass of electrons and holes.]

Return to the experimental set up shown in Figure 3.4 and rewrite the Lorentz force in unit vector form:

$$\mathbf{F}_L = eE\mathbf{i} + ev_xB_z(\mathbf{i} \times \mathbf{k}) \tag{3.31}$$

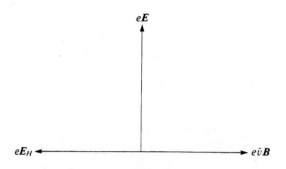

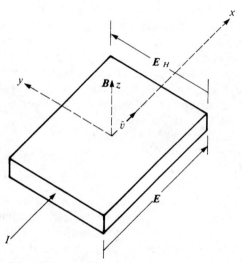

FIGURE 3.4 Experimental configuration for determining the Hall effect.

This reduces to:

$$\mathbf{F}_L = eE\mathbf{i} + ev_x B_z(-\mathbf{j}) \tag{3.32}$$

The applied electric field $E\mathbf{i}$ produces a current I with electron velocity along the x-axis (v_x). The applied magnetic field $\mathbf{B}$ is along the z-axis. The electrons respond to the second term in the Lorentz force by moving in a curved path in the negative y-axis ($-\mathbf{j}$). Thus, charge builds up on the right-hand side of the block. This build up of charge creates an electric field, E_H, the Hall effect. The forces on the electrons must come to equilibrium. Therefore, the Lorentz force balances the Hall force in the y-axis.

$$F_L \mathbf{j} = F_H \mathbf{j} \tag{3.33}$$
$$ev_x B_z(-\mathbf{j}) = eE_H(\mathbf{j}) \tag{3.34}$$

Rearranging and solving for the Hall field:

$$E_H = -v_x B_z \quad (3.35)$$

which is the y-direction. The current I is defined by

$$I \equiv Nev_x \quad (3.36)$$

where N is the number of charge carriers. If we solve for v_x we obtain:

$$v_x = \frac{I}{Ne} \quad (3.37)$$

This can now be substituted into the definition for the Hall field E_H:

$$E_H = -\left(\frac{I}{Ne}\right)B_z \quad (3.38)$$

The *Hall constant* R_H is defined by

$$R_H \equiv \frac{1}{Ne} = \frac{E_H}{IB_z} \quad (3.39)$$

By measuring the physical quantities on the right-hand side of the definition of the R_H, the number of charge carriers can be determined to great accuracy. The sign of the Hall constant indicates whether electrons, $R_H > 0$, or holes, $R_H < 0$, dominate the conduction process.

3.4 ELECTRONIC CONDUCTIVITY

When an electron is excited across the band gap, its motion in the conduction band can be represented by a quasi-free electron in a potential well. The hole on the valence side of the band gap also acts like a particle of positive charge. The hole and the electron can each be assigned with an effective mass and drift velocities can be calculated for each of them.

Calculation of n, the concentration of charge carriers, and μ, the mobility, is required in order to calculate electronic conductivity σ, since:

$$\sigma = |e|(n_e \mu_e + n_h \mu_h) \quad (3.40)$$

where n_e = concentration of electrons
μ_e = electron mobility
n_h = concentration of holes
μ_h = hole mobility

The mobilities have the form of velocity

$$v = at \tag{3.41}$$

where a = acceleration
t = time

but the acceleration is equal to force divided by the mass, thus

$$v = \frac{F}{m}t \tag{3.42}$$

Likewise, the mobilities of each type of charge carrier have the form

$$\mu_e = \frac{e}{m_e^*}t \tag{3.43}$$

and

$$\mu_h = \frac{e}{m_h^*}t \tag{3.44}$$

The effective mass of the charge carrier reflects the electric field of the lattice. Thus the ratio

$$\frac{e}{m^*} = \text{effective acceleration} \tag{3.45}$$

and

$$t = \text{relaxation time between collisions in the lattice} \tag{3.46}$$

Substituting the equations for effective mass into equation (3.40) for electronic conductivity yields:

$$\sigma = e^2 t \left[\frac{n_e}{m_e^*} + \frac{n_h}{m_h^*} \right] \tag{3.47}$$

3.5 BOLTZMANN ENERGY DISTRIBUTION

From the kinetic theory of gases, the ideal gas law and the average kinetic energy of a molecule of a gas can all be described by the *Boltzmann–Maxwell energy distribution*. In an ideal gas, each particle is free to move independent of the state of any other particle. As shown in Chapter 2 the electrons in a quantum mechanical square well potential also can have a distribution of energies. However, the electron energies (eigenvalues) are limited by the boundary conditions imposed. No two electrons can have exactly the same state or energy. This is not true for an ideal gas. Any of the molecules of a gas will have energies near or exactly equal to the average energy:

$$E_{gas} = \tfrac{3}{2}kT \tag{3.48}$$

By analyzing the permutations of available states and the energy of each particle, a distribution of particle energies for an ideal gas is

$$n(E) = g(E)\frac{N}{kT}\exp\left[-\frac{E}{kT}\right] \tag{3.49}$$

where $g(E)$ = density of states
N = total number of molecules
T = temperature, °K
k = Boltzmann's constant = 8.62×10^{-5} eV/°K
E = energy of the molecule

Figure 3.5 shows the Boltzmann distribution function for two different

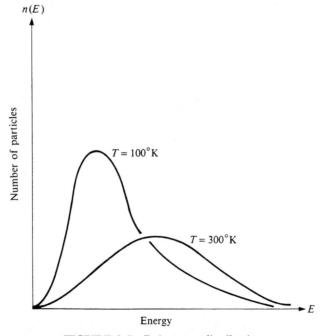

FIGURE 3.5 Boltzmann distribution.

temperatures. This characteristic distribution has a maximum value at some energy greater than zero. However, electrons in a semiconductor behave much differently than molecules in a gas. The description of the behavior of electrons in a solid requires use of the Fermi–Dirac distribution.

3.6 FERMI–DIRAC DISTRIBUTION

The requirement that electrons cannot have the exact same state is known as the *Pauli exclusion principle*. If electrons cannot have the same state, then their distribution will be significantly different than that of an ideal gas.

This distribution has the following form:

$$n(E) = \frac{g(E)}{\exp[(E - E_f)/kT] + 1} \quad (3.50)$$

where

$$g_c(E) = \frac{m_e^*[2m_e^*(E - E_c)]^{1/2}}{\pi^2 \hbar^3} \quad (3.51)$$

for $E \geq E_c$, the conduction band, or

$$g_v(E) = \frac{m_h^*[2m_h^*(E_v - E)]^{1/2}}{\pi^2 \hbar^3} \quad (3.52)$$

for $E \leq E_v$, the valence band, and

$$E_f = \text{Fermi level or Fermi energy} \quad (3.53)$$

Analysis of the Fermi distribution can give insight into the energy dependence of the electrons. At very low temperature where

$$T \approx 0°K \quad (3.54)$$

the quantity

$$\frac{(E - E_f)}{kT} \to -\infty \quad (3.55)$$

for all energies where

$$E < E_f \quad (3.56)$$

Also,

$$\frac{(E - E_f)}{kT} \to +\infty \quad (3.57)$$

for all energies

$$E > E_f \quad (3.58)$$

Thus,
$$n(E < E_f) = 1 \qquad (3.59)$$
and
$$n(E > E_f) = 0 \qquad (3.60)$$

and is represented by the step function in Figure 3.6.

At temperatures greater than absolute zero, the plot of $n(E)$ takes on the curved step function shown in Figure 3.6. The maximum value is at an energy of zero which is significantly different than the Boltzmann distribution.

At an energy of
$$E = E_f \qquad (3.61)$$
the distribution becomes
$$n(E_f) = \tfrac{1}{2}g(E_f) \qquad (3.62)$$

This indicates that the Fermi level lies halfway between the conduction band and the valence band if the density of states is the same for both of these

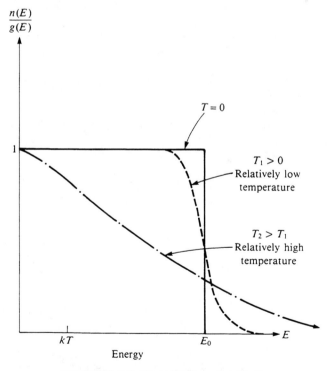

FIGURE 3.6 Fermi–Dirac distribution.

bands. This condition is typical of *intrinsic semiconductors* where the number of electrons is equal to the number of holes.

The band structures for a metal, semiconductor, or an insulator differ by the value of the band gap E_g. For metals

$$E_g \approx 0 \tag{3.63}$$

The schematic structure is shown in Figure 3.7. The work function, $e\phi$, is the energy required to free electrons from the metal.

In Figure 3.8, an intrinsic semiconductor band structure is shown. This band gap energy satisfies the condition where

$$kT \approx 0.02 \text{ eV} \leq E_g \leq 3 \text{ eV} \tag{3.64}$$

It is between kT at room temperature and about 3 eV

In Figure 3.9 the band structure of MgO, an insulator, is shown. The

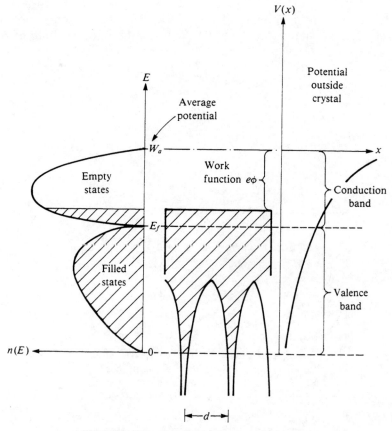

FIGURE 3.7 Fermi distribution for a metal.

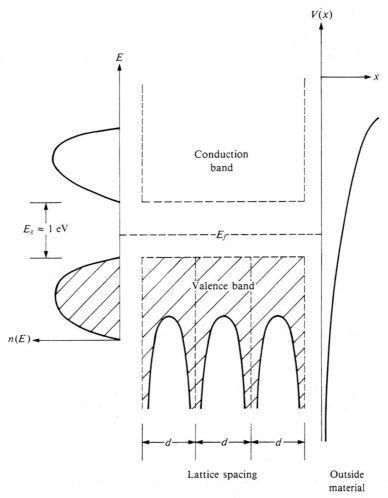

FIGURE 3.8 Intrinsic semiconductor (n = p).

band gap E_g is 8 eV. The dividing line between semiconductors and insulators is generally

$$E_g \geq 3.0 \text{ eV} \tag{3.65}$$

Table 3.1 shows the band gaps of various materials with the distribution between insulators and semiconductors set arbitrarily at 3 eV. Semiconductors with band gaps between 2 to 3 eV are often termed *high-band-gap* semiconductors. If a transition metal is exchanged for the alkaline earth cation, Mg, in MgO, then the band gap in general decreases due to the larger density of states.

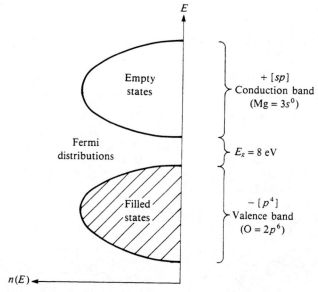

FIGURE 3.9 MgO insulator band structure.

3.7 FERMI LEVEL

We have seen that the number of electrons in an energy state E is the product of the state function $g(E)$ and the Fermi–Dirac distribution $f(E)$:

$$N(E) = g(E)f_e(E) \tag{3.66}$$

The total number of electrons would be integral for all energies above the conduction edge:

$$N_c = \int_{E_g}^{\infty} g(E)f_e(E)\,dE \tag{3.67}$$

This is above the Fermi level such that

$$(E - E_f) \gg kT \tag{3.68}$$

Then the Fermi distribution function $f_e(E)$ can be approximated

$$f_e(E) \approx \exp\left[(E_f - E)/kT\right] \tag{3.69}$$

Substituting this into the integral for the total number of electrons, we

TABLE 3.1 Band Gap of Electronic Materials

Semi-conductor		Band gap E_g (eV)	Insulator		Band gap E_g (eV)
KBr		0.185	Fe_2O_3		3.100
PbTe		0.275	ZnO		3.200
PbS	Galena	0.350	AsCl		3.200
PbSe		0.400	TiO_2	Rutile	3.400
Ge		0.740	ZnS		3.600
Si		1.120	CoO		4.000
GaAs		1.400	NiO		4.200
CdTe		1.450	PbF_2		4.276
CdSe		1.850	Ga_2O_3		4.600
Cu_2O		2.100	BN		4.800
CdO		2.100	UO_2		5.200
GaP		2.250	C	Diamond	5.330
CdS		2.420	CdF_2		6.200
TlBr		2.480	NaF		6.667
ZnSe		2.600	KCl		7.000
$BaTiO_3$		2.800	NaCl	Salt	7.300
AsI		2.800	MgO	Periclase	7.800
AgBr		2.800			
alpha-SiC		2.900	Al_2O_3 (parallel)	Sapphire	8.000
			SiO_2	Fused Silica	8.000
			BaF_2		8.857
			Al_2O_3 (perpendicular)		8.857
			SrF_2		9.538
			MgF_2		10.973
			LiF		12.000
			CaF_2		12.000
			MnF_2		15.500

obtain:

$$N_e = \int_{E_g}^{\infty} g(E) \exp(E_f - E) \, dE \qquad (3.70)$$

$$N_e = \frac{1}{2\pi^2} \left(\frac{2m_e^*}{\hbar^2}\right)^{3/2} \exp\left[\frac{E_f}{kT}\right] \int_{E_g}^{\infty} (E - E_g)^{1/2} \exp\left[-\frac{E}{kT}\right] dE \qquad (3.71)$$

The integration for the total number of electrons in the conduction band becomes:

$$N_e = 4\pi \left(\frac{m_e^* kT}{h^2}\right)^{3/2} \exp[(E_f - E_g)/kT] \qquad (3.72)$$

In order to calculate the number of holes in the valence band, we can logically define a hole as the absence of an electron. Therefore the Fermi distribution for the hole becomes

$$f_h(E) = 1 - f_e(E) \tag{3.73}$$

If the valence band edge E_v is low enough in energy such that

$$(E_f - E) \gg kT \tag{3.74}$$

for all energies

$$E < E_v \tag{3.75}$$

then

$$f_h(E) = 1 - \frac{1}{\exp\left[(E - E_f)/kT\right] + 1} \tag{3.76}$$

This can be simplified by using the binomial expansion

$$(1 + x)^{-1} \approx 1 - x + \cdots \tag{3.77}$$

and using the approximation that the valance band will be well below the Fermi level:

$$f_h(E) \approx \exp\left[(E - E_f)/kT\right] \tag{3.78}$$

This distribution of holes can then be used to calculate the total number of holes in the valence band, just like in the case for electrons:

$$N_h = \int_{-\infty}^{0} g_h(E) f_h(E) \, dE \tag{3.79}$$

Evaluating

$$N_h = 4\pi \left(\frac{m_h^* kT}{h^2}\right)^{3/2} \exp\left[-E_f/kT\right] \tag{3.80}$$

For an intrinsic semiconductor, the number of holes must equal the number of electrons:

$$N_h = N_e \tag{3.81}$$

By substituting the expressions for N_h and N_e, we can obtain an expression for the Fermi level in a semiconductor:

$$4\pi \left(\frac{m_e^* kT}{h^2}\right)^{3/2} \exp\left[(E_f - E_g)/kT\right] = 4\pi \left(\frac{m_h^* kT}{h^2}\right)^{3/2} \exp\left[-E_f/kT\right] \tag{3.82}$$

Simplifying this, we obtain:

$$\exp\left[\frac{2E_f}{kT}\right] = \left(\frac{m_h}{m_e}\right)^{3/2} \exp\left[\frac{E_g}{kT}\right] \qquad (3.83)$$

Taking the natural log of both sides, we obtain:

$$\frac{2E_f}{kT} = \frac{E_g}{kT} + \ln\left(\frac{m_h^*}{m_e^*}\right)^{3/2} \qquad (3.84)$$

Simplifying this, we obtain:

$$E_f = \frac{1}{2}E_g + \frac{3}{4}kT \ln\left(\frac{m_h^*}{m_e^*}\right) \qquad (3.85)$$

This equation gives us the general relation that simplifies to

$$E_f = \tfrac{1}{2}E_g \qquad (3.86)$$

if and only if

$$m_h^* = m_e^* \qquad (3.87)$$

We can see that if the effective masses are different, then the Fermi level moves linearly with temperature. If

$$m_h^* > m_e^* \qquad (3.88)$$

Then the Fermi level moves toward the valence band with increasing temperature. Likewise if

$$m_e^* > m_h^* \qquad (3.89)$$

Then increasing temperature moves the Fermi level nearer to the conduction band.

3.8 EXTRINSIC SEMICONDUCTION

As an electron is excited from the valence band to the conduction band, a hole is left behind. If the Fermi level is close to the conduction band:

$$E_f > \tfrac{1}{2}E_g \qquad (3.90)$$

Then the product of the Fermi–Dirac distribution $f(E)$ and the density of states $g(E)$ gives a larger concentration of carriers $n(e)$, in the conduction

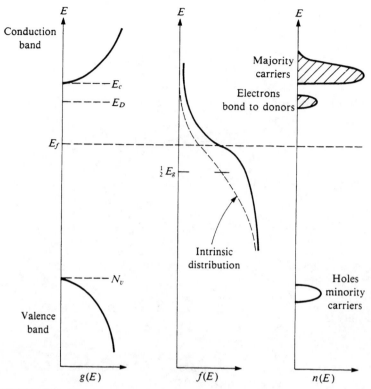

FIGURE 3.10 Charge carrier structure for extrinsic semiconductors (n-type).

band than in the valence band. As the energy increases above the conduction edge E_c, the electron concentration falls off exponentially. Figure 3.10 shows the structure as described above, this type of semiconductor is known as an *extrinsic semiconductor*.

In order to raise the Fermi level near the conductor band, electron donor atoms must be added to the semiconductor. When silicon is doped with a group V element such as As^{5+} or P^{5+}, it creates an *n-type extrinsic semiconductor*. The donor level is usually 0.05 eV below the conduction band edge E_c.

When trivalent elements such as B, Al, Ga, or In are used as a dopant in silicon, one of the covalent bonds cannot be satisfied. Therefore, a nearby silicon atom must transfer a valence electron and the dopant is said to have accepted a valence electron. This process adds holes in the valence band and lowers the Fermi level nearer to the conduction edge. The acceptor level and energy band schematic is shown in Figure 3.11 where

$$E_f < \tfrac{1}{2} E_g \qquad (3.91)$$

and the material is termed a *p-type semiconductor*.

3.8 EXTRINSIC SEMICONDUCTION

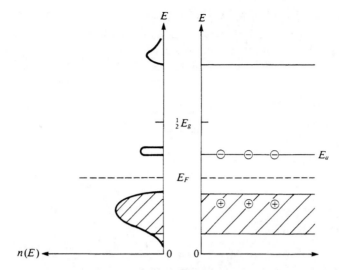

FIGURE 3.11 Acceptor impurities and band structure (p-type).

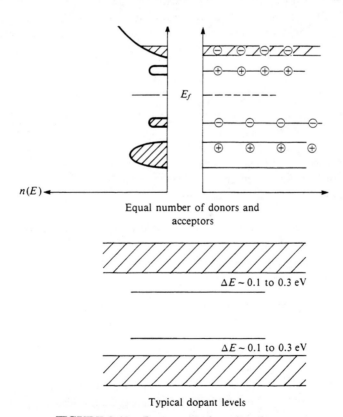

FIGURE 3.12 Compensated semiconductor.

When equal levels of both donor and acceptor dopants are added to a silicon crystal, a *compensated semiconductor* is formed. The band structure for a compensated semiconductor is shown in Figure 3.12. A concentration of acceptors and donors at room temperature of $10^{14}/cm^3$ yields energy levels approximately 0.2 eV above and below the band edges.

3.9 CONCENTRATION OF CARRIERS

The increase in electrons in the conduction level cannot occur without a corresponding reduction in holes. In fact the product of the number of electrons n and number of holes p is a constant:

$$np = \text{constant} \tag{3.92}$$

This relation holds for all doped semiconductors that are in equilibrium and nondegenerate. A *degenerate semiconductor* is defined when the Fermi level is within $3\,kT$ (~ 0.06 eV) of either band edge.

The Fermi level is strongly influenced by the concentration of dopants. Increasing the dopant pushes the Fermi level closer to the band edge. Increasing the temperature moves the Fermi level away from the band edges. These relationships are schematically shown in Figure 3.13.

3.10 POLARONS

The electron freed by thermal or other means of excitation causes polarization in the surrounding lattice. This polarization also occurs around the parent (or ionized) atom. There are generally two types of polarization. The first is electron displacements occurring within 10^{-15} s and the second is lattice displacements occurring in approximately 10^{-13} s. These displacements create energy levels just below the conduction band (see Figure 3.14).

The polarization reduces the potential energy of the excited electrons. The energy levels of the electrons are below the bottom of the conduction band. This level in the band gap is ~ 0.1 eV less than E_c in ionic crystals. The electron takes its state of polarization with it as it moves. New regions polarize while others return to normal. The movement of these charges in a polarized medium are called *polarons*.

3.11 EXCITONS

An electron is transferred to an excited state by a photon or by thermal means. The resulting hole and electron may behave as a single paired system, called an exciton. Less energy is needed than is required to form a free

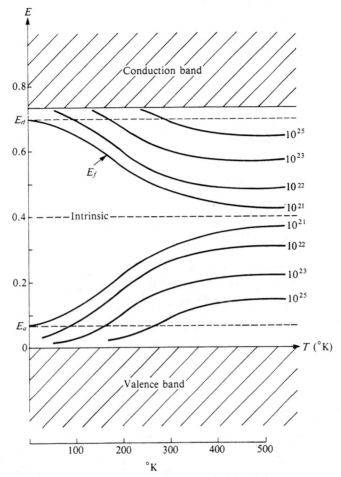

FIGURE 3.13 Fermi level as a function of charge carrier concentration per cm^3 and temperature.

electron and a free hole. The electron remains bound to the hole much like a hydrogen atom and they move together until one becomes bound or the electron and hole recombine releasing light (see Figure 3.15). *Excitons* are discussed in detail in Chapter 10.

3.12 SEMICONDUCTOR SURFACES

At the surface of an n-type semiconductor there exist irregularities where the crystal lattice has been terminated. Bonds that would normally be bridging are exposed and are very active sites. Oxides or nitrides are formed at

98 CHAPTER 3: SEMICONDUCTORS

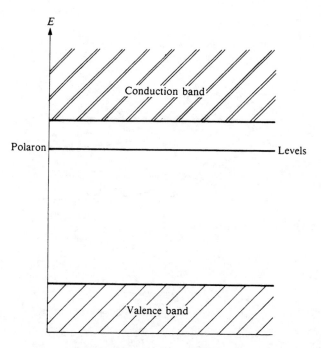

FIGURE 3.14 Polaron levels.

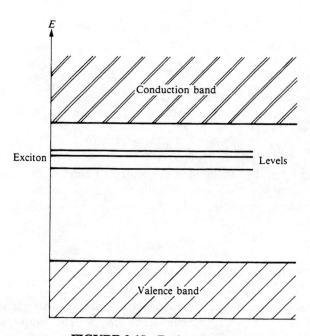

FIGURE 3.15 Exciton levels.

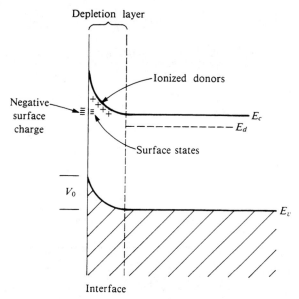

FIGURE 3.16 Semiconductor surfaces (n-type).

the surface which change the band structure. Surface states are created as shown in Figure 3.16 that are basically negative. They form by ionizing donors which lower the edge of the conduction band. A negative surface charge forms on the surface to maintain charge neutrality. The bands are lowered by an energy potential V_0 and this region near the surface is called the *depletion layer*.

When the potential energy V_0 at the surface is greater than the bulk conduction band width, an *inversion layer* is formed. This can cause an electric field to be built up across the interface. The field is due to the large concentration of negatively charged ions adsorbed onto the surface. In order to maintain charge neutrality a correspondingly large concentration of positively ionized donors are formed. This lowers the conduction band so far as to create a p-type surface on an n-type bulk semiconductor (see Figure 3.17a).

The exact opposite happens for a p-type bulk semiconductor if the surface has a large concentration of positive charge (Figure 3.17b). The thickness of the inversion layer that can be created is on the order of one micrometer.

3.13 SEMICONDUCTOR CONTACTS

Nature does not like discontinuities. If a dam breaks, the water does not stay put, but rapidly seeks a new energy level that is continuous and at

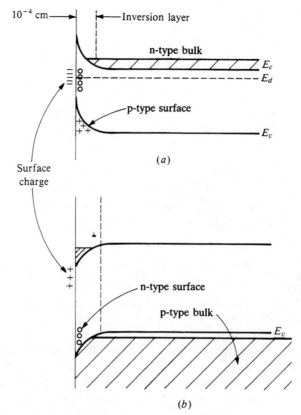

FIGURE 3.17 Inversion layer.

equilibrium. Much the same thing happens when a metal is placed in contact with a semiconcuctor. The energies of the electronic states are different, and adjustments or redistribution of charge must take place before the interface is at equilibrium. The primary measure of the electronic energy of two materials is their Fermi levels or energies. Discontinuities in this energy are not allowed for quantum-mechanical reasons.

Thus, the Fermi levels at a junction must match. Figures 3.18 and 3.19 show the lowering and raising of the Fermi level in n-type and p-type semiconductors, respectively. At the interface the original band edges remain. As we go into the semiconductor the band edges change continuously to their new equilibrium value.

For n-type semiconductors, E_f is near the conduction band and therefore must be lowered to match the Fermi level of the metal (see Figure 3.18). For a p-type semiconductor, E_f is near the valence band and must be raised to come into equilibrium with the metal (see Figure 3.19). The difference in

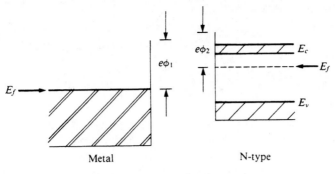

$e\phi \equiv$ Work function

This becomes:

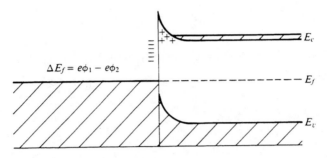

FIGURE 3.18 Rectifying contacts (n-type).

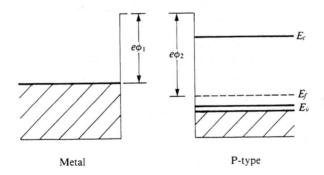

This becomes:

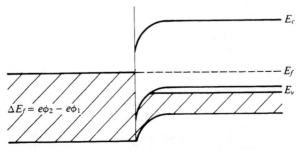

FIGURE 3.19 Rectifying contacts (p-type).

work functions $e\phi$ gives a measure of the difference in E_f that has to occur.

$$\Delta E_f = e\phi_{metal} - e\phi_{semi} \tag{3.93}$$

An ohmic contact can be created between a semiconductor and a metal. This contact allows resistive currents to freely move in either direction across the interface. This may occur if the semiconductor work function is greater than that of the metal. When equilibrium occurs at the interface, the conduction band of the metal overlaps the conduction band of the semiconductor:

$$\Delta E_f < 0 \tag{3.94}$$

If no overlap exists, then a *rectifier* has been created. Figure 3.20 illustrates the condition where

$$\Delta E_f > 0 \tag{3.95}$$

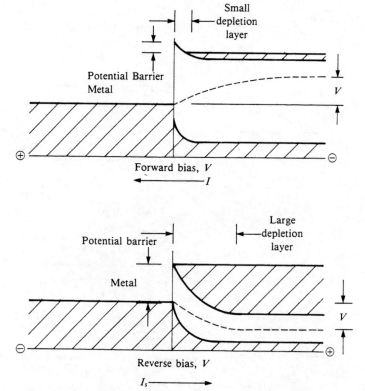

FIGURE 3.20 n-Type rectifier.

and electric current can flow only in one direction, i.e., rectification occurs.

If a potential V is applied on a forward bias (n-type semiconductor negative) as shown in Figure 3.20, a current I will flow easily. The potential barrier is small and the electrons fall to a lower energy level after crossing the barrier. If the bias potential is reversed, then the electrons from the metal must jump a large potential barrier. Only a small current I_s can be maintained for a given reverse bias.

The current in a rectifier obeys the following exponential equation:

$$I = I_s\left[\exp\left(\frac{eV}{kT}\right) - 1\right] \tag{3.96}$$

For the forward bias,

$$V > 0 \tag{3.97}$$

the current increases exponentially with bias voltage (see Figure 3.21).

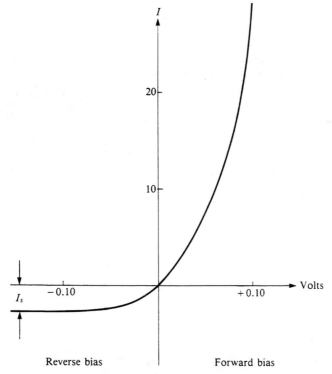

FIGURE 3.21 Rectifier equation.

104 CHAPTER 3: SEMICONDUCTORS

However, in the case of reverse bias, where

$$V < 0 \tag{3.98}$$

The factor $\exp(eV/kT)$ varies as

$$0 \le \frac{1}{\exp\left(\dfrac{eV}{kT}\right)} \ge 1 \tag{3.99}$$

for all values of V. The reverse bias current I_s is the limiting current in this case.

3.14 P–N JUNCTIONS

Consider a junction between a high-conductivity p-type semiconductor where there is a large excess of acceptors in contact with a low-conductivity n-type semiconductor (Figure 3.22). This results in the band diagram shown in Figure 3.23 where the space charge layer originates because of the concentration gradient. The holes tend to diffuse into the n-region and electrons into the p-region (acceptors).

Any net flow of carriers leaves behind a space charge consisting of the impurity ions of the carriers, positive for donor states, negative for acceptors. The space charge builds up to a value where the effects of diffusion are balanced out by the electric field V_0 which is equivalent to a

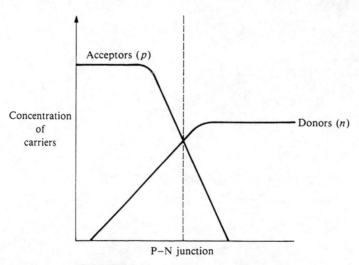

FIGURE 3.22 Linear graded junction.

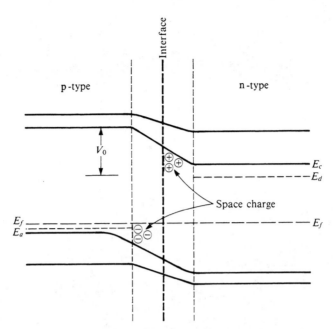

FIGURE 3.23 Band structure.

distortion of the energy levels. As discussed previously, the condition for equilibrium is that the Fermi levels match. The energy bands shift to provide the match. The magnitude of V_0 depends on the width of the forbidden energy gap E_g, impurity concentration of both types, and temperature. At high temperature, both regions become intrinsic, V_0 disappears and the p–n junction disappears. This is called the exhaustion region.

At equilibrium, there are few holes in the n-region, and few electrons in the p-region, but they are important because they provide a current through the junction when small voltages are applied.

3.15 DIODES

A *p–n junction diode* is shown in Figure 3.24. There is only a small flow of holes from the p-region into the n-region balanced by holes in the n-region diffusing thermal fluctuations into the p-region. Therefore,

$$I_r \text{ (recombination current)} \approx I_g \text{ (generation current)} \quad (3.100)$$

When a forward voltage V_a is applied, the holes move into the n-type semiconductor, and electrons into the p-type semiconductor, as shown in Figure 3.25. The number of carriers is high, because E_f moves up due to the

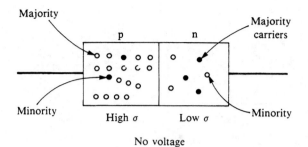
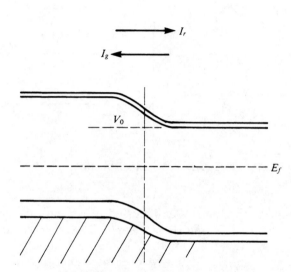

FIGURE 3.24 Diode: p–n Junction.

potential applied. The current flowing from the n- to p-regions is proportional to the equivalent hole concentration P_n in the n-region:

$$I_r = C_1 P_n \tag{3.101}$$

I_r is independent of the applied voltage.

The current from p to n is proportional to the hole concentration P_p in the p-region with sufficient energy to climb the barrier V_0

$$I_g = C_1 P_p \exp\left[-(V_0 - eV_a)/kT\right] \tag{3.102}$$

When

$$V_a = 0 \tag{3.103}$$

Then

$$I_g = I_r \tag{3.104}$$

also

$$P_n = P_p \exp\left(\frac{-V_0}{kT}\right) \tag{3.105}$$

3.15 DIODES

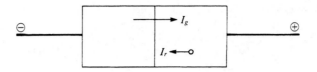
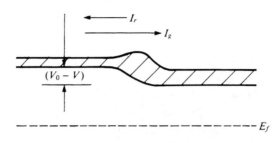

FIGURE 3.25 Forward bias (voltage).

The net hole current I_h is

$$I_h = I_g - I_r = C_1 P_p \exp\left[-(V_0 - eV_a)/kT\right] - C_1 P_p \exp\left(\frac{-V_0}{kT}\right) \quad (3.106)$$

$$I_h = C_1 P_p \left[\exp\left(\frac{eV_a}{kT}\right) - 1\right] \exp\left(\frac{-V_0}{kT}\right) = C_1 P_n \left[\exp\left(\frac{eV_a}{kT}\right) - 1\right] \quad (3.107)$$

The net electron current I_e is

$$I_e = C_2 N_p \left[\exp\left(\frac{eV_a}{kT}\right) - 1\right] \quad (3.108)$$

where N_p is the electron concentration in the p-region. Therefore, the total current I_T becomes

$$I_T = I_e + I_h = I_0 \left[\exp\left(\frac{eV_a}{kt}\right) - 1\right] \quad (3.109)$$

which is the rectifier equation presented earlier (equation 3.96) without derivation, where

$$I_0 = C_1 P_n + C_2 N_p \text{ is the saturation current}$$

with C_1 and C_2 being proportionality constants.

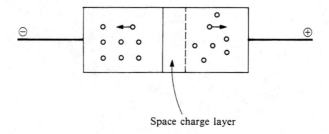

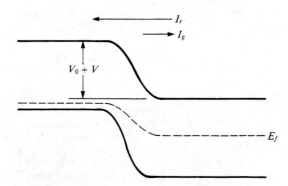

FIGURE 3.26 Reverse bias (nonconduction).

$$I = I_0\left[\exp\left(\frac{eV}{kT}\right) - 1\right]$$

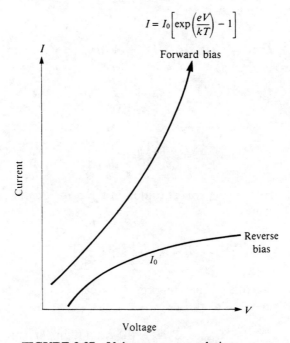

FIGURE 3.27 Voltage–current relation.

The field tends to pull both types of carriers away from the junction and broaden out as well as deepen the space charge layer. The potential difference between the two layers is increased and few holes are excited over the potential barrier (see Figure 3.26).

The reverse current consists only of electrons in the p-type material which diffuse to the junction holes in the n-type material. These come from thermally generated carriers that are pulled away from the junction by the field before they can recombine (Figure 3.27).

3.16 JUNCTION TRANSISTOR

Two parallel p–n junctions in the same crystal separated by less than a minority carrier diffusion length compose a *junction transistor,* as shown in Figures 3.28 and 3.29. In the operation of a transistor, the holes injected

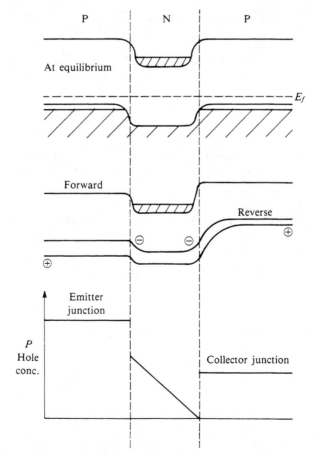

FIGURE 3.28 Junction transistors.

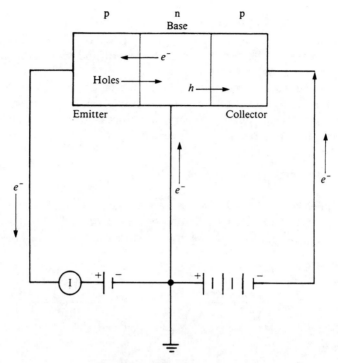

FIGURE 3.29 Transistor.

into the n-type base region at the emitter junction diffuse across to the collector junction where they are collected by the collector field. Variation of the emitter–base voltage changes the injected current correspondingly and this signal is observed at the collector junction.

A forward-biased emitter represents a small resistance and the reverse-biased collector a large resistance. Nearly the same current flows through both and a large power gain results. The input signal is amplified. The ratio of collector current to emitter current is the current gain factor of the transistor.

3.17 GALLIUM ARSENIDE SEMICONDUCTORS

Referring back to Table 3.1, we see that the band gap for silicon is less than that of GaAs (1.1 eV vs. 1.4 eV). Table 3.2 shows various types of semiconductors with their corresponding electron mobilities. GaAs semiconductor devices are more stable with respect to temperature because of the larger band gap. The electron mobility in GaAs is more than five times that of silicon and the electron mobility in InSb is 50 times that of silicon.

Although silicon currently dominates the commercial semiconductor

TABLE 3.2 Mobilities of Various Semiconductors

Semiconductor	Type	Electron mobility (m^2/Vs)	Hole mobility (m^2/Vs)
α-SiC	IV–IV	0.04	~0.02
Si	Elemental	0.15	0.049
Ge	Elemental	0.39	0.18
GaAs	III–V	0.85	0.30
InAs	III–V	3.30	0.02
InSb	III–V	8.00	0.17
Diamond	IV	0.18	0.12
PbS	IV–VI	0.06	0.02
GaSb	III–V	0.30	0.065
CdS	II–VI	0.04	—
CdSe	II–VI	0.065	—
CdTe	II–VI	0.12	0.005

market, the superior electrical properties of compound semiconductors challenge silicon for certain applications. For example, $GaAs_xP_{1-x}$ is used in devices because of electronic transitions which are in the visible range of light. *Light-emitting diodes* (LEDs) are of primary importance. *GaAs lasers* are discussed in Chapter 11.

The crystalline structure of III–V semiconductors is similar to the diamond lattice of silicon except that Ga takes the place of the face and corner atoms while As is substituted for the four interior atoms. This type of structure is called a *zinc-blende lattice*.

The reason for the increased mobility in GaAs is directly related to the reduced effective mass of the electrons.

$$\frac{m_e^*}{m_e}(GaAs) = 0.068 \qquad (3.110)$$

While for silicon

$$\frac{m_e^*}{m_e}(Si) = 1.1 \qquad (3.111)$$

Because of the larger band gap of intrinsic GaAs, the number of charge carriers is two orders of magnitude less than that of intrinsic silicon. This is generally true from room temperature to over 700°C. This improves the performance of GaAs over silicon for high-temperature applications. For comparison, silicon devices can operate at $\approx$200°C while germanium devices are limited to ~100°C. This relates directly to the differences in the band gap of the intrinsic semiconductors.

3.18 SILICON CARBIDE SEMICONDUCTORS

Silicon carbide (SiC) has a high band gap ranging from 2.4 eV to 5.1 eV compared with 1.1 eV for silicon. Consequently, SiC is good for high temperature electronic applications where thermal generation of electron–hole pairs is low. The exhaustion range is much higher, and doped devices can operate at much higher temperature than Si or GaAs. Silicon carbide is chemically inert and stable and can be chemically doped to make p–n junctions. It is superior for microwave applications at 1 GHz, since devices can operate in a velocity activation regime and at high breakdown fields before avalanche breakdown can occur. SiC is also a natural superlattice, compared with the artificial superlattices of III–V compounds, and therefore can potentially be used for molecular quantum well devices. However, SiC devices are more expensive.

Because of its high band gap, SiC can also be used for electro-optic applications, where Si and GaAs are not suitable. For example, blue light-emitting diodes have been developed with a peak emission at 475 mm (see Chapter 10). Also, very high breakdown voltages can be achieved for SiC avalanche photodiodes (APD) as discussed in Chapter 10.

However, silicon carbide has several disadvantages as a semiconductor, primarily the difficulty in achieving inexpensive, reproducible single crystals. This is because of the crystallography of the SiC lattice and a structural phenomenon called *polytypism*.

Silicon carbide exists in three Bravais lattices: closed-packed cubic (β, or zinc-blende structure), hexagonal (α, or wurtzite structure), and rhombohedral. The bonding is tetrahedral, and in real space, each Bravais lattice corresponds to two interpenetrating sublattices of silicon and carbon atoms. The stacking along the c-axis determines the structure that prevails. In the zinc-blende structure, there are three alternating layer pairs of Si and C, so the regular succession of alternating layers in the Bravais lattice is indicated

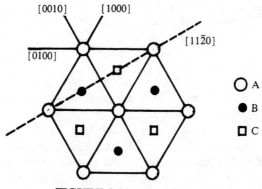

FIGURE 3.30 Crystal lattice.

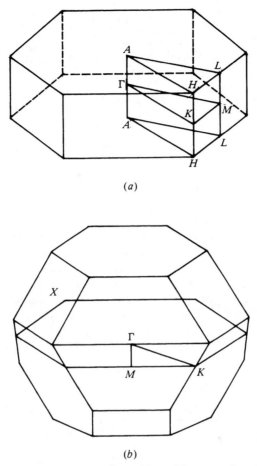

FIGURE 3.31 Brillouin zones for cubic and hexagonal configurations.

by ABCABC.... In the wurtzite structure, there are two alternating layers, ABAB.... For the hexagonal structure, see Fig. 3.30. These basic forms are indicated by 3C and 2H, the number giving the layer pairs of the succession period and C, H, and R referring to the cubic, hexagonal, and rhombohedral forms. The Brillouin zones for the cubic and hexagonal configurations are given in Figure 3.31.

Besides the basic structures indicated above, many other stacking arrangements along the c-axis can occur. Thus, SiC exhibits the phenomenon of polytypism. In the plane perpendicular to the c-axis these polytypes are all similar but in the stacking sequence and cycle along the c-axis they differ. For example, the 15R polytype (Ramsdell notation) has the sequence ABCACBCABACABCB. The 6H stacking, one of the most important technical polytypes in present technology, is ABCACB.

Band structure calculations have been made for only the simplest

TABLE 3.3 Summary of Band Structure for SiC Polytypes

	Polytypes			
	3C	6H	4H	2H
Direct band gaps (ev)	5.14	4.4	4.6	4.46
Indirect band gaps (eV)				
Experimental values	2.39	3.0	3.26	3.35
Theoretical values	2.4 (Γ–X)	2.4 (Γ–M)	2.8 (Γ–M)	3.35 (Γ–K)

polytypes. Such calculations become very cumbersome when the number of atoms in the unit cell increases. Table 3.3 lists some results for four common SiC polytypes.

More than 45 different polytypes are known and there is no theoretical limit to exclude new findings. A schematic arrangement of Si and C atoms in the [11$\bar{2}$0] plane for five different polytypes is depicted in Figure 3.32. The zigzag sequence of sublattices along the c-axis is also often used to characterize the polytypes. Different notations are compared in Table 3.4.

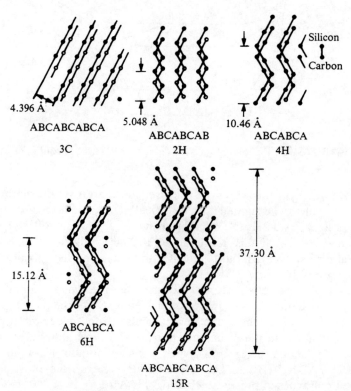

FIGURE 3.32 Silicon carbide polytype structures.

TABLE 3.4 Nomenclature of the Polytypes of Silicon Carbide

ABC notation	Ramsdell	Zigzag sequence	Sequence of inequivalent layer
AB	2H	(11)	h
ABC	3C	(∞)	c
ABCACB	6H	(33)	hcc
ABCACBCABACABCB	15R	(323232)	hcchc
ABCB	4H	(22)	hc

Another complication is that the large period of the potential field in the crystal symmetry of the more complex polytypes suggests that they should be regarded as superlattices. The elementary Brillouin zone has an energy-momentum structure characterized by the reciprocals of the large periods. The E vs. **k** curves have breaks within the Brillouin zone boundaries leading to minibands. For the simple polytypes, the width of the mini-Brillouin zones is sufficiently large that the three-dimensional parabolic band model is still approximately valid. So the transport properties are not much affected by the miniband structure. On the contrary, for complex polytypes with long periods the miniband widths are quite small and the band discontinuities at miniband zones control the scattering and transport properties.

Some of the more simple polytypes can be obtained as single crystals or ingots, thus enabling the fabrication of devices of (e.g.,) 2H or 6H polytypes. Doping is also possible. Silicon carbide is one of the few near-insulating (e.g., large band gap) materials that can be doped equally well as n-type, usually through the introduction of nitrogen donors, or as p-type, usually through overcompensation by the introduction of aluminum acceptors. For a discussion of preparation of SiC crystals and devices see Van Vliet, Bosman, and Hench in the reading list.

The n-type SiC polytypes affect carrier mobility perpendicular to the c-axis, (i.e.,) $\mu_{6H} < \mu_{15R} < \mu_{4H} < \mu_{3C}$. Ionized impurity scattering occurs up to 300°K with $T^{-2.4}$ dependence between 300°K and 800°K due to a mixture of ionized impurity scattering and polar optical phonon scattering. The hole effective mass (m_h^*) varies from $3.0\,m_e$ to $4.4\,m_e$ (where m_e is the free electron mass) whereas the electron effective mass m_e^* varies only $1.0 \pm 0.2\,m_e$.

Table 3.5 compares the electrical properties at 300°K of several n-type SiC polytypes with the nitrogen concentration adjusted to $6 \times 10^{16}\,\text{cm}^{-3}$. The polytypes greatly affect the mobility, ionization energy of the nitrogen donor (E_D), and the aniostropy of the effective mass of the carriers. This is due to the variation in quantum boundary conditions on the electrons that result from the different stacking sequences of the Si and C bonds (see Chapter 2 and Problems 3.7, 3.8, and 3.9).

TABLE 3.5 A Summary of Electron Mobilities, Ionization Energies of Nitrogen Donor (E_D), and Effective Masses of Electrons in n-type SiC ($N_D = 6 \times 10^{16}$ cm^{-3}) at Room Temperature

Polytype	Electron Mobility (cm^2/V.s)	E_D (meV)	$m_\parallel^*/m_e$	$m_\perp^*/m_e$
4H	700	33	0.19	0.21
15R	500	47	0.27	0.25
6H	330	95	1.39	0.35

In contrast to the above data, the temperature dependencies of the conductivity in equally doped p-type 4H, 6H, and 15R SiC is independent of polytype. This is because p-type conductivity is usually direct while n-type conductivity is a consequence of a degree of overcompensation.

Silicon p–n junctions have been used as photovoltaic cells for visible and infrared radiation. However, they show a drop towards the ultraviolet region due to the strong increase of the absorption coefficient, causing the electron–hole pairs to be generated too far from the junction. On the contrary, silicon carbide, with a band gap near 3.0 eV has an absorption coefficient which is several orders of magnitude less than for silicon at blue light of 4000 Å. The maximum response of SiC photovoltaic cells is usually in the near-ultraviolet region.

Silicon carbide structures with linearly graded p–n junctions have been found useful as alpha-particle detectors, neutron counters, and fission product indicators. Using heavy doping tunnel diodes with useful characteristics from 77°K to 673°K have also been manufactured.

For metal semiconductor field effect transistors (MESFETs) a mesa technology is used. An oxide layer of about 400 Å is grown on the carbon face (00$\bar{1}$) basal plane of the p-type epitaxial layer by exposure to wet oxygen. The mesa regions are defined photolithographically. (For a description of SiC MESFET processing see Van Vliet et al., 1988.)

The transconductance of SiC MESFET devices is abut 3.0 mA/V, that is, an order of magnitude lower than for GaAs. This is mainly due to the higher channel resistance. A better g_m is obtained if highly doped regions are used for the ohmic regions of the device. The pinchoff voltage is high, near 10 V, as expected for a SiC device. Table 3.6 gives comparative data for SiC and GaAS MESFETs.

Power devices (output 1 W or more) have also been fabricated. For this purpose, SiC is preferable over GaAs (where a construction of many FETs in parallel with the gate must be used) since natural drain–source breakdown voltages of 80 V for a single FET can be obtained.

TABLE 3.6 Electrical Characteristics of GaAS and SiC MESFETs (dc performance)

Symbol	Parameters and test conditions		Units	GaAs	SiC
ρ/d	Sheet resistance		Ohm/cm^2	461	536
I_{DSS}	Saturated drain current	$V_{DS} = 5, 10$ V $V_{GS} = 0$ V	mA	12	12
V_P	Pinchoff voltage	$V_{DS} = 5, 10$ V $I_{DS} < 100\ \mu$A	V	2.0	10
V_{sat}	Saturation voltage		V	0.8	8.0
R_{SD}	Source–drain resistance		Ohm	20	500
U_{DS}	Source–drain voltage	$I_{DS} < 100\ \mu$A	V	12	>80
U_{GD}	Gate–drain voltage	$I_{DG} < 100\ \mu$A	V	8	80
g_m	Transconductance	$V_{DS} = 5, 10$ V $V_G = 0$	mA V^{-1}	24	3.0

The microwave characteristics of a SiC MESFET diode are quite attractive. Two figures of merit, indicative for microwave behavior, have been suggested in the past:

$$Z_J = (4\pi)^{-1} E_m v_{sat}^2, \qquad (3.112)$$

$$Z_K = \lambda (c v_{sat}/4\pi\epsilon)^{1/2} \qquad (3.113)$$

Here E_m is the breakdown field (3×10^6 V/cm for SiC), v_{sat} the saturated drift velocity, ϵ the dielectric constant, c the speed of light, and λ the thermal conductivity. The second by Keyes (1974) does not include, E_m. These quantities, listed in Table 3.7 show that SiC has very favorable properties as indicated by Z_J and Z_K. A drawback, however, is the smaller lifetime and diffusion length of about 1 μm. This requires the fabrication of devices of smaller dimensions than have been achieved up till now.

3.19 AMORPHOUS SEMICONDUCTORS

Understanding the electronic structure of amorphous materials is extremely difficult due to the lack of periodic boundary conditions as used in Chapter 2. As a result most calculations are based upon one-dimensional models such as a disordered Kronig–Penney model. Disorder is invariably

TABLE 3.7 Figures of Merit for Microwave Semiconductors

		Si	GaAs	SiC
V_{sat}	cm/s	$1 = 10^7$	2×10^7	2×10^7
E_m	V/cm	2×10^5	3×10^5	3×10^6
ϵ	(relative)	12	11	9.7
λ	W/cm.°C	1.5	0.5	3.5
Z_J	V^2/s^2	3.2×10^{23}	2.9×10^{24}	2.9×10^{26}
Z_K	W/s.°C	6.7×10^7	3.5×10^7	2.4×10^8
T_{max}	°C	250	420	>500

assumed to be a strictly random nature which is appropriate to a gas but not to a glass.

Figure 3.33 shows the development of the Kronig–Penney linear model of an amorphous semiconductor. The ordered model potential is represented

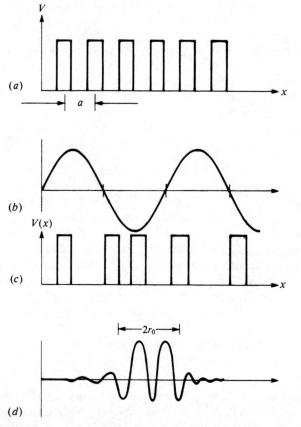

FIGURE 3.33 Ordered and disordered Kronig–Penney Model.

by Figure 3.33a. It is a simply a regular series of square-well potentials with finite walls. This has a characteristic solution

$$\psi \propto e^{ikx} \qquad (3.114)$$

shown in Figure 3.33b.

The amorphous linear model potential is represented in Figure 3.33c. The solution is a wave-packet represented by

$$\psi \propto e^{-\gamma|x|} \sin kx \qquad (3.115)$$

where $2r_0$ is the characteristic dimension of the packet.

This model was originally developed by Professor Sir Nevill Mott of the University of Cambridge. He argued that the observed sharp band edges of crystalline materials are the result of the long-range periodic structure. Thus, in amorphous materials these sharp edges should disappear. Consequently, the Fermi distributions would have extended tails. The overlapping of conduction and valence bands would generally describe the model of a metal. However, Mott's argument continued by assuming that these overlapping bands occurred only on a local scale. The local areas of overlap would be completely analogous to the impurities in crystalline extrinsic semiconductors. The perturbations to a mildly disordered structure cause band tails as shown in Figure 3.34. In a strongly disordered solid, localized tails overlap as shown in Figure 3.35. This leads to localized charge traps where charge carriers can concentrate.

The localized states are separated from the extended states in the main

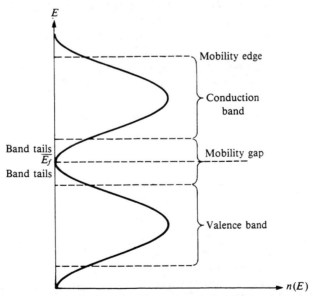

FIGURE 3.34 Extended distribution tails without overlap.

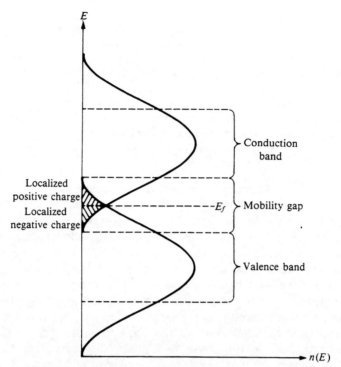

FIGURE 3.35 Extended distribution tails with overlap.

part of the bands by *"mobility edges."* The region lying between the mobility edges of the valence and conduction bands is termed the *"mobility gap."* Under an applied potential, this leads to a redistribution of electric charge by *"hopping,"* as electrons move from one localized state to another, trying to lower their energy. The result is a high density of positively and negatively charged carriers in the traps. This decreases the mobility of the carriers and makes the material less sensitive to efforts to modulate its conductivity by adding impurities.

Once the traps are filled, the mobility increases significantly and conduction occurs. This effect is exactly analogous to the increased conduction of doped crystalline semiconductors above a specific applied potential.

3.20 RADIAL DISTRIBUTION FUNCTION

The atomic structure of glasses and amorphous semiconductors is described by the relative probability of finding an atom a certain distance r_i from a central reference atom. This is termed a *radial distribution function* (RDF). An RDF is obtained by X-ray or neutron, or electron diffraction methods. It is impossible, however, to reconstruct the three-dimensional distribution of atoms from the RDF. Any given RDF can result from an essentially infinite number of three-dimensional structures. Furthermore, when more than one type of atom is present, the assignment of peaks in the RDF to specific atomic correlations is impossible without additonal information. However, qualitative information on the three-dimensional structure can be obtained from the RDF by regarding the atoms as nearly hard spheres and then using geometrical arguments to construct a reasonable mathematical approximation of the lattice.

3.21 AMORPHOUS GERMANIUM

For a particularly useful example, let us consider the work done on *amorphous germanium*. It is found from the area under the first peak of the RDF, which corresponds to nearest neighbor correlations, that the fourfold crystalline coordination and spacing are preserved in the amorphous states. This suggests that the fundamental building block in amorphous Ge is a Ge_4 tetrahedron. These tetrahedra may be connected by having common corners. The fourfold coordination of Ge may be satisfied on an extended scale if the tetrahedra are oriented in either a staggered or an eclipsed configuration. Repeating only the staggered configuration leads to the diamond crystal lattice. Repeating only the eclipsed configuration (with some small distortion) leads to a regular dodecahedron each of whose sides consists of a ring of five atoms. The lack of a periodic structure is attributed to the appearance of these elements with fivefold symmetry. By mixing the staggered and eclipsed configuration, it is found that a structure may be obtained which reproduces the actual RDF fairly well. A configuration midway between the staggered and eclipsed configurations is also possible.

3.22 ONE-DIMENSIONAL MODEL

Because of the difficulty of obtaining an accurate description of lattice sites for amorphous materials, the most reliable calculations of electronic properties are performed on one-dimensional models.

With a perfectly periodic structure, all states are the normal band states (Figure 3.33). As a small amount of disorder is introduced, the number of band states decreases and localized states are formed, having energies at both the top and bottom of the band. As the amount of disorder increases, more states are split off from the band and form localized states. A definite energy E_c (Figure 3.36) limits all states that are localized. If the disorder is sufficient, all states will be localized.

This is considered the tight-binding model, that is, one in which the electronic wavefunctions may be approximated reasonably well as a sum of

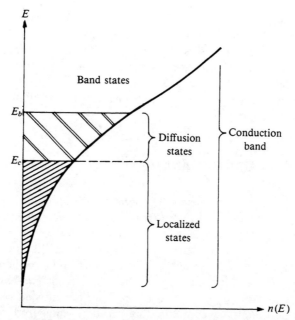

FIGURE 3.36 Distribution tail in partially disordered lattice. E_b = highest energy of diffusion states; E_c = higher energy of localized states.

atomic or molecular wavefunctions. If the quantity

$$\frac{J_{ij}}{\Delta E} \equiv \text{Anderson parameter} \qquad (3.116)$$

is sufficiently small (of the order 1/5), then all states are localized. Here J_{ij} is defined as:

$$J_{ij} = \sum_{n \neq i} \int \psi_i^* V_n \psi_j \, dt \qquad (3.117)$$

with ψ_i the wavefunction of the ith atom or molecule, V_n the potential of the nth atom or molecule, and ΔE the mean spread of energy levels of the electron on each of the atoms or molecules isolated from each other.

The essential reason for this result is that, due to the scattering of the wave from the various centers in which the phase undergoes a random shift, there will be, at best, few sites in the lattice where the scattered waves interfere constructively. If the energy of such a site is just right, then the phases can be adjusted and the wavefunction can blow up. The probability of finding such a site increases much less rapidly with increasing distance than the amplitude of the wavefunction decreases, provided that Anderson's criterion is met. The wavefunction therefore goes to zero at infinite distances.

Now let us consider a partially disordered solid, such as in Figure 3.33c. Considering first the sites which have the smallest energy, we find that these are the least numerous (at least for random disorder), since they fall at the extremity of the energy distribution. Hence, electrons localized on these sites must be the most localized (i.e, the radius of the electron cloud is the smallest), since the probability of finding another site anywhere near the proper energy is smallest because of the low density of these sites. This condition is analogous to the problem of describing impurity states in a semiconductor. If we have a semiconductor with a single impurity (a donor for simplicity), the energy of the donor state decreases as the charge increases and the radius of the state decreases.

Now let us suppose that we are in an energy region E near E_c, the critical energy for electron localization, where band states begin. Again we can identify critical sites. That is, the probability of finding an electron is large in a certain region. We can identify a certain site which causes this modulation of the wavefunction. By this, we mean that if the energy of the site is changed appreciably, the state disappears from our range of energy interest.

We now proceed to calculate the wavefunction of an electron on each site with all other sites so altered as to be outside the range of interest. We can use the wavefunction so generated as a new set of "molecular" functions and apply them in Anderson's theory. We will find that if $\langle J/\Delta E \rangle$ is sufficiently small, our states will be localized. Here J is the overlap integral, defined in equation (3.117), using our "molecular" wavefunctions. By considering sites above and below E_c (the Anderson parameter for these E_c will be exceeded), we can pin down E_c.

We need to consider one additional topic before leaving the structure problem. This is the effect of short-range order on the electronic structure. As we have previously mentioned, disorder in any real glass or amorphous semiconductor is not completely random. If we again limit our attention to the conduction band of an n-type semiconductor, it may be possible to split off a relatively high density of states below the conduction band. By limiting the type of disorder, fairly delocalized wavefunctions may be formed. As an extreme example of this, we can imagine starting with a monatomic crystal and then fractionally increasing every other nuclear charge while decreasing the remaining charges by the same amount. The mean deviation of such a process would be large, however, the result will be eventually to split the conduction band in two. Some related structure may arise in an amorphous semiconductor. In amorphous silicon and germanium, there are probably eclipsed and staggered configurations of tetahedra, and comparatively small distortions of these. If the electronic energies of electrons in these configurations differ moderately, there may result a splitting of the energies into two groups, about which the deviation is small.

3.23 TRANSPORT

We next consider the problem of *transport* in a disordered lattice. Three regimes have been recognized: band conduction, as in crystalline materials, for electrons with energy above E_c; thermally activated hopping among localized states for electrons with energy below E_c; and a diffusion mechanism for electrons with energy just slightly greater than E_c.

The hopping mechanism may be understood by considering a localized electron on one site, and a second site nearby having a slightly different energy. (In order for the electron to hop it will need to gain or lose energy from thermal vibrations.) Due to its localized nature, however, the electron will polarize the lattice. Movement of the lattice distortion at high temperature (in the classical region) also requires energy.

Let us first consider the so-called nonadiabatic hopping. The classical activation energy for a hop when the energy of both sites is the same has been derived from electrostatics by Mott. First, we compute the binding energy of the localized electron. This is the energy lost by the system when the electron polarizes the lattice.

The energy of the electron in polarizing the lattice will be lowered by

$$E_e = -\frac{e^2}{r_0}\left(\frac{1}{k_\infty} - \frac{1}{k_s}\right) \quad (3.118)$$

where k_s is the static dielectric constant (more precisely, to avoid many polaron effects, the dielectric constant measured at 10^{10} Hz or so), k_∞ the infinite-frequency dielectric constant (see Chapter 5), and r_0 the radius of the electronic state (see Figure 3.33). The energy required to polarize the lattice is

$$E_l = \frac{1}{2}\frac{e^2}{r_0}\left(\frac{1}{k_\infty} - \frac{1}{k_s}\right) \quad (3.119)$$

So that the binding energy, the sum of these two terms will be

$$-E_b = E_e + E_l = -\frac{1}{2}\frac{e^2}{r_0}\left(\frac{1}{k_\infty} - \frac{1}{k_s}\right) \quad (3.120)$$

Note that since the binding energy is inversely proportional to its radius, the localized states further from the conduction band will have a greater binding energy.

In order for the electron to move classically, the electron on each of the two sites must have the same energy. The minimum energy required to do this is depicted in Figure 3.37.

Let the electronic energy of the electron on site 1 be zero and on site 2 be

(a) Original site energies

(b) Lattice polarization on site

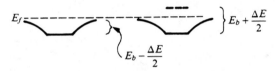

(c) Configuration mid-hop

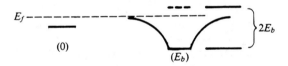

(d) Configuration after hop

FIGURE 3.37 Localized electron hopping model.

ΔE (Figure 3.37a). Then when we allow the electron to polarize the lattice on site 1, the electronic energy is reduced by $2E_b$, while the lattice stores an elastic energy e_b, which means that the total energy of the system is reduced by E_b (Figure 3.37b). In order for the electron to jump to site 2, the electronic energies must be made equal. The minimum energy required to do this is shown in Figure 3.37c. In determining this energy, we have assumed that the change in the electronic energy is proportional to the displacement of the ions, while the energy stored in the lattice is proportional to the square of this displacement.

The energy stored in the lattice site 1 during midhop (Figure 3.37c) is

$$E_1 = \frac{E_b}{4} - \frac{\Delta E}{4} + \left(\frac{\Delta E}{4E_b}\right)^2 \tag{3.121}$$

Likewise, the energy stored in the lattice site 2 is

$$E_2 = \frac{E_b}{4} + \frac{\Delta E}{4} + \left(\frac{\Delta E}{4E}\right)^2 \tag{3.122}$$

If we calculate the difference in energy between this configuration, and the ground state in Figure 3.37b, we find that it is

$$E_H = \frac{\Delta E}{2} + \frac{E_b}{2} + \frac{(\Delta E)^2}{8E_b} \tag{3.123}$$

This will be the classical activation energy for hopping.

With the energy on the two sites equal, the electron will be able to tunnel from site 1 to site 2. The probability for tunneling will be proportional to:

$$\text{Probability} \approx \exp(-2\alpha d) \tag{3.124}$$

where α is the tunneling constant and d the distance between the two sites. Once the electron has moved, the system will return to the configuration shown in Figure 3.37d, and the transition will have been accomplished.

The overall transition rate can be written as:

$$W = v \exp(-2\alpha d) \exp[-(\Delta E + E_b + \Delta E^2/4E_b)/2kT] \tag{3.125}$$

where v is a phonon frequency. For a quantum-mechanical treatment, see Schaake and Hench (1970).

The third term in the activation energy will be significant if $\Delta E > 2E_b$. It merely reflects the fact that if the electron is loosely bound to the lattice, the probability of its gaining large amounts of energy is small.

Equation (3.125) is valid only over a limited range. In particular, consider the case when two sites are close together (adiabatic hopping). Under these conditions the electron may tunnel back and forth between the two sites when their energies are similar; the energy of the electron and hence the halfway configuration is lowered by a resonance process, yielding for the transiton rate:

$$W = v \exp\left[-\left(\frac{\Delta E}{2} + \frac{E_p}{2} + \frac{\Delta E^2}{8E_b} - \tfrac{1}{2}J\right)/kT\right] \tag{3.126}$$

The criteria for this equation is that $J > \hbar\omega_0$ where ω_0 is the optical phonon frequency. J is as defined in equation (3.117).

When the temperatures are sufficiently low, neither of the above equations is valid. This results from the fact that a single phonon process may be favored over a multiphonon process such as considered above. A calculation of the transition rate for single acoustical phonons in the nonadiabatic case has been made by Schaake and Hench (1970), who find

$$W = \frac{E_1^2 \Delta E J^2}{\pi \rho h^4 c^5} \exp[-2S(T)] \coth(-\Delta E/kT) \quad (3.127)$$

where E_1 is the deformation potential, ρ the density, c the velocity of sound, and $\exp[-2S(T)]$ is a complex function of temperature which decreases more rapidly with increasing T as the binding energy is increased.

Physically, this process would correspond to the electron gaining or losing the necessary energy from a single phonon, with the electron and its distortion then tunneling to the new site.

Where the binding energy is very small, such as impurity states in silicon, the exponental term goes to unity, and the results become identical to earlier theories. Also of interest is the fact that equation (3.127) is valid only if

$$\frac{E_b}{2h\omega_0} \coth \frac{h\omega_0}{2kT} \gg 1 \quad (3.128)$$

If the binding energy is very small (again as in Si), equation (3.127) will form a good approximation through an extensive temperature range, although interactions with single optical phonons also need to be considered.

A consideration of the single phonon process for the adiabatic case should make the dependence on the distance between the two sites disappear. This is implied by the distance dependence of the overlap integral:

$$J \propto \exp[-2\alpha d] \quad (3.129)$$

A third hopping region, also adiabatic, may exist. If the sites are extremely close, then the electron may desire to localize itself on two sites with the lattice polarization then surrounding both sites. Under these conditions, transitions would take place between bonding and antibonding orbitals. The transition rate activation energy should be given by an expression in the high-temperature region such as

$$W_H = \nu \exp\left[\left(-\frac{\Delta E}{2} - \frac{\Delta E^2}{8E_b}\right)\bigg/kT\right] \quad (3.130)$$

Here ΔE is the energy difference between the bonding and antibonding orbitals.

Let us summarize this discussion of electron transport in an amorphous semiconductor. For hopping of an electron between two sites, we find that the transition rate will increase as the sites come closer together. If other factors are kept the same the rate will level off at some critical distance. The transition rate increases exponentially with temperature and the activation energy also increase. If the binding energy is small, this increase may not come until quite high temperatures. Finally, in the high-temperature region, the activation energy will be smaller for closer sites than for more distant sites.

Regarding transport in the band states, electrons in them obey laws similar to conventional semiconductors, but with a much lower mobility, due to the large amount of scattering caused by the disorder.

For band states just above the critical energy for localization, E_c, the electronic wavefunction will be highly modulated, with comparatively large probabilities of finding the electron in some regions and small probabilities in the regions connecting them. In this diffusion region, a mobility equation such as

$$\mu = \frac{ea^2}{kT} \nu_{el} \qquad (3.131)$$

is appropriate, where a is the distance between the maxima in the probability, and ν_{el} is an electronic frequency, i.e., the frequency of the electron in a maxima region.

An additional factor must be taken into consideration in these diffusion states. If the dielectric constant of the material is high and localization small then there may be appreciable interaction between the lattice and the electron. This leads essentially to polaron states. From this, we conclude that as the temperature increases, E_c increases, i.e., more states become localized.

Finally, if the semiconductor is not degenerate, that is, if the Fermi level is located in a band gap, then an additional factor of $\exp(-E_g/2kT)$ will enter into the conductivity expression, reflecting the population of carriers at the bottom of the conduction band. The meaning of ΔE for a rapidly changing density of localized states with increasing energy is complicated, however.

The ac conductivity of this model can be considered a function of relaxation time τ. Each hop will contribute to the conductivity through the factor

$$\sigma_{hop} \propto \frac{\omega^2 \tau}{1 + \omega^2 \tau^2}, \quad \text{with } \tau = \frac{1}{W} \qquad (3.132)$$

Hops can occur between distant neighbors with a large τ, and between near neighbors with a small τ. On summing all possible hops, it is found that

$$\sigma_{ac}(\omega) \propto \omega^{0.8} \quad (3.133)$$

The above conductivity is due to nonadiabatic hopping and is additive to the dc conductivity. An additional contribution may come from adiabatic hopping. Since all such hops occur with approximately the same transition rate, if $\sigma < W$, we find a relation where

$$\sigma_{ac}(\omega) \propto \omega^2 \quad (3.134)$$

The temperature dependence occurs through τ, and through a site population factor, $\exp[(-E - E_F)/kT]$.

If the glass is heterogeneous, that is, consists of two or more phases, then additional mechanisms are possible (see Schaake and Hench). The conductivity is given by a complex mixture equation. If the conductivity of the isolated (precipitated) phase is greater than the matrix phase at high temperatures and smaller at low temperatures, then the activation energy of the composite is found to obey a simple mixture formula. When the conductivities are equal the activation energy of the matrix phase dominates at high and low temperatures.

Two new ac effects may be introduced by the heterogeneities. Maxwell–Sillars heterogeneous dielectric losses occur (see Chapter 5) when the ratios of the dielectric constants and conductivities of the two phases differ. The distribution of the electric field at high frequencies is determined primarily by the electrical conductivities and dielectric constants of the two phases. If the ratios of these terms differ, a relaxation region going from one regime to the other will be observed. If the concentration of the precipitate is very small, the real ac conductivity will obey an equation such as equation (3.133).

The second mechanism occurs if the precipitated phase has greater conductivity than the matrix phase, and if it is surrounded by a barrier through which the carriers must tunnel. If these conditions are met then the dc path conductivity will be dependent on frequency. Above a critical frequency the carriers will be less and less able to complete their tunneling through the barrier. An appropriate expression is

$$\sigma_{dc\,path}(\omega) = \sigma_{composite} + (\sigma_{matrix} - \sigma_{composite})(1 + \omega^2 \tau_0^2)^{-1} \quad (3.135)$$

where τ_0 is the tunneling time. This equation predicts a decreasing conductivity with increasing frequency when $\omega > 1/\tau_0$.

The complex part of the conductivity has not been considered, since for each of these mechanisms it is implied to follow from the Kramers–Kronig relations (see Chapter 8).

3.24 MATERIALS

The electrical conductivity of most amorphous materials is insensitive to small amounts of cation impurities (several percent). This is in marked contrast with the situation in crystalline semiconductor where impurities in the parts per million range cause orders of magnitude change in the conductivity. This is attributed to the ability of the amorphous structure to satisfy the bond requirements of the impurities. It may also be a result of the effects of the impurities being swamped by defects produced by the amorphous state.

The most frequently studied amorphous materials which exhibit semiconducting properties fall into one of three categories: elemental glasses (Ge, Si, As, Te, C, B, Sb, Se), chalcogenide glasses (compounds containing S, Se, and/or Te), and transition-metal-oxide glasses (V, Ti, and Fe oxides in particular).

Some conductivity data on several amorphous semidonductors are shown in Figures 3.38 and 3.39. The high temperature dc activation energy in the elemental and chalcogenide glasses is attributed mainly to the excitation of carriers across the forbidden gap. These activation energies correspond reasonably well to half the optically measured gap energy. In amorphous Ge and Si, the measured activation energy drops with decreasing temperature. This drop is generally attributed to a hopping of carriers in

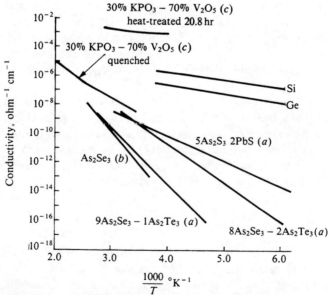

FIGURE 3.38 DC conductivity of some amorphous semiconductors. (*a*) Owen and Robertson (1970). (*b*) Lang (1968). (*c*) Schaake and Hench (1970).

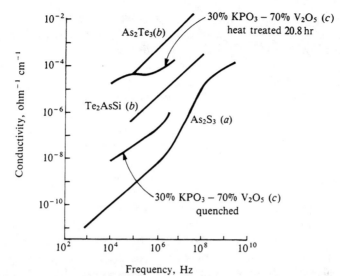

FIGURE 3.39 AC room temperature conductivity of some amorphous semiconductors. (a) Owen and Robertson (1970). (b) Rockstad (1970). (c) Schaake and Hench (1970).

localized states near the center of the gap. These states arise from overlap in the localized tails from the top of the valence band and the bottom of the conduction band.

The ac conductivity in the elemental and chalcogenide glasses is frequently found to have two regions. One is proportional to ω and at higher frequencies it is proportional to ω^2 (see data for As_2S_3, Figure 3.39). The linear region is evidence of nonadiabatic hopping. The ω^2 region is probably due to adiabatic hopping to nearby sites.

Optical measurements on V_2O_5–P_2O_5 glasses indicate a band gap width of 2–2.5 eV. The high temperature dc activation energy is found to be from 0.2 to 0.6 eV, indicating extrinsic conduction, i.e., conduction by carriers from impurities.

In the V_2O_5–P_2O_5 glasses, the ac conductivity shows linear and quadratic regions but it is at most very slightly dependent on temperature. This is indicative of a very small binding energy of the adiabatic hopping of the polaron for most of the localized states. It is of the order of a few hundredths of an electron volt at most. If we go to a heterogeneous structure where we have a matrix in which the binding energy is high and it controls overall conduction, we would need to introduce considerable disorder in the matrix to accomplish such localization.

Since the activation energy also decreases to a few hundredths of an electron volt at low temperatures, this latter explanation is unsatisfactory. Furthermore, since the ac conduction is only slightly dependent on temperature, the majority of the activation energy cannot be attributed to the thermal population of localized states. Therefore, it appears that the amorphous V_2O_5–P_2O_5 semiconductor is degenerate and that the activation energy is due to thermal activation over barriers. It may also be due to regions in the glass having band conduction states a few tenths of an electron volt above the Fermi level. At low temperatures, hopping in the neighborhood of the Fermi level would dominate in these regions. These two cases would be particularly attractive if a large number of states are split off from the conduction band as previously described. Optical absorption evidence is inconclusive on this.

Thermopower results favor a hopping mechanism. The effect of a heterogeneous electronic structure as postulated above on the thermopower has not been considered theoretically, however.

Heat treatment of a quenched 30% KPO_3–70% V_2O_5 glass has been found to produce a marked phase separation. The electrical properties of these glasses are found to obey the properties discussed under the conduction mechanisms of heterogeneous glasses (see Schaake and Hench 1970).

The low frequency ac conductivity (Figure 3.39c) of the heat-treated glass is found to fit a Debye relaxation process with a single well-defined relaxation time. This relaxation time is found to have an activation energy

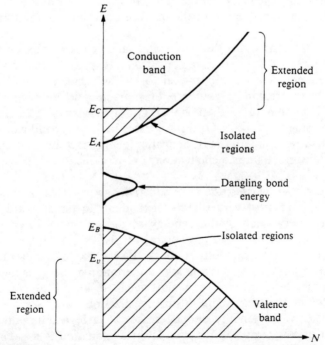

FIGURE 3.40 Amorphous heterogeneous semiconductor band structure.

equal to the activation energy of dc conduction. This is strong evidence of Maxwell–Wagner losses (see Chapter 5). Presumably, these losses mask the nonadiabatic hopping which is probably also present.

At high temperatures, the dc activation energy has a well-defined break. At temperatures above this break, the dc path conductivity is found to decrease with increasing frequency. It decreases to a value at 4 MHz equal to that expected by a continuation of the low-temperature activation energy. This is in agreement with equation (3.135) where the break in the dc activation energy is attributed to the onset of the conductivity in the isolated phase exceeding the conductivity in the matrix phase. At the same time the decreasing conductivity with increasing frequency is due to a tunneling process through a barrier surrounding the isolated phase.

Figure 3.40 shows a schematic energy band structure for a heterogeneous amorphous semiconductor that is consistent with the above experimental results and theoretical development. The frequency and temperature dependence of transport properties is dependent upon the dimensions of the isolated regions shown in Figure 3.40 and the relative magnitude of energy barriers between the regions.

PROBLEMS

3.1 (a) What temperatures would correspond to the maximum band gap considered to be the upper limit of intrinsic high band gap and low band gap semiconductors?

(b) What is the wavelength of light that corresponds to these upper limits?

3.2 (a) Why must Fermi–Dirac statistics be used to describe electron behavior of solids rather than Maxwell-Boltzmann statistics?

(b) Plot the density of states $g(E)$ for silicon and GaAs for $E_c < E < 4\,eV$.

(c) Discuss the effect of $g(E)$ for Si and GaAs on device function.

3.3 (a) Plot the energy, E, of a free electron for values of k between zero and 3.

(b) What is the range of electron de Broglie wavelength for the above range of k.

3.4 (a) Assume that the effective mass of a hole in intrinsic silicon is 0.59 of the mass of the electron. What is the Fermi level at 300 K?

(b) At 3 K?

(c) At 373 K?

3.5 Using equations (3.80) and (3.70) prove that equation (3.92) is true, i.e., the product of the number of holes, N_h, and the number of electrons, N_e, in a semiconductor is constant for a specific band gap and temperature.

3.6 (a) Silicon doped with phosphorous, P^{5+}, creates a semiconductor. What is the effective mass ratio of holes to electrons at room temperature?

(b) Compare this with intrinsic silicon.

(c) What type of semiconductor has this become?

3.7 (a) What are polytypism boundaries in SiC?

(b) Sketch the change in stacking sequence between 3C and 2H boundaries.

3.8 (a) Draw a diagram comparing the free electron density of states curve (i.e., 3D gas) with the density of states for electrons restricted in one dimension L (i.e., 2D gas).

(b) How large must L be before it becomes insignificant?

(c) Relate L to the behavior of SiC semiconductors.

3.9 (a) For the boundary described in Problem 3.8, sketch a diagram that represents the density of states for SiC with polytypism boundaries of distance $L = 100$ Å.

(b) For $L = 1000$ Å.

3.10 At room temperature:

(a) For silicon the number of charge carriers $n_i = n_e = n_p = 1 \times 10^{10}$ per cc. What is the conductivity?

(b) Calculate the conductivity for Germanium if $n_i = 1 \times 10^{13}$ per cc.

(c) Calculate the conductivity for GaAs if $n_i = 1 \times 10^7$ per cc.

(d) Calculate the conductivity for alpha-SiC if $n_i = 6 \times 10^{16}$ per cc.

(e) Compare these calculated values with the literature.

READING LIST

Adler, D., (1977). *Sci. Amer.*, **236**(5), 36.

Azaroff, L. and Brophy, J. J., (1963). *Electronic Processes in Materials*, McGraw-Hill, New York, Chapters: 3, 5, 6, 7, 8, 9, 10.

Blatt, F. J., (1968). *Physics of Electronic Conduction in Solids*, McGraw-Hill, New York, Chapters: 1, 3, 4, 8, 9.

Elliott, S. R., (1983). *Physics of Amorphous Materials*, Longman, London, Chapters: 3, 5.

Hench, L. L. and Dove, D. B. (1971). *Physics of Electronic Ceramics*, Part A, Dekker, New York, Chapters: 2–10.

Hummel, R. E., (1985). *Electronic Properties of Materials*, Springer-Verlag, Berlin, Chapters: 7, 8.

Keyes, R. W., Marshall, R. C., Faust, Jr., J. W., and Ryan, C. E., (1974). University of South Carolina Press, Columbia, SC. 534.

Kittel, C., (1986). *Introduction to Solid State Physics,* 6th ed., Wiley, New York, Chapters: 11, 12, 13, 14.

Lang, J. E., (1968). Ph.D Thesis, Rensselaer-Polytechnic Institute.

Owen, A. E. and Robertson, J. M., (1970). *J. Non-crystalline Solids,* **2,** 40.

Rockstad, H. K., (1970). *J. Non-crystalline Solids,* **2,** 86.

Schaake, H. F. and Hench, L. L., (1970). *J. Non-crystalline Solids,* **2,** 292.

Slater, J. C., (1968). *Quantum Theory of Matter,* 2nd. ed., McGraw-Hill, New York.

Van Vliet, C. M., Bosman, G., and Hench, L. L., (1988). "New Perspectives on Silicon Carbide: An overview with special emphasis on noise and space-charge-limited flow," *Ann. Rev. Mater. Sci.,* **18,** 381–421.

4

IONIC AND DEFECT CONDUCTORS

4.1 INTRODUCTION

In order for a current to flow in either a crystalline ceramic or an amorphous glass, charge must somehow be transported. Electrons may move through the lattice in one of two ways: either by moving with nuclei, in which case ionic conductivity (σ_i) occurs, or by becoming detached from one atom, and moving to another, in which case electronic conductivity (σ_e) occurs. It is also possible, of course, that both electronic and ionic conduction processes take place simultaneously. The *transference number*, $t_x = \sigma_x/\sigma_T$, expresses the fraction of total conductivity (σ_T) contributed by each type of charge carrier. Thus, if $t_i = 0.9$ then 90% of the charge is by ionic transport and 10% is electronic since $t_i + t_e = 1.0$. The transference number of cations t_i^+, anions t_i^-, and electrons t_e^-, or holes t_h^-, in several compounds is given in Table 4.1. The table shows that for most crystal compounds or glasses, one type of charge carrier dominates the electrical conductivity of the material.

As shown in Chapter 3, the electrical conductivity σ of any material is defined as the ratio of the current density $\mathbf{j}$, to the electric field $\mathbf{E}$, required to produce the density:

$$\sigma = \frac{\mathbf{j}}{\mathbf{E}} \tag{4.1}$$

TABLE 4.1 Transference Numbers of Cations t_{i+}, Anions t_{i-}, and Electrons or Holes $t_{e,h}$ in Several Compounds

Compound	Temperature (°C)	t_{i+}	t_{i-}	$t_{e,h}$
NaCl	400	1.00	0.00	
	600	0.95	0.05	
KCl	435	0.96	0.04	
	600	0.88	0.12	
KCl + 0.02% CaCl$_2$	430	0.99	0.01	
	600	0.99	0.01	
AgCl	20–350	1.00		
AgBr	20–300	1.00		
BaF$_2$	500		1.00	
PbF$_2$	200		1.00	
CuCl	20	0.00		1.00
	366	1.00		0.00
ZrO$_2$ + 7% CaO	>700	0	1.00	10^{-4}
Na$_2$O·11Al$_2$O$_3$	<800	1.00(Na$^+$)		$<10^{-6}$
FeO	800	10^{-4}		1.00
ZrO$_2$ + 18% CeO$_2$	1500		0.52	0.48
+ 50% CeO$_2$	1500		0.15	0.85
Na$_2$O·CaO·SiO$_2$ glass		1.00(Na$^+$)		
15% (FeO·F3$_2$O$_3$)·CaO·SiO$_2$·Al$_2$O$_3$ glass	1500	0.1(Ca^{2+})		0.9

While in the most general case σ is a *second-rank tensor*,† since both **j** and **E** are vectors, because glass and polycrystalline ceramics are, to first order, isotropic in their properties, we need only consider it as a scalar. The current density j is the electric charge crossing a unit area per second. The charge q on a carrier is the product of its valence times unit charge e: $q = Ze$. For electrons q is -1, for holes it is $+1$. For ions, q typically can vary from -1 to -2 or $+1$ to $+4$ depending upon the valence of the anion $-Z$ or cation $+Z$.

If the velocity of the carrier is v and there are n identical carriers per unit volume then the current density will be

$$j = nqv \qquad (4.2)$$

We now examine how the velocity v is related to the electric field **E**. For electrons in a conduction band, the classical free-electron model can be used

---

† A "zeroth"-rank tensor is a scalar, while a first-rank tensor is a vector. A second-rank tensor either (1) relates one vector linearly to another, as in the case of electrical conductivity, where an applied electric vector field induces a vector current density which is proportional to the applied electric field (i.e., doubling the applied field without changing its direction results in a doubling of the magnitude of the current density) but which is not necessarily in the same direction as the electric field; or (2) relates a second-rank tensor (such as strain) to a scalar such as temperature. An example of the latter is the thermal expansion tensor. Second-rank tensors are generally represented by 3 × 3 matrices (three rows and three columns) and are manipulated by matrix algebra.

with an argument that the electron will be accelerated by the field (see Chapter 2) and therefore its velocity at any time t will be changed from its initial random value v_0 by an amount

$$\Delta v = (\text{acceleration} \times t) = \frac{q\mathbf{E}}{m}t \tag{4.3}$$

where m is the mass of the electron (or more properly, the effective mass, m^* in which factors introduced by quantum mechanics are taken into account). However, the electron will undergo this acceleration only for a very short increment of time before it collides with an atom and its random velocity is restored. If the mean time between collisions is τ, the average velocity of an electron is independent of time and is given by

$$\langle v \rangle = \langle v_0 + v(t) \rangle = 0 + \frac{q\mathbf{E}\tau}{2m} \tag{4.4}$$

Combining equations (4.1), (4.2), and (4.4) we find that

$$\sigma = \frac{nq^2\tau}{2m} \tag{4.5}$$

Equation (4.5) is generally written as

$$\sigma = nq\mu \tag{4.6}$$

where μ is the mobility defined by

$$\mu \equiv \frac{q\tau}{2m} \tag{4.7}$$

While equation (4.7) is based on a simple model, a full scale quantum-statistical mechanical treatment yields similar results.

Another case of interest is that of ions which hop from site to site. A relation between the ordinary diffusion constant D_i and the mobility μ can be derived from the Boltzmann transport equation. This yields the so-called *Einstein relation*:

$$\mu = \frac{eD_i}{kT} \tag{4.8}$$

The essential argument in deriving equation (4.8) is that the electric field only slightly perturbs the thermally activated diffusion process. This leads to a slightly larger number of ions moving in the direction of the field than against it (or vice versa, depending on the sign of the charge of the ion). The T^{-1} dependence of equation (4.8) is very slight, and is usually swamped out by the exponential temperature dependence of the diffusion coefficient, $D = D_0 \exp[-Q_D/kT]$. Consequently, a general expression for ionic conduction will be of the form

$$\sigma = nq \frac{e}{kT} D_0 \exp\left[-\frac{Q_D}{kT}\right] = \sigma_0 \exp\left[-\frac{Q_D}{kT}\right] \quad (4.9)$$

4.2 THEORY OF IONIC CONDUCTIVITY

Ionic conductivity involves the long-range migration of ionic charge carriers through the material under the driving force of an applied electric field. The charge carriers will be the most mobile ions in the material. Therefore for silicate glasses, the mobile ions are usually +1 valence cations moving in an immobile SiO_2-based matrix. Alkali ion migration in grain boundaries is often the dominant ionic conduction mechanism in polycrystalline ceramics. In fast ion conductors, it is also alkali ions that move through the ordered ionic defect lattice. The electric force on the cation serves to perturb its random thermal motion by increasing the probability of a transition in the direction of the applied field. This behavior can be described easily by considering the two schematic potential energy wells shown in Figure 4.1. The potential barrier V represents the maximum barrier in the path of least resistance in the glass or ceramic.

Consider that the ions move in one dimension parallel to the x-axis jumping over the potential barrier V. The probability that an ion will move

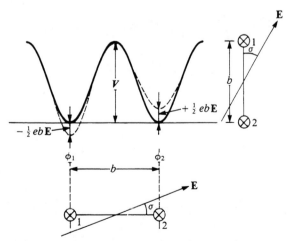

FIGURE 4.1 Potential well energy configuration in a glass or ceramic. Solid lines indicate energy without field; dashed lines indicate energy with field **E** applied.

to either the right or the left is

$$P = \alpha \frac{kT}{h} \exp\left[-\frac{V}{kT}\right] \qquad (4.10)$$

where α is an accommodation coefficient related to the irreversibility of the jump; kT/h is the vibrational frequency of the ion in the well, k is Boltzmann's constant, and T is absolute temperature. When a field $\mathbf{E}$ is applied the ion coordinations are slightly distorted within the glass structure and the potential barrier to the ion motion is slightly shifted as shown in Figure 4.1. The field will lower the potential barrier on the side in the direction of the field and raise it on the other side by an equivalent amount. If the distance between the two wells is b, the potential on the right will now be smaller by an amount $\frac{1}{2}ze\mathbf{E}b = \frac{1}{2}Fb$, where F is the force on the ion and z is the valence on the ion. Therefore, the probability of motion to the right is

$$P^+ = \tfrac{1}{2}\alpha \frac{kT}{h} \exp\left[-\frac{V - \tfrac{1}{2}Fb}{kT}\right]$$

By using equation 4.10, we obtain

$$P^+ = \tfrac{1}{2}P \exp\left[+\frac{Fb}{2kT}\right] \qquad (4.11)$$

The probability of motion to the left is thus

$$P^- = \tfrac{1}{2}\alpha \frac{kT}{h} \exp\left[-\frac{V + \tfrac{1}{2}Fb}{kT}\right] \qquad (4.12)$$

or

$$P^- = \tfrac{1}{2}P \exp\left[-\frac{Fb}{2kT}\right] \qquad (4.13)$$

Consequently, the positive transitions will be more frequent than the negative. Therefore there will be an average drift velocity to the right in the direction of the field. The mean velocity of the drift motion is

$$\bar{v} = b(P^+ - P^-) = \tfrac{1}{2}bP\left[\exp\left(+\frac{Fb}{2kT}\right) - \exp\left(-\frac{Fb}{2kT}\right)\right] \qquad (4.14a)$$

$$\bar{v} = bP \sinh \frac{Fb}{2kT} \qquad (4.14b)$$

4.2 THEORY OF IONIC CONDUCTIVITY

It is worthwhile to repeat that this result is based on the electric field merely directly the random diffusion of ions that occur spontaneously at any temperature T_i. The field does not pull the ions out of the wells. In fact, as long as the field strength is much smaller than kT, (i.e., $\frac{1}{2}Fb \ll kT$) then sinh can be replaced by the argument and the drift velocity can be expressed as

$$\bar{v} \approx \frac{b^2 PF}{2kT} \quad (4.15)$$

In the case where very large field strengths exist, the first term in sinh of equation (4.14) will dominate and

$$\bar{v} \approx \text{Const.} \exp\left[\frac{bF}{2kT}\right] \quad (4.16)$$

At room temperature bF is small compared with kT up to field strengths of 10 V/cm. Therefore, the current will be proportional to the field strength, in agreement with Ohm's law.

$$j = nze\bar{v} \quad (4.17)$$

where n is the number of ions per cubic centimeter, and $q = ze$. So

$$j = \frac{nzeb^2 PF}{2kT} = \frac{nz^2 e^2 b^2 P\mathbf{E}}{2kT} \quad (4.18)$$

from substituting 4.10 and $F = ze\mathbf{E}$ and substituting:

$$P = \frac{\alpha kT}{h} \exp\left[-\frac{V}{kT}\right] \quad \text{and} \quad V = \frac{\Delta F_{dc}}{N} \quad (4.19)$$

where ΔF_{dc} is the change in free energy for dc conduction in units of kilocalories/mole and N is Avogadro's number $= 6.023 \times 10^{23}$ atoms/mole. This yields:

$$j = \frac{n\alpha z^2 e^2 b^2 \mathbf{E}}{2h} \exp\left[-\frac{\Delta F_{dc}}{RT}\right] \quad (4.20)$$

where R = gas constant = 8.31 joule/mole °K. Therefore, the electrical resistance is

$$R_{(ohm)} = \frac{\mathbf{E}}{j} = \frac{2h}{n\alpha z^2 e^2 b^2} \exp\left[+\frac{\Delta F_{dc}}{RT}\right] \quad (4.21)$$

or

$$\log R_{(ohms)} = \log \frac{2h}{n\alpha z^2 e^2 b^2} + \frac{\Delta F_{dc}}{RT}$$

or conversely,

$$\log G_{(ohm^{-1})} = \log \frac{n\alpha z^2 e^2 b^2}{2h} - \frac{\Delta F_{dc}}{RT} \qquad (4.22)$$

which is similar to the *Rasch–Hinrichsen law of electrical resistivity* of glass, discovered empirically many years ago:

$$\log \rho = A + \frac{B}{T} \qquad (4.23)$$

(Note, σ^{-1} (ohm cm)$^{-1}$ = G^{-1} (ohm^{-1}) · (thickness) · (area^{-1}); and ρ (ohm cm) = R(ohm) · (area) · (thickness)$^{-1}$.)

Many glass compositions behave according to equation (4.22), as shown in Figure 4.2. The ionic conductivity of crystalline oxide ceramics is also described by equations (4.22) and (4.23), as illustrated in Figure 4.3.

The free energy of activation for dc conduction is shown in Figure 4.3 for the various ceramics. Table 4.2 compares the diffusion activation energies with conduction activation energies for several glasses. The close comparison gives credence to the formulation of ionic conductivity theory, as based on field-directed diffusion of charged ions.

It is also possible to distinguish between enthalpy and entropy changes

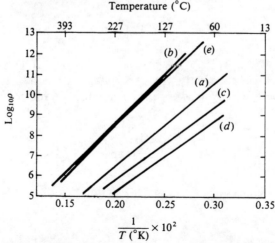

FIGURE 4.2 Resistivity of some ionic glasses. (*a*) 18Na$_2$O·10CaO·72SiO$_2$. (*b*) 10Na$_2$O·20CaO·70SiO$_2$. (*c*) 12Na$_2$O·88SiO$_2$. (*d*) 24Na$_2$O·76SiO$_2$. (*e*) Pyrex.

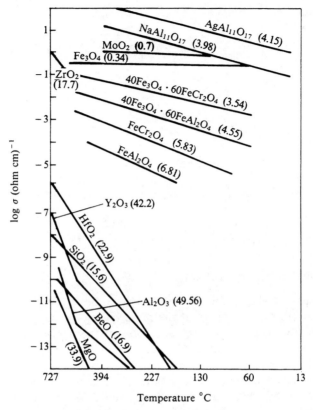

FIGURE 4.3 Temperature dependence of conductivity for several oxides with activation energy in Kcal/mol shown in brackets ( ).

TABLE 4.2 Activation Energies for Diffusion and dc Conductivity of Several Glasses

Composition, mole%					$E_a \left(\dfrac{\text{Kcal}}{\text{mole}} \right)$	
Na_2O	CaO	Al_2O	SiO_2	GeO_2	Diffusion (Na^+)	Conduction
33.3	—	—	66.7	—	13–14	14–16
25.0	—	—	—	75.0	17–18	16–18
15.7	—	12.1	72.2	—	16.4	15.6
11.0	—	16.1	72.9	—	15.6	15.1
15.9	11.9	—	72.2	—	22.0	20.8
14.5	12.3	5.8	67.4	—	20.2	19.5

involved in the dc conduction by using *second law of thermodynamics*

$$\Delta F_{dc} = \Delta H_{dc} - T \Delta S_{dc} \tag{4.24}$$

which yields a final expression for the electrical conductance:

$$\log G = \log \left[\frac{n\alpha z^2 e^2 b^2}{2h} \exp\left(\frac{\Delta S_{dc}}{R}\right) \right] - \frac{\Delta H_{dc}}{RT} \tag{4.25}$$

Therefore, the enthalpy and entropy can be calculated from the slope and intercept of the $\log G$ vs. $1/T$ plot. Some interpretations of ΔS values have been determined although they are largely speculative in nature.

4.3 IONIC CONDUCTIVITY

4.3.1 Electrode Polarization

Ion transport within the glass or ceramic under a dc potential eventually leads to a build-up of charge at the material–electrode interface if the ions are not replenished at the electrode. This phenomenon is termed *electrode polarization* and it can drastically affect the magnitude of dc conduction measured. Three regions of ionic conductivity can be seen in Figure 4.4 which is an experimental measurement of the conductivity or current after a voltage of 1 V had been applied to a 30% Li_2O–70% SiO_2 glass for the times

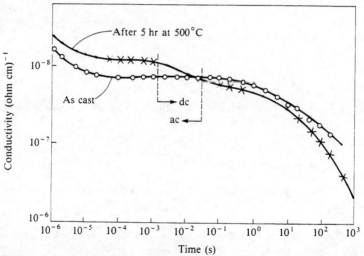

FIGURE 4.4 Ionic conductivity as a function of time for a $30Li_2O:10SiO_2$ glass as-cast and heat treated. After Kinser and Hench (1969).

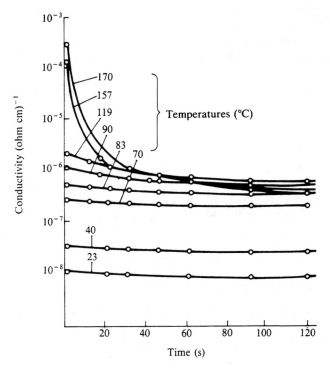

FIGURE 4.5 Conductivity as a function of time for a $33Na_2O:67SiO_2$ glass at various temperatures (°C). After Kinser and Hench (1969).

indicated. The conductivity over very short time intervals increases due to various dielectric contributions discussed in Chapter 5. Over long time intervals, the conductivity decreases indefinitely and only over intermediate intervals is the conductivity time independent. It is within this time-independent region that conductivity can be correctly termed σ_{dc}.

The reason for the continued decrease in σ over long time intervals is that the mobile cations pile up at the electrodes which results in an induced back field of potential. The back field concentrates the potential drop in the electrode region (Figure 4.5). Since the potential across the bulk of the sample is decreased to a fraction of the applied potential, the force F on the mobile ions in the bulk is decreased proportionally and the conductivity decreases, as predicted from equation (4.20). The back field builds up indefinitely as time progresses resulting in a continuous decrease in conductivity.

It is thus evident that dc conductivity is dependent on the time interval at which the measurement is made, and care must be taken to conduct measurements in the time-dependent region of behavior. This may be especially difficult at higher temperatures when the conductivity of most ceramics and glasses is in the range of $>10^{-5}$ ohm^{-1} cm^{-1} because electrode polarization occurs so rapidly (<10 s).

There are three possible solutions to the problem of accurate experimental measurement of ionic σ_{dc} posed by electrode polarization. The first is to construct σ vs. t curves, such as Figure 4.4, for each temperature, to ensure that measurements are made in the time-independent regime. Secondly, it is possible to use non-blocking electrodes, which provide a source and sink for the mobile ions, (e.g., sodium amalgam electrodes for Na-containing glasses and ceramics). However, the alloy composition must be adjusted such that the chemical potential across the interface is equal before the electrode is completely nonblocking.

The third approach is to calculate the current flow required to produce the back potential necessary to decrease the conductivity to a given fraction of its true nonblocked value. The back potential ϕ_b, due to the polarization charge per unit area q, is simply

$$\phi_b = \frac{qd}{2\epsilon\epsilon_0} \tag{4.26}$$

where ϵ is the permittivity of the insulator, d is the thickness, and ϵ_0 is the permittivity of free space (see Chapter 5). The charge q_b to form the back potential can be related to the current flow as

$$q_b = \int_0^t j\, dt = jt \tag{4.27}$$

The current j required to form the back charge is directly related to the applied potential ϕ_a and the conductivity of the material as $j = \phi_a\sigma$. Therefore substitution into equation (4.27) yields

$$q_b = \phi_a \sigma t \tag{4.28}$$

So, if there is a critical response time for the measuring equipment, t_x, and an acceptable level of error in the conductivity measurement of 0.1, for example, then the ratio ϕ_b/ϕ_a is equal to 0.9 and the limit of conductivity that can be accurately measured within t_x is

$$\phi_b/\phi_a = \frac{0.9(2\epsilon\epsilon_0)}{dt_x} = \sigma_x \tag{4.29}$$

Since σ increases rapidly with temperature, equation (4.29) implies that there will be a critical temperature, T_c, where the conductivity error becomes greater than 10% when T_c is exceeded. Such a behavior is shown in Figure 4.5. By using equation (4.29) and an estimated value of ΔF_{dc}, this problem can be eliminated.

4.3.2 Glass Composition and Structure Effects

The electrical conductivity of alkali silicate glasses changes with the replacement of silica by alkali ions. In nearly all cases, a rapid increase in conductivity is observed until a range of 20–30 mole% modifier is reached, after which the effect becomes progressively less. The range of conductivity is from $10^{-16}\,\text{ohm}^{-1}\,\text{cm}^{-1}$ to $10^{-7}\,\text{ohm}^{-1}\,\text{cm}^{-1}$ at room temperature (see Figure 4.2). The relative effect of alkali type on conductivity is: $Na^+ > Li^+ > K^+$. The addition of only parts per million of Na^+ ions is sufficient to drastically increase the conductivity of fused silica and the conductivity is directly related to the Na^+ content.

Additions of divalent modifiers to a soda silica glass have consistently been observed to decrease the electrical conductivity with the relative effect being: $Ba^{2+} > Pb^{2+} > Sr^{2+} > Ca^{2+} > Mg^{2+} > Zn^{2+} > Be^{2+}$. Additions of a 20 weight% Ca to an 18% Na_2O–82% SiO_2 glass will decrease the room temperature conductivity to $10^{-15}\,\text{ohm}^{-1}\,\text{cm}^{-1}$, whereas Zn^{2+} additions up to 40% will be of no influence. The effectiveness of the divalent oxides in decreasing the conductivity has been found to increase as the radius of the metal ion increases. This correlation is shown in Figure 4.6. The interpreta-

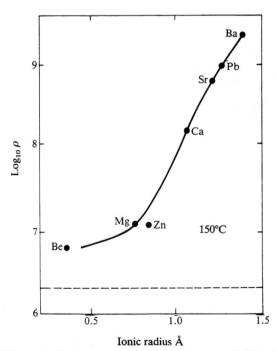

FIGURE 4.6 Effect of divalent ion radii on resistivity of $20Na_2O \cdot 20MO \cdot 60SiO_2$ glass. Dashed line is the resistivity of $20Na_2O:80SiO_2$. After Owen (1963).

tion of the effect is that the divalent ions block the conduction paths of the alkali ions and as the divalent ion size increases, the blocking becomes more effective. The divalent and monovalent ions form a coordination complex with mutual oxygen ions, thus maximizing the local charge distribution in the glass. The coordination complex has a higher activation energy barrier to migration. Consequently, the ionic conductivity decreases.

Partial substitution of one alkali-metal oxide for another in silicate, borosilicate and borate glasses produce such large increases in electrical resistance that this behavior is termed the *mixed alkali effect*. This effect is illustrated for three systems in Figure 4.7. Early work interpreted the mixed effect largely in terms of the motion of "coupled ions." The evidence suggesting the formation of a cation complex was that the activation energy for conduction also increases by almost a factor of two as the substitution occurs. Studies have shown that the substitutions induce a glass-in-glass phase separation in the Li_2–Na_2O–Cs_2O silicate glasses. The smaller, more mobile ions are isolated in the dispersed phase with the larger ions in the

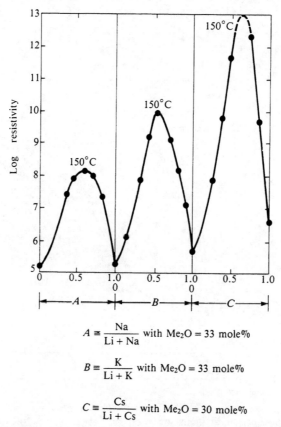

$A \equiv \dfrac{Na}{Li + Na}$ with $Me_2O = 33$ mole%

$B \equiv \dfrac{K}{Li + K}$ with $Me_2O = 33$ mole%

$C \equiv \dfrac{Cs}{Li + Cs}$ with $Me_2O = 30$ mole%

FIGURE 4.7 Resistivity of mixed alkali silicate glasses. After Charles (1965).

matrix phase thus decreasing the total dc conductivity. However, in order for the conductivity to be suppressed to the magnitude demonstrated, the larger cations must also be largely coordinated within the dispersed phases. This yields a nearly alkali-free silica-rich matrix which determines the very low dc conductivity.

Substitution of Al_2O_3 for SiO_2 in sodium silicate glasses is especially interesting. As the Al_2O_3 content increases, the electrical conductivity is also increased and the activation energy for conduction is reduced. This behavior is associated with the aluminum ion entering the network in fourfold coordination reducing the number of nonbridging oxygen ions (see Chapter 1). The nonbridging oxygen coordination shell about the sodium ions must be decreased, which increases the sodium ion mobility.

The electrical properties of glasses also can depend on the morphology of the glass-in-glass phase separation. A rapid quench of a low, 7.5 mole%, lithia silicate glass changes the phase-separated structure from that of a dispersed phase to a tabular, interconnected phase, characteristic of spinodal decomposition. The dc conductivity is increased by nearly an order of magnitude by the rapid quenching due to the presence of the interconnected Li_2O-rich phase.

The onset of crystallization in silicate glasses also produces large changes in electrical conductivity. The dc conductivity of both Li_2O–SiO_2 and Na_2O–SiO_2 glasses decreases by over two orders of magnitude concurrent with the appearance of the equilibrium crystal phases in the glass. This

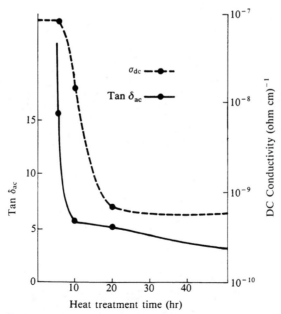

FIGURE 4.8 Conductivity and loss tangent for a $30Li_2O \cdot 70SiO_2$ glass as a function of heat treatment at 500°C.

behavior is illustrated in Figure 4.8. The major change in the conductivity is associated with the presence of only a few percent of crystals in the glass. The more energetic alkali cations which usually participate in the thermally activated dc conduction are the first removed from the glass during the crystallization process. The conductivity decreases as the least mobile cations are left behind in the matrix. Dielectric effects (e.g., frequency dependence) associated with structural and compositional changes in glasses are discussed in Chapter 5.

4.3.3 Molten Silicates

The production of glass by means of electric furnaces provides a commercial impetus for understanding the nature of the electrical conductance of molten glasses. Such furnaces operate by I^2R heating of the glass by ionic conduction. Thus, it is important to understand the effect of composition on the conductance of glasses in the molten state. Obviously, from a commercial standpoint, the effect of temperature on electrical conductivity is also of extreme importance.

The electrical conductivity of molten silicates depends upon the concentration of alkali ions. For concentrations of R_2O greater than 22%, the

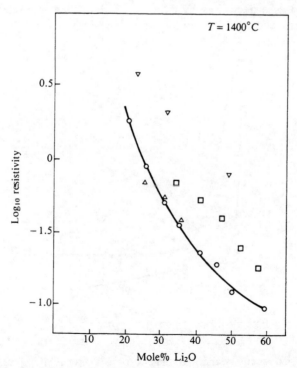

FIGURE 4.9 Resistivity of molten lithia–silica glasses as a function of Li_2O content.

4.3 IONIC CONDUCTIVITY

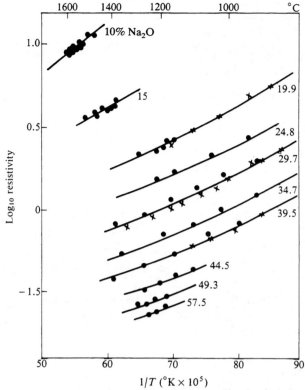

FIGURE 4.10 Resistivity of molten soda–silica glasses as a function of temperature. After Tickle (1967).

order of equivalent conductances is $Li_2O > Na_2O > K_2O$, but below about 22% R_2O, the order is $Na_2O > Li_2O > K_2O$.

A comparison of resistivity at 1400°C for various Li_2O–SiO_2 glasses, measured by different workers, is shown in Figure 4.9. The resistivity of a series of soda glasses is shown as a function of reciprocal temperature in Figure 4.10 over a temperature range of 900°C to 1600°C. The resistivity decreases by nearly three orders of magnitude as the Na_2O content decreases from 10 to 57.5% in the binary $x(Na_2O)$–$(100-x)SiO_2$ compositions.

The numbers on the curves in Figure 4.10 denote the mole% of Na_2O in the binary glasses. A curvature of the plots is clearly seen, indicating a departure from the simple exponential temperature dependence of the Rasch–Hinrichsen law (equation 4.23).

The conductivity of the alkaline earth silicate glasses is, in general, about an order of magnitude lower than that of alkali-containing glasses of equivalent concentration. The general order of equivalent conductances for the various alkaline oxides is $CaO > MgO > SrO > BaO$. The differences between the resistivities of the high and low members of this series is very

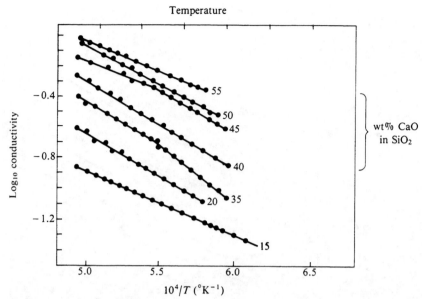

FIGURE 4.11 Conductivity of molten calcia–silica glasses as a function of temperature.

small, with the resistivity of calcium-containing glasses being approximately one-half of barium-containing glasses of equivalent concentration.

The change in conductivity of calcium oxide silicate glasses is shown as a function of reciprocal temperature in Figure 4.11.

Binary liquid silicates containing iron and manganese modifier ions possess electrical conductivities nearly equivalent to that of the alkali-silicate system. Binary liquid silicates containing a second network former, such as Al_2O_3 or TiO_2, possess electrical conductivities that are as much as three to four orders of magnitude lower than those of the alkali silicate liquids.

The electrical conductivity of glasses and liquid melts containing two alkali modifier ions are important because of the so-called mixed alkali effect mentioned above. The mixed alkali effect increases as the temperature is decreased. The conductivity of glass systems containing alkali modifier ions at high temperatures is nearly equivalent to linear additions of the conductivities of the two end members, while at lower temperatures, the conductivity is decidedly less. The mixed alkali effect may decrease the conductivity by as much as a factor of four at a temperature of 800°C for soda potassium silicate glasses, whereas at 1200°C the decrease in conductivity in the mixed soda potassium glass is only a factor of two.

Understanding the effects of composition and temperature on the electrical conductivity of molten glasses of the alkali–alkaline earth–silicate system is of the greatest importance from a commercial viewpoint since most commercial glasses are based upon combinations of alkali and alkaline earth modifiers with silica. The RO modifiers that have been studied in the ternary $Na_2O-RO-SiO_2$ systems include CaO, MgO, PbO, ZnO, BaO, BeO, and CdO. In the range of 1100°C to 1400°C, the conductivity of these glasses obey equation (4.23). The conductivity is proportional to the concentration of sodium oxide present in the ternary systems. The variation of resistivity of $Na_2O-RO-SiO_2$ glasses with the alkali ion concentration at 1200°C is shown in Figure 4.12. It can be seen from this figure that the resistivity is directly a function of concentration of sodium present in the glasses regardless of the nature and concentration of the divalent cation present. The temperature dependence and sodium ion concentration dependence of ternary glasses in the $Na_2O-RO-SiO_2$ system is:

$$\log \sigma = 1.508 + 0.0204C - \frac{4836 + 128C}{T} \qquad (4.30)$$

where C is the Na_2O content in weight% and T is the temperature in °K. Equation (4.30) is valid over a temperature range of 1100°C to 1450°C for the composition ranges: 12–20% Na_2O, 0–6% MgO, 0–11% CaO, 0–11% Al_2O_3 and 68–78% SiO_2 when the sodium concentration is >10 mole%. When the sodium ion concentration decreases to <10% or the divalent cation concentration increases to >10%, the relationship of conductivity to concentration no longer fits equation (4.30).

Bivalent cations appear to have very little influence on the mobility of monovalent ions in the liquid silicate melts. The electrical conductivity of

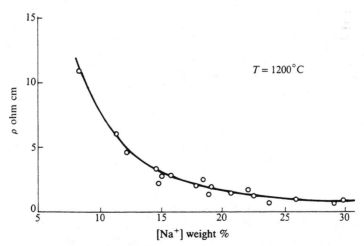

FIGURE 4.12 Dependence of resistivity on sodium ions of $Na_2O-RO-SiO_2$ glasses on weight% of Na^+ ions. Data points represent MgO-, CaO-, BaO-, BeO- and CdO-containing glasses with 68–78% SiO_2.

sodium borate glasses is also a function of sodium ion concentration, however, sodium borate glasses have ~50% higher values of conductivity than sodium glasses of equivalent alkali concentration.

A change in the conductance will be noticed if either of the terms on the right-hand side of equation (4.25) are changed. The first term may be changed if the ion vibrational frequency (see equation 4.10), the connectivity of hopping sites (i.e., the number of nearest neighbor equivalent hopping sites), the cationic valence or the hopping distance b are changed. The second term may be altered if the height of the energy barrier is changed. This energy barrier consists of two terms: (1) the electrostatic energy required to move a point charge, $1 = Ze$, to a midpoint between the two sites, and (2) the elastic energy required to distort the amorphous lattice so that the ion will "jump" through to the new site. Many of these factors can be responsible for the difference between the borate and silicate glasses.

The conductivity of complex compositions of calcia alumina silicate slags is a maximum in the region 38%–43% SiO_2, 44%–50% CaO, and 10%–15% Al_2O_3. This maximum is close to the composition of the ternary eutectic in the $CaO-Al_2O_3-SiO_2$ system, located at approximately 41% SiO_2, 47.3% CaO, and 11.7% Al_2O_3. The conductivity is a minimum at concentrations of 46% SiO_2, 37% CaO, and 17% Al_2O_3. Other than these regions of maximum and minimum conductivity, the conductivity generally increases with increasing basicity of the melt.

4.4 SOLID ELECTROLYTES AND FAST ION CONDUCTORS

Many crystal compounds show exceptionally high ionic conductivities. These compounds fall into three groups:

1. Halides and chalcogenides of silver and copper where the metal atoms are bonded to differing sites in a relatively random fashion,
2. Oxides with the β-alumina structure with highly mobile monovalent cations, and
3. Oxides of the CaF_2 fluorite structure with large concentrations of defects caused by either a variable valence cation or a solid solution with a second cation of lower valence. Examples include $CaO \cdot ZrO_2$ or $Y_2O_3 \cdot ZrO_2$.

The conductivities of solid electrolytes are several orders of magnitude higher than normal ionic compounds. Consequently they are often called *fast ion conductors* (FIC), *superionic conductors*, or *optimized ionic conductors*.

Conductivities of FICs are in the range of the ionic conductivity of molten salts where free motion of cations or anions are possible, e.g.,

4.4 SOLID ELECTROLYTES AND FAST ION CONDUCTORS

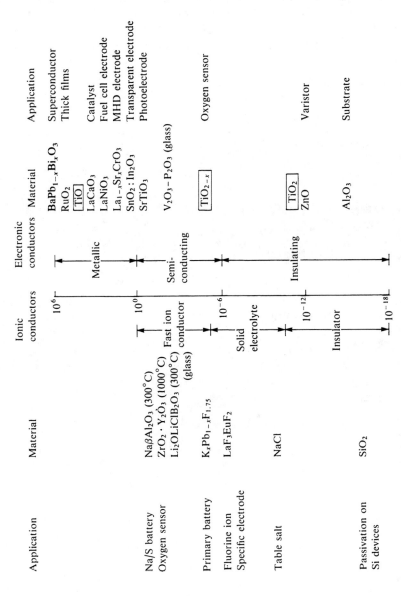

FIGURE 4.13 Fast ion conductors. Logarithmic scale of conductivity for ionic vs. electronic materials.

CHAPTER 4: IONIC AND DEFECT CONDUCTORS

($\sigma > 10^{-2}\,\text{ohm}^{-1}\,\text{cm}^{-1}$). However, for FICs a conductivity of 10^{-2} $(\text{ohm cm})^{-1}$ occurs at a temperature very much lower than the temperature of melting.

Figure 4.13 shows that FICs have a range of conductivity that overlaps that of semiconducting ceramics and metallic-like, electronic-conducting ceramics.

Some FICs have purely cationic conduction, e.g., Na^+ in β-alumina.

TABLE 4.3 Some Types of Fast Ionic Conductors

Material	Conductivity $\sigma(\text{ohm}^{-1}\,\text{cm}^{-1})$	Activation $\Delta H_{dc}(\text{eV})$	Enthalpy (kcal/mole)	Comments
Cationic conductors				
α-AgI (146–555°C)	1(150°C)	0.05	1.15	bcc iodine skeleton Tridimensional conductivity
$RbAg_4I_5$ (−155–+200°C)	0.12(20°C)	0.07	1.61	
Ag_2S (>170°C)	3.8(200°C)	0.05	1.15	
CuS (>91°C)	0.2(400°C)	0.25	5.75	Cu-deficient phase
$AgAl_{11}O_{17}$	0.1(500°C)	0.18	4.14	
β-alumina	0.35(300°C)	0.01	0.23	Effective formula: $11Al_2O_3(1+x)M_2O$ Very stable bidimensional conductivity
$NaAl_{11}O_{17}$	0.1(500°C)	0.17	3.91	
$Na_3Zr_2Si_2PO_{12}$	0.2(25°C)	0.27	6.21	Tridimensional structure with intersecting tunnels
$NaSbO_3$	0.06	0.35	8.05	Cubic structure stabilized by NaF
$K_{2x}Mg_xTi_{8-x}O_{16}$ (hollandite)	0.02(25°C)	0.22	5.06	Tunnel structure
$Li_xV_2O_5$	0.01(25°C)	0.07	1.61	Bronze structure monodimensional conductivity
Li_3N	0.0005	0.19	4.37	
Li_2SO_4	1.04(600°C)	0.4	9.20	
$Li_{5-x}Al_{1-x}SiO_4$	0.001(300°C)	1	23.00	
Anionic conductors				
ZrO_2 12% CaO	0.055(1000°C)	1.1	25.30	Fluorite fcc 9% structure
9% Y_2O_3	0.12(1000°C)	0.8	18.40	
10% Sc_2O_3	0.25(1000°C)	0.65	14.95	
$ThO_2 \cdot 8\%\ Y_2O_3$	0.0048	1.1	25.30	
$La_2O_3 \cdot 15\%$ CaO	0.0024	0.88	20.24	
$Bi_2O_3 \cdot 25\%\ Y_2O_3$	0.16(700°C)	0.6	13.80	
CaF_2	0.01(700°C)	1	23.00	Stabilized δ-Bi_2O_3 fcc lattice
PbF_2	0.001(200°C)	0.45	10.35	
$LaF_3 \cdot 5\%\ SrF_2$	0.01(400°C)	1	23.00	

Thus, β-alumina with $t_{Na^+} = 1$, is useful in sodium–sulfur batteries at 300°C. Other FICs, e.g., the ferrite $KFe_{11}O_{17}$ which also has the β-alumina structure, has a mixed ionic and electronic conductivity due to a mixture of Fe^{2+} and Fe^{3+} ions. Therefore it can be useful as an electrode constituent in batteries where both cation and electronic conductivity is useful. For other FICs, e.g., $MM'O_{2-x}$ fluorites such as CaO-stabilized ZrO_2, the conductivity is entirely by O^{2-} anions. Table 4.3 lists examples of both cationic and anionic FICs, their magnitude of conductivity, and activation energy of conduction.

Structural features of the crystal lattices determine the type of ionic charge carrier and the magnitude of conduction. In general, fast ion transport can occur where sublattices are disordered and partially occupied. For example, three-dimensional disorder and ion transport can occur in compounds such as $Na_3Zr_2PSi_2O_{12}$. For β-alumina ($Na_2O \cdot 11Al_2O_3$) the Na^+ conduction is via two-dimensional defects and for $LiAlSiO_4$ the transport is one-dimensional.

In all cases there are four important structural features:

1. The skeleton of the structure is formed by one set of ions which occupy fixed sites,
2. Large voids are present in the network with the number of voids being considerably higher than the number of mobile ions. Therefore there are always empty sites on the disordered lattice into which mobile ions can move,
3. The sublattice sites have nearly equivalent energies and relatively low activation energies between the sites (see Table 4.3), and
4. There is interconnection between the sites such that translational motion is possible along preferred pathways.

For some FICs and especially for stoichiometric compounds there is ordering of the conducting species at low temperature. At higher temperatures the disorder occurs on the sublattice with ionic motion as easy as in a liquid-like state. For defect compounds, disorder occurs even at low temperatures.

4.5 CRITERION FOR FAST ION CONDUCTION

Referring to equation (4.25):

$$\text{Log } G = \log \left[\frac{n \alpha z^2 e^2 b^2}{2h} \right] \exp \left[\frac{\Delta S_{dc}}{R} \right] - \left(\frac{\Delta H_{dc}}{RT} \right) \quad (4.31)$$

the characteristic of fast ion conduction arises from the large value of n, the concentration of the charge carriers, with respect to n_0, the total number of potentially mobile ions.

CHAPTER 4: IONIC AND DEFECT CONDUCTORS

In normal ionic solids, the number of charge carriers arising from Frenkel or Shottky defects is determined by thermal generation, e.g.,

$$n = n_0 \exp(-\Delta E_f/kT) \tag{4.32}$$

$$n = n_0 \exp\left[\frac{\Delta S_f}{R} - \frac{\Delta H_f}{RT}\right] \tag{4.33}$$

where ΔE_f is the energy of defect formation. (See the following section for a discussion of Frenkel and Shottky defects and defect theory.) The conductivity equation (4.25) thus becomes:

$$\text{Log } G = \log\left[\frac{n_0 \alpha z^2 e^2 b^2}{2h}\right] \exp\left[\frac{\Delta S_m + \Delta S_f}{R}\right] - \left[\frac{\Delta H_m + \Delta H_f}{RT}\right] \tag{4.34}$$

where ΔS_m, ΔH_m are the entropy and enthalpy of ionic migration or motion under an applied field and ΔS_f, ΔH_f are the entropy and enthalpy of formation of the charge carrier or defect associated with motion.

In FICs, the number of charge carriers n is a large fraction β of the potentially mobile ions n_0. This is due to the existence of structurally inherent vacant sites in the lattice. Therefore

$$n = \beta n_0 \tag{4.35}$$

Since the lattice vacancies are structurally inherent their concentration is not thermally dependent and therefore $\Delta S_f = 0$ and $\Delta H_f = 0$. Consequently, for FICs the equation describing ionic conductivity, equation (4.31) can be expressed as

$$\text{Log } G = \log\left[\frac{\alpha \beta n_0 z^2 e^2 b^2}{2h}\right] \exp\left[\frac{\Delta S_m}{R}\right] - \left[\frac{\Delta H_m}{RT}\right] \tag{4.36}$$

where ΔH_m is the activation enthalpy for the ion motion (see Table 4.3).

Thus, for normal ionic compounds there are a limited number of charge carriers which depend entirely on thermal energy. Their activation energy is a sum of the energy of defect formation ($E = \sim 2 - 4$ eV) and the energy for defect migration ($E \approx 0.05 - 1$ eV), e.g., (2 eV + 0.5 eV) = 2.5 eV. However, for FICs there are a large number of carriers in the sublattice and only the energy barrier for migration must be overcome for conduction to occur:

$$E_m = 0.05 \text{ to } 1 \text{ eV} \tag{4.37}$$

Other factors which contribute to the high conductivity of FICs include:

1. *Anharmonic lattice vibrations.* Atomic vibrations with a large anharmonic character can decrease the bulk modulus of the lattice with increasing temperature and thereby decrease $\Delta H_f + \Delta H_m$, and $\Delta S_f + \Delta S_m$. Examples include: Li in Li_3N, Ag + Hg in $HgAg_2I_4$, X^- or X^{2-} in oxides of fluorides with fluorite structure, and F in PbF_2.
2. *Lattice polarizability.* A large amount of local lattice polarization (see Chapter 5) is unfavorable because a migrating cation or anion has to carry its polarization with it, e.g., the ionic polaron behavior. Thus, the highest ionic conductivity occurs when the migrating species is easily polarized, such as Ag^+ and Cu^+, in a deformable framework. Consequently, compounds involving I, Pb^{2+}, or Bi^{3+} give higher conductivities than less easily polarized lattices.
3. *Correlation effects.* When migrating ionic species interact during transport, e.g., cation–cation correlations, the effective activation energies can be significantly lowered. Na^+ transport in β-alumina is an example.

4.6 β-ALUMINA AND β″-ALUMINA

Both β and β'' alumina FICs have been extensively investigated for use in sodium–sulfur batteries, due to the very high conductivity of Na^+ in the material (Table 4.3 and Figure 4.13). The generalized chemical formulation of β-alumina has the form

$$\beta\text{-alumina} = AM_{11}O_{17} \qquad (4.38)$$

The A cation has a single valence electron and is usually the most mobile. Examples include: Na^{+1}, K^{+1}, Rb^{+1}, Ag^{+1}, Tl^{+1}, or Li^{+1}. The cation M has three valence electrons, (e.g., Al^{3+}, Fe^{3+}, or Ga^{3+}). Since β-alumina is a nonstoichiometric aluminate, its chemical formula can also be expressed as

$$\beta\text{-alumina} = (1+x)\,Na_2O \cdot 11Al_2O_3 \qquad (4.39)$$

where $0 < x < 0.3$.

Table 4.3 shows the range of conductivities for β-alumina with various A cations. The conductivities are in the range of liquid chemical electrolytes. The size of the A cation is the dominant factor in the conductivity such as

$$G\,(Na\,\beta\text{-alumina}) > G\,(K\,\beta\text{-alumina}) \qquad (4.40)$$

The conductivity is also a function of the anisotropic crystal structure. Conductivity parallel to conduction channels is considerably greater than

TABLE 4.4 Occupancy Factors on the Available Sites of β-Alumina

Type of site	Conducting ion				
	Na$^+$	Ag$^+$	K$^+$	Ti$^+$	NH$_4^+$
BR	1.55(5)	1.54(3)	1.56(1)	1.76(2)	1.85(5)
mO	0.97(6)	0.09(2)	0.75(1)	0.51(2)	0.55(6)
aBR	0	0.89(2)	0.28(1)	0.21(2)	0.18(2)
Total	2.52(11)	2.52(7)	2.59(3)	2.48(6)	2.56(13)

Number of M^+ ions in ( ).

conductivity perpendicular to these channels. The structure of β-alumina is composed of two spinel-like blocks containing Al^{3+} and O^{2-} ions separated by a mirror plane in which loosely packed M^+ cations and one O^{2-} anion are located (Figure 4.14). The space group is P6$_3$/mmc with cell constants of $a = 5.59$ Å and $c = 22.53$ Å for Na β-alumina.

In the mirror plane there are seven alternative positions for the Na$^+$ ions. The lower energy, and therefore most probable position theoretically, is called the *Bevers–Ross* or *BR position* (Figure 4.15). However, mid-oxygen positions (mO) or anti-BR positions can also be occupied. The occupancy of the various sites and the total number of M^+ ions depends on the size of the cation (Table 4.4).

Charge compensation to accommodate the nonstoichiometry is via Frenkel defects, as illustrated in Figure 4.16. The Frenkel defect involves an interstitial oxygen ion in the conducting plane, e.g., on mO sites, with an Al vacancy/Al interstitial pair within the spinel blocks (Figure 4.16).

Therefore, in the conducting plane there will be a variety of sites filled with a large fraction of the cations displaced from the theoretical BR, mO, and aBR positions. The net effect is a quasi-liquid-like state for the M^+ ions within the defect lattice.

Other phases of alkali β-aluminas are β' and β". The general chemical formulae for these materials are:

$$\beta' - A_{(1-x)}M_7O_{11} \tag{4.41}$$
$$\beta'' = A_{(1-x)}M_5O_8 \tag{4.42}$$

where A and M are defined above.

β'-Alumina is easily stabilized by additions of MgO with a formula close to Na$_2$O·MgO·5Al$_2$O$_3$. The β'-alumina structure has rhombohedral symmetry of R3 which consists of three spinel blocks separated by loose-packed layers containing mobile M^+ cations. Since the structure is deficient in M^+ there is no need for interstitial oxygen anions as is required in β-alumina for charge compensation. Therefore, conductivity values can be quite high; (e.g., $\sigma = 10^{-2}$ ohm^{-1} cm^{-1} at 25°C for H$_3$O$^+$) and is extremely anisotropic with respect to the c-axis:

$$\sigma_{\perp c} \gg \sigma_{\parallel c} \tag{4.43}$$

4.6 β-ALUMINA AND β″-ALUMINA   161

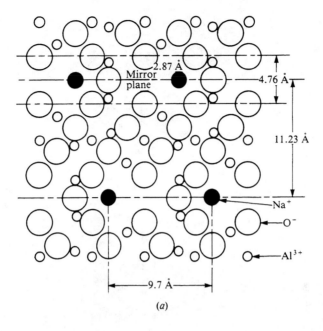

(a)

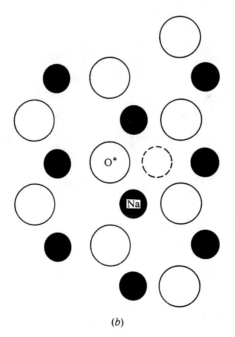

(b)

FIGURE 4.14 β-Alumina. (a) Crystal structure. This section, a plane parallel to the c-axis, does not show the closest sodium–sodium distance. (b) The arrangements of atoms in the mirror plane of β-alumina.

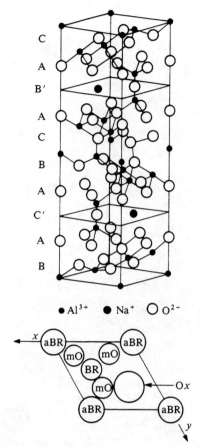

FIGURE 4.15 Structure of β-alumina; site occupation in the conducting plane.

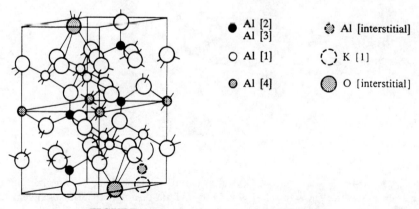

FIGURE 4.16 Frenkel defect in β-alumina.

4.7 GRAIN BOUNDARY EFFECTS

The fast ion conduction paths within a crystal are often impeded at a grain boundary due to the lattice mismatch. Consequently, grain boundaries in polycrystalline fast ion conductors decrease the conductance and give rise to ac frequency effects (see Chapter 5). Figure 4.17 shows that the product of specific conductivity times absolute temperature for polycrystalline β-alumina is more than 10 times lower than the crystal phase. As expected from equation (4.31), the activation energy of grain boundary migration is higher. This is because $\Delta H_f \neq 0$ for grain boundary migration, whereas $\Delta H_f = 0$ for the crystal.

An important consequence of the higher resistance of the grain boundaries is that a large fraction of the applied voltage of an electrochemical device appears across the grain boundaries. It can be calculated that the local electric field can be as large as 10^5 V/cm when 1 V is applied across a typical β-alumina ceramic of 1 mm thickness with 10^3 grain boundaries/cm, and a grain boundary thickness of 10^{-7} cm. Such high local fields can lead to electro-chemical deterioration of grain boundary phases, dielectric breakdown, and cracking.

4.8 DEFECT SEMICONDUCTORS

Because ceramic crystal compounds, $M_a N_b X_c$, consist of more than one element, their defect nature is much more complex than metals or elemental semiconductors, such as Si or Ge, or even III–V semiconductors such as

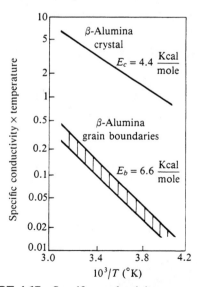

FIGURE 4.17 Specific conductivity × temperature.

GaAs. The major differences are:

1. Defects can occur on either M, N, or X lattice sites,
2. Various combinations of cation (M, N sites) defects and anion (X-sites) defects are possible,
3. Defects and defect concentrations will be atmospheric dependent, e.g., partial pressure of O_2 for oxides,
4. Atomic defects require electronic defects to maintain charge balance, and
5. Atomic and electronic defects can be either associated (charge coupled) *or* disassociated (ionized) depending upon temperature.

Thus, in order to understand ceramic defect semiconductors it is necessary to:

1. Describe the various types of atomic defects in crystal compounds,
2. Describe the electronic changes induced by the atomic defects,
3. Formulate expressions for defect energies, and
4. Describe the thermodynamics of defect formation using the law of mass action.

The nomenclature we will use is as follows:

Cation	M	Anion	X
Cation interstitial	M_i	Anion interstitial	X_i
Cation vacancy	V_m	Anion vacancy	V_x
Cation on anion site	M_x	Anion on cation site	X_m

Cation vacancy–anion vacancy pair ($V_m X_m$)
Electron in conduction band e^-
Electron hole in valence band e^+

To develop the equations describing the relative concentrations of atomic and electronic defects we use the *law of mass action* which is: "The rate of any reaction is at each instant proportional to the concentrations of the reactants with each concentration raised to a power equal to the number of molecules of each species participating in the process."

For a general reaction:

$$lL + mM + \cdots = qQ + rR + \cdots \qquad (4.44)$$

the free energy change at equilibrium is

$$\Delta F^0 - \Delta F = -RT \ln(K) \qquad (4.45)$$

where

$$K \equiv \left(\frac{a_Q^q a_R^r}{a_L^l a_M^m}\right) \qquad (4.46)$$

and

$$a \equiv \frac{P_i}{P_{i_0}} \qquad (4.47)$$

with

P_{i_0} = equilibrium pressure of a gas i with its pure solid
P_i = equilibrium pressure of gas i with its solid in the reaction mixture

The development of mass action theory of defects can be evaluated by writing the equilibrium constant as

$$K = \frac{[Q]^q [R]^R}{[L]^l [M]^m} \qquad (4.48)$$

For ionization of defects we consider as a typical reaction:

$$V = V^+ + e^- \qquad (4.49)$$

which yields therefore, using the law of mass action:

$$K = \frac{[V^+][e^-]}{[V]} \qquad (4.50)$$

This expression is valid, however, only as long as classical statistical mechanics is obeyed, i.e., when $T \gg 0°K$. The donor level must be close to the conduction band and the donor solubility is not too large. Otherwise, K is a function of solute concentration as well as a function of temperature. We consider only cases where this is obeyed, e.g., $(E_i - E_f) \gg 4kT$ where E_f is the Fermi level. Therefore, nearly complete ionization of the vacancy has occurred.

4.9 FRENKEL DISORDER IN A STOICHIOMETRIC CRYSTAL

The combination of a cation vacancy and a cation interstitial is termed a *Frenkel defect*. As illustrated in Figure 4.18a, a trapped electron at the metal vacancy (V_m) contributes holes to the valence band. An electron (or two electrons if the cation is divalent) will be missing from the valence shell of one of the anions. This effectively provides an empty energy site or

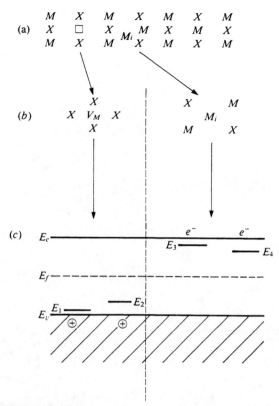

FIGURE 4.18 Frenkel disorder in a stoichiometric crystal.

"hole" associated with the cation vacancy. Ionization of the hole occurs at energy E_1. Ionization of the second hole occurs at E_2. The ionization of the vacancy–hole defect yields an electron acceptor state for the cation vacancy and a dissociated hole. Thus, ionization of the cation vacancy can be written as

$$V_m = V_m^- + e^+(E_1) \tag{4.51}$$

$$V_m^- = V_m^{2-} + e^+(E_2) \tag{4.52}$$

and leads to p-type semiconduction.

The trapped electron at the metal interstitial atom (M_i) contributes electrons to the conduction band, i.e., M_i serves as a donor with the following equations for ionization:

$$M_i^0 = M_i^+ + e^-(E_3) \tag{4.53}$$

$$M_i^+ = M_i^{2+} + e^-(E_4) \tag{4.54}$$

4.9 FRENKEL DISORDER IN A STOICHIOMETRIC CRYSTAL

Thus, for Frenkel defects, removal of a fraction (δ) of M atoms leaves an equivalent number of vacancies (δV_m) and metal interstitial atoms (δM_i). Therefore, we can write

$$MX = M_{(1-\delta)}X + \delta V_m + \delta M_i \tag{4.55}$$

with an equilibrium constant associated with reaction (4.50):

$$K' = \frac{(a_{M_{(1-\delta)}})(a_{V_m})^\delta (a_{M_i})^\delta}{a_{MX}} \tag{4.56}$$

By assuming that $\delta \ll 1$, the defects will obey *Henry's law*, i.e., no interaction between defects, and the crystal will obey *Raoult's law*, i.e., ideal behavior, so that

$$a_{M_{(1-\delta)}} \approx a_{V_m} \equiv 1 \tag{4.57}$$

and

$$(a_{V_m})^\delta \approx [V_m] \tag{4.58}$$

where

$$[V_m] \text{ is } \frac{\text{number of defects}}{\text{cm}^3 \text{ of crystal}} \tag{4.59}$$

and

$$(a_{M_i})^\delta \approx [M_i] \tag{4.60}$$

where

$$\delta_{M_i} = \frac{M_i}{M} = \frac{\text{number of interstitial defects}}{\text{number of } M \text{ sites}/(\text{cm}^3 \text{ of crystal})} \tag{4.61}$$

Consequently, there is a constant

$$K_1 = [V_m][M_i] = A_1 k' \tag{4.62}$$

where

$$k' = \exp\left[-\frac{\Delta G_1^0}{kT}\right] = \exp\left[\frac{\Delta S_1^0}{k}\right] \exp\left[-\frac{\Delta H_i^0}{kT}\right] \tag{4.63}$$

where ΔH is the enthalpy associated with reaction (4.63) and ΔS is the entropy.

Now, since

$$\Delta H^0 = \Delta H_{V_m} + \Delta H_{M_i} \tag{4.64}$$

the enthalpy of formation of the cation vacancy (ΔH_{V_m}) and the enthalpy of formation of a cation interstitial (ΔH_{M_i}), we can write for the equilibrium constant of the Frenkel defect:

$$K_1 = A_1 \exp\left[\frac{\Delta S_1^0}{k}\right] \exp\left[-\frac{\Delta H_{V_m}}{kT}\right] \exp\left[-\frac{\Delta H_{M_i}}{kT}\right] \qquad (4.65)$$

Thus, if ΔH_{V_m} or ΔH_{M_i} is large, K_1 will be small and $[V_m]$ and $[M_i]$ will be small.

As we saw earlier in equations 4.51–4.54, some of the defects may be ionized, and at $T \gg 0°K$ which we assume, nearly all will be ionized. Consequently, we can write for the divalent Frenkel metal-vacancy defects the following based upon equations (4.46) and (4.47)

$$K_2 = \frac{[e^+]^2[V_m^{2-}]}{[V_m]} = A_2 \exp\left[-\frac{\Delta E_2}{kT}\right] \qquad (4.66)$$

Since for a solid $\Delta H \equiv \Delta E$ due to the $\Delta(PV)$ term being so small in $\Delta H = \Delta E + \Delta(PV)$.

Thus, when ΔE_2 is small, K_2 is large and ionization is complete. Complete ionization means that 2 holes are created for every V_m^{2-} defect, i.e.,

$$[e^+] = 2[V_m] \qquad (4.67)$$

and for ionization of divalent metal interstitials we can write

$$K_3 = \frac{[e^+]^2[M_i^{2-}]}{[M_i]} = A_3 \exp\left[-\frac{\Delta E_3}{kT}\right] \qquad (4.68)$$

Examples of oxides that exhibit Frenkel defects are: FeO, NiO, CoO, FeS, and Cu_2O.

Other types of defect reactions include:

1. Frenkel defects on the anion sublattice:

$$MX = MX_{(1-\delta)} + \delta V_x + \delta X_i \qquad (4.69)$$

$$K_4 = A_4 \exp\left[\frac{\Delta S_4}{k}\right] \exp\left[-\frac{\Delta H_{V_x}}{kT}\right] \exp\left[-\frac{\Delta H_{X_i}}{kT}\right]$$

$$\times \exp\left[-\frac{\Delta E_{V_x}}{kT}\right] \exp\left[-\frac{\Delta E_{X_i}}{kT}\right] \qquad (4.70)$$

2. Shottky disorder (equal $V_m = V_x^-$):

$$MX = M_{(1-\delta)}X_{(1-\delta)} + \delta V_m + \delta V_x \qquad (4.71)$$

$$K_5 = A_5 \exp\left[\frac{\Delta S_5}{k}\right] \exp\left[-\frac{\Delta H_{V_m}}{kT}\right] \exp\left[-\frac{\Delta H_{V_x}}{kT}\right]$$

$$\times \exp\left[-\frac{\Delta E_{V_m}}{kT}\right] \exp\left[-\frac{\Delta E_{V_x}}{kT}\right] \qquad (4.72)$$

Examples are: NaCl, KCl, KBr.

3. Substitutional disorder:

$$MX = M_{(1-\delta)}X_{(1-\delta)} + \delta X_m + \delta M_x \qquad (4.73)$$

$$K_6 = A_6 \exp\left[\frac{\Delta S_6}{k}\right] \exp\left[-\frac{\Delta H_{X_m}}{kT}\right] \exp\left[-\frac{\Delta H_{M_x}}{kT}\right]$$

$$\times \exp\left[-\frac{\Delta E_{X_m}}{kT}\right] \exp\left[-\frac{\Delta E_{M_x}}{kT}\right] \qquad (4.74)$$

In this case an X ion on an M site will reduce the electro-static repulsion energy of negatively charged surroundings by losing electron charge, i.e., it acts as a donor. Thus, we can write

$$X_m^- = X_m^0 + e^- \qquad (4.75)$$
$$X_m^0 = X_m^+ + e^- \qquad (4.76)$$

whereas an M ion on an X site yields holes and is an acceptor. Thus,

$$M_x^+ = M_x^0 + e^+ \qquad (4.77)$$
$$M_x^0 = M_x^- + e^+ \qquad (4.78)$$

In summary, defects that are electron donors include:

1. Anion vacancy (see Figure 4.19).
2. Cation interstitial (see Figures 4.18, 4.20).
3. Anion on a cation site (see Figure 4.21).

Defects that are acceptors include:

4. Cation vacancy (see Figures 4.18, 4.22).
5. Anion interstitial (see Figure 4.23).
6. Cation on an anion site (see Figure 4.24).

4 IONIC AND DEFECT CONDUCTORS

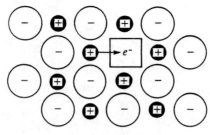

Donor site

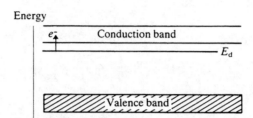

FIGURE 4.19 Anion vacancy.

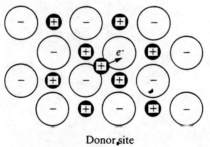

Donor site

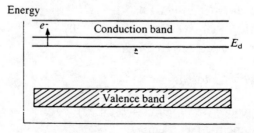

FIGURE 4.20 Cation interstitial.

4.9 FRENKEL DISORDER IN A STOICHIOMETRIC CRYSTAL

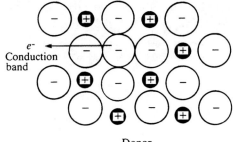

Donor

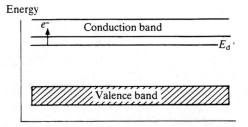

FIGURE 4.21 Anion on a cation site.

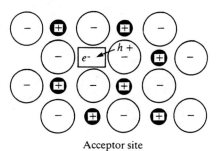

Acceptor site

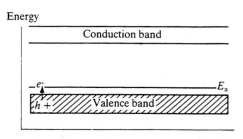

FIGURE 4.22 Cation vacancy.

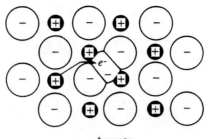

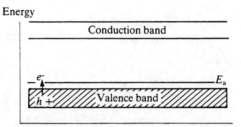

FIGURE 4.23 Anion interstitial.

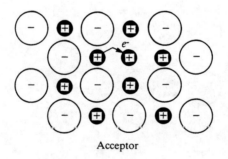

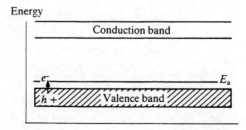

FIGURE 4.24 Cation on anion site.

4.9 FRENKEL DISORDER IN A STOICHIOMETRIC CRYSTAL

For each crystal, the enthalpies of formation of the various defects are considerably different. As a result, only one type of atomic defect will usually predominate where K is the largest. Consequently, a total of 11 possible equilibrium equations, such as $K_1, K_2, K_3, \ldots$ will usually reduce to three equilibrium constants. (See Problem 4.4 for demonstration of K_7 to K'.) In addition, the following equations, which represent physical constraints on the defect equilibrium, can be imposed, i.e., intrinsic holes = intrinsic electrons

$$K_{12} = [e^-][e^+] \tag{4.79}$$

$$K_{12} = A_{12} \exp\left[-\frac{E_g}{kT}\right] \tag{4.80}$$

For electrical neutrality:

$$[V_m^-] + [e^-] + [X_i^-] + [M_x^-] = [V_x^+] + [e^+] + [M_i^+] + [X_m^+] \tag{4.81}$$

If the crystal is stoichiometric:

$$[M_i] + [M_i^+] + [M_x] + [M_x^-] - [V_m] - [V_m^-]$$
$$= [X_i] + [X_i^-] + [X_m] + [X_m^+] - [V_x] - [V_x^+] \tag{4.82}$$

For a Shottky type of defect, characteristic of alkali halides, at ambient temperature there are usually equivalent concentrations of cation vacancies and anion vacancies with both being singly ionized. For MgO and other cubic oxides the concentration of singly vs. doubly ionized defects is much more temperature dependent.

Thus, we can write for a crystal with Shottky defects:

$$M_x = M_{(1-\delta)}X_{(1-\delta)} + \delta V_m + \delta V_x \tag{4.83}$$

$$K_5 = [V_m][V_x] \tag{4.84}$$

$$V_m = V_m^- + e^+ \tag{4.85}$$

$$K_2 = \frac{[e^+][V_m^-]}{[V_m]} \tag{4.86}$$

$$V_x = V_x^+ + e^- \tag{4.87}$$

$$K_7 = \frac{[V_x^+][e^-]}{[V_x]} \tag{4.88}$$

Imposing neutrality conditions:

$$[V_m^-] + [e^-] = [V_x^+] + [e^+] \tag{4.89}$$

and requiring maintenance of stoichiometry,

$$[V_m] + [V_m^-] = [V_x] + [V_x^+] \tag{4.90}$$

and finally electron–hole balance

$$K_{12} = [e^+][e^-] \tag{4.91}$$

In principle, each of these equilibrium constants from K_1 to K_{12} can be obtained from experimental measurements. Usually, associated data from lattice parameter and density measurements as a function of M or X, along with electrical conductivity and Hall constants, are required to analyze the defect chemistry of a particular crystal.

4.10 EXAMPLE OF A DEFECT SEMICONDUCTOR: LEAD SULFIDE

PbS is an intrinsic semiconductor with a band gap of 0.40 eV. The crystal structure is that of NaCl rock salt. Therefore, Shottky defects are the most likely type of defect. The concentration of defects can be altered significantly by varying the partial pressure of S_2 above the material while heating it. Quenching from the heat treatment temperature freezes in the high temperature defect.

Measurement of the concentration and type of charge carriers as a function of the partial pressure of S_2 overpressure at 1000°K and 1200°K produces results such as shown in Figure 4.25. The electrical conductivity and Hall effect (see Chapter 3) were used to obtain the data.

The electron concentration and therefore the conductivity decrease as the partial pressure of S_2 increases. An abrupt change in the curve (discontinuity) occurs when the electron and hole concentration are equal. This is at 10^{-3} atmospheres of S_2 overpressure at 1000°K. A further increase in S_2 partial pressure causes the electron concentration to increase. The conductivity is n-type before the discontinuity occurs. It is p-type after the discontinuity. As the temperature increases, the discontinuity occurs at high S_2 partial pressures, e.g., at 10^{-1} atmosphere at 1200°K.

If we assume that the PbS defect is of the Shottky type, then

$$PbS = Pb_{(1-\delta)}S_{(1-\delta)} + \delta V_{Pb} + \delta V_S \tag{4.92}$$

where

$$K_1 = [V_{Pb}][V_S] \tag{4.93}$$

The cation vacancies may act as acceptors.

$$V_{Pb}^0 = V_{Pb}^- + e^+ \tag{4.94}$$

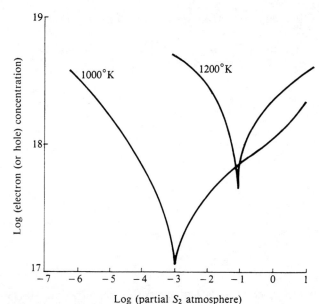

FIGURE 4.25 Pure PbS defect concentrations. (After Bloem, 1956).

where

$$K_2 = \frac{[V_{Pb}^-][e^+]}{[V_{Pb}^0]} \quad (4.95)$$

The formation of the crystal involves the reaction:

$$Pb + \tfrac{1}{2}S_2 = PbS_{(solid)} \quad (4.96)$$

where

$$K_4 = P_{Pb} \cdot (P_{S_2})^{1/2} \quad (4.97)$$

when P_{S_2} (the partial pressure of S_2) increases, the sulfur S will tend to be moved into the sulfur vacancies. The concentration of the vacancies, $[V_S^+]$, then decreases. For each $[V_S^+]$ removed, a conduction electron is also removed. When P_{Pb} increases, the Pb will tend to be moved into the Pb vacancies. Therefore the concentration of the Pb vacancies, $[V_{Pb}]$, will decrease and thereby an electron hole is also removed.

The condition of electrical neutrality must be maintained regardless of the pressure variation:

$$[V_S^+] + [e^+] = [V_{Pb}^-] + [e^-] \quad (4.98)$$

A cation vacancy and an electron hole must be formed for each anion vacancy filled and electron removed. The reverse is likewise true. Therefore,

$$\tfrac{1}{2}S_2(g) = S_{(S)} + V_{Pb}^- + e^+ \quad (4.99)$$

where

$$K_5 = \frac{[V_{Pb}^-][e^+]}{P_{S_2}^{1/2}} \quad (4.100)$$

and

$$Pb(g) = Pb_{(S)} + V_S^+ + e^- \quad (4.101)$$

where

$$K_6 = \frac{[V_S^+][e^-]}{P_{Pb}} \quad (4.102)$$

The product of electron hole and electron concentrations define a constant:

$$K_7 = [e^-][e^+] \quad (4.103)$$

Using equations (4.88)–(4.92) we can solve for K_1:

$$K_1 = [V_{Pb}^-][V_S] \quad (4.104)$$

$$K_1 = \frac{[V_{Pb}^-][e^+]}{K_2} \cdot \frac{[V_S^+][e^-]}{K_3} \cdot \frac{1}{[e^+][e^-]} \quad (4.105)$$

$$K' = \frac{K_1 K_2 K_3}{K_7} \quad (4.106)$$

$$K' = [V_{Pb}^-][V_S^+] \quad (4.107)$$

The experimental results can now be interpreted. Using equation (4.97)

$$P_{Pb} \cdot P_{S_2}^{1/2} = K_4 \quad (4.108)$$

we see that as the partial pressure of $S_2^{1/2}$ goes down, the partial pressure of Pb must increase in order for K_4 to remain constant. Using equation (4.102):

$$\frac{[V_S^+][e^-]}{P_{Pb}} = K_6 \quad (4.109)$$

we see that at large values of the partial pressure of Pb the sulfur vacancies

4.10 EXAMPLE OF A DEFECT SEMICONDUCTOR: LEAD SULFIDE

and ionized electrons must also be high. For this case, a high n-type conductivity results. Conversely, when P_{S_2} increase the conductivity decreases.

The trend continues until

$$[e_i^-] = [e_i^+] = \sqrt{K_7} \tag{4.110}$$

which corresponds to the discontinuity in the slope of conductivity vs. P_S in Figure 4.25. Above the discontinuity at $P_{S_2} = 10^{-3}$ we have high concentration of $[V_{Pb}^-]$ and $[e^+]$ which corresponds to p-type conduction.

Since K_7 increases with temperature the values of

$$[e_i^-] = [e_i^+] \tag{4.111}$$

increases as the temperature increases. The discontinuity then corresponds to larger values of the partial pressure P_{S_2} at higher temperatures.

We can evaluate these constants from Figure 4.25. At a temperature of 1000°K the discontinuity occurs at

$$[V_{Pb}^-] = [V_S^+] = [e^-]_i = [e^+]_i = 10^{17} \text{ cm}^{-3} \tag{4.112}$$

then

$$K_7 = 10^{34} \tag{4.113}$$

and

$$K' = 10^{34}$$

Likewise at a temperature of 1200°K

$$K_7 = 10^{35.6} \tag{4.114}$$

and

$$K' = 10^{35.6} \tag{4.115}$$

Using linear interpolation, an equation for $\log K'$ as a function of inverse temperature can be extracted:

$$\Delta \log K' = 1.6 \tag{4.116}$$

for

$$T = 1000°K \rightarrow 1200°K \tag{4.117}$$

Likewise

$$\Delta\left(\frac{1}{T}\right) = \frac{1}{1000} - \frac{1}{1200} \tag{4.118}$$

$$= 1.667 \times 10^{-4} \tag{4.119}$$

178 CHAPTER 4: IONIC AND DEFECT CONDUCTORS

The slope m of the log K' vs. $1/T$ relationship is

$$m = \frac{\Delta \log K'}{\Delta \left(\frac{1}{T}\right)} = \frac{1.6}{1.667 \times 10^{-4}} \qquad (4.120)$$

$$m = 9600 \qquad (4.121)$$

But we must convert the slope to eV/°K: thus

$$\frac{k}{k}m = \frac{k(9600)}{k} = \frac{1}{k}\frac{\Delta \log K'}{\Delta \left(\frac{1}{kT}\right)} \qquad (4.122)$$

where

$$k = 1.38 \times 10^{-23} \text{ J/°K} \qquad (4.123)$$

or

$$k = \frac{1.38 \times 10^{-23} \text{ J/°K}}{1.602 \times 10^{-19} \text{ J/eV}} \qquad (4.124)$$

$$k = 0.826 \times 10^{-5} \text{ eV/°K} \qquad (4.125)$$

Then the slope in units of eV becomes

$$m = 1.934 \text{ eV/°K} \qquad (4.126)$$

which represents the energy required to form $[V_{Pb}^-]$ and $[V_S^+]$.
The linear equation now has the form

$$y = mx + b \qquad (4.127)$$

where

$$y = \log K' \qquad (4.128)$$

and

$$x = \frac{1}{kT} \qquad (4.129)$$

The y intercept b can be found at $x = 0$ by using the slope equation (4.121):

$$m = 9600 \qquad (4.130)$$

4.10 EXAMPLE OF A DEFECT SEMICONDUCTOR: LEAD SULFIDE

The intercept is

$$\Delta b = \Delta\left(\frac{1}{T}\right) \cdot m \tag{4.131}$$

$$\Delta b = \frac{1}{1200} \cdot 9600 = 8 \tag{4.132}$$

$$b = 35 + \Delta b \tag{4.133}$$

$$b = 43 \tag{4.134}$$

Therefore the final equation is found from equation (4.127) to be

$$10^y = 10^{(mx+b)} \tag{4.135}$$

This becomes

$$K' = 10^{(m/kT + b)} \tag{4.136}$$

which is approximately

$$K' \approx 10^{44} \exp\left[-0.83/kT\right] \tag{4.137}$$

We can compare this experimental result with what is expected from theory:

$$K = \exp\left[\Delta S^0/k\right] \exp\left[-\Delta H^0/kT\right] \tag{4.138}$$

ΔS^0 is usually very small therefore we can consider the first factor:

$$e^{\Delta S^0/k} \approx 1 \tag{4.139}$$

Then we have

$$K = e^{-\Delta H^0/kT} \tag{4.140}$$

$$K = [X_{V_{\overline{Pb}}}][X_{V_S^+}] \tag{4.141}$$

where $[X]$ is the concentrations of vacancies expressed as atom fractions:

$$[X_{V_{Pb}}] = \frac{[V_{\overline{Pb}}]}{N_{Pb}} \tag{4.142}$$

$$[X_{V_S}] = \frac{[V_S^+]}{N_S} \tag{4.143}$$

where N is the number of atoms per cubic centimeter.

From equation (4.102)

$$K' = [V_{Pb}^-][V_S^+] \quad (4.144)$$

$$K' = N_{Pb}N_S[X_{V_{Pb}}][X_{V_S^+}] \quad (4.145)$$

Substituting equations (4.140) and (4.141):

$$K' = N^2 K \quad (4.146)$$

$$K' = N^2 \exp[-\Delta H^0/kT] \quad (4.147)$$

where

$$N = N_{Pb} = N_S = 2 \times 10^{22} \quad (4.148)$$

Thus

$$K' = 10^{44.6} \exp[-\Delta H^0/kT] \quad (4.149)$$

which compares favorably with the experimental equation (4.137).

4.11 PROTONIC CONDUCTION IN GLASSES

As discussed earlier, the charge carriers in most glasses are mobile monovalent cations, such as Na^+ in silicate glasses, or electrons in glasses containing multivalent transition metal ions. Proton (H^+) conduction is generally very low for silicate glasses because O–H bonding is very strong, e.g., $v_{OH} = 3700$ cm^{-1}. However, O–H bonding in phosphate glasses is weak due to hydrogen bonding where the counter oxygen is of a bridging type (see Abe et al., 1988 for details).

The electrical conductivity in various alkaline earth phosphate glasses is proportional to the square of the proton concentration

$$\sigma - A_0[H^+]^2 \quad (4.150)$$

where $[H^+]$ is proton concentration in mol/l and A_0 is a constant depending on the host glass. The constant A_0 is a measure of proton mobility. Abe (1988) has shown that the proton mobility increases linearly with proton concentration in the phosphate glasses and assuming that all protons are mobile charge carriers:

$$\mu = \frac{A_0[H^+]}{e} \quad (4.151)$$

Figure 4.26 shows the relation between electrical conductivity and proton concentration $[H^+]$ in 45MO–55P$_2$O$_5$ glasses where M = Mg, Ca, and Ba.

4.11 PROTONIC CONDUCTION IN GLASSES

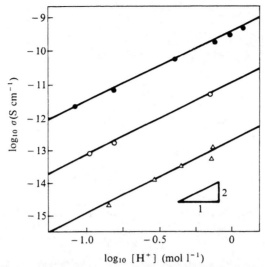

FIGURE 4.26 Relationships between electrical conductivity (σ at 417 °K) and proton concentration [H$^+$] in 45MO–55P$_2$O$_5$ glasses [after Abe et al. (1988)].

Experimentally it was found that the activation energy for conduction E is also a function of proton concentration [H$^+$] (Figure 4.27). For the phosphate glasses the following relations were obtained:

$$45\text{MgO–}55\text{P}_2\text{O}_5; \quad E = 110 - 12 \log_{10} [\text{H}^+] \quad (4.152)$$

$$45\text{CaO–}55\text{P}_2\text{O}_5; \quad E = 101 - 14 \log_{10} [\text{H}^+] \quad (4.153)$$

$$50\text{CaO–}50\text{P}_2\text{O}_5; \quad E = 105 - 16 \log_{10} [\text{H}^+] \quad (4.154)$$

$$45\text{BaO–}55\text{P}_2\text{O}_5; \quad E = 95 - 16 \log_{10} [\text{H}^+] \quad (4.155)$$

Consequently, E can be expressed as

$$E = E_0 + E_1 \quad (4.156)$$

where E_0 is an activation energy at unit concentration of the proton ([H$^+$] = 1) and E_1 is an activation energy dependent on carrier concentration. Figure 4.28 is a plot of E_0 vs. ν_{OH} with E_0 increasing linearly with the O–H vibrational frequency. (For a discussion of optical vibrational frequencies see Chapter 8.) Thus, E_0 and E_1 can be expressed as:

$$E_0 = B_0 + B_1 \nu_{\text{OH}} \quad (4.157)$$

$$E_1 = -B_2 \log_{10} [\text{H}^+] \quad (4.158)$$

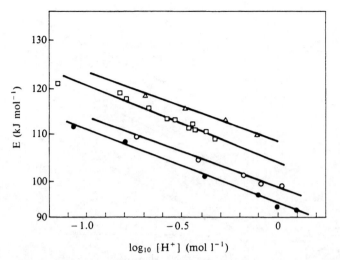

FIGURE 4.27 Activation energy for electrical conduction E vs. $\log_{10}[H^+]$ in glasses: △ = 45MgO–55P_2O_5; □ = 50CaO–50P_2O_5; ○ = 45CaO–55P_2O_5; ● = 45BaO–55P_2O_5. [After Abe et al. (1988).]

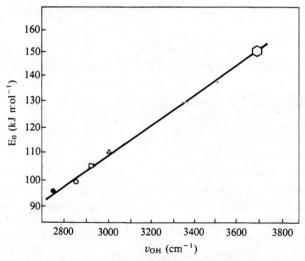

FIGURE 4.28 Plot of E_0 vs. V_{OH} in glasses: △ = 45MgO–55P_2O_5; □ = 50CaO–50P_2O_5; ○ = 45CaO–55P_2O_5; ● = 45BaO–55P_2O_5; ⬡ = SiO_2 by extrapolation. [After Abe et al. (1988).]

where B_0 and B_1 are constants ($B_0 = -66$, $B_1 = 5.89 \times 10^{-2}$) and B_2 is a value depending on host-glass compositions.

This analysis shows that protonic conduction is controlled by two elemental processes; (1) breaking of the O–H bond; controlled by E_0 and (2) transport between neighboring sites; controlled by E_1.

These findings are potentially important for the development of fast proton conductive glasses for a solid electrolyte in H_2–O_2 fuel cells and H_2 sensors.

PROBLEMS

4.1 For a given diatomic defect semiconductor similar to PbS, the concentration of electrons at 800°K shows a distinct discontinuity of $10^{16.5}$ at a specific partial pressure of the anion. At 1150°K a similar discontinuity occurs for an electron concentration of $10^{17.2}$ at a higher partial pressure. Plot the discontinuity concentrations as a function of temperature.

4.2 For the condition in problem 4.1, calculate the activation energy in eV per °K and in joules per °K for the formation of the vacancies.

4.3 For problem 4.1, calculate the number of cation and anion pairs N per cubic centimeter.

4.4 Express the equation for K' for problem 4.1.

4.5 Which compounds in Table 4.1 have mixed conduction? Why?

4.6 Explain the effect of temperature on mixed conduction in a transition metal oxide.

4.7 Calculate the conductivity of the ternary $(75\% - X\%) SiO_2 + 6\%$ $MgO + X\%$ Na_2O glass system:

   (a) Plot the conductivity as a function of temperature from 1000°C to 1400°C for $X = 20\%$.

   (b) Plot the conductivity as a function of $[Na_2O]$ for $12\% < X < 20\%$.

4.8 (a) Compare the results in problem 4.7 with Figure 4.12. Is the model above or below the experimental results?

   (b) Using least square analysis, modify the coefficients in equation (4.30) to match Figure 4.12 more closely.

   (c) What is the mobile ionic species in this case?

4.9 Consider the glass in problem 4.7 with 12% Na_2O that is 4 cm long:

   (a) What is the activation energy in eV?

   (b) Compare this with the band gap for this type of material?

   (c) Plot the activation energy in eV as a function of sample length in centimeters.

4.10 Consider the fast ion conductor β-alumina. From values in Table 4.3 construct an equation like equation (4.30) for this material.
  (a) Plot the conductivity as f(T) for $25°C < T < 1000°C$.
  (b) Compare this result with experimental values and determine the useful temperature range for this model.

READING LIST

Abe, Y., Hosono, H., Ohta, Y., and Hench, L. L., (1988). *Physics Review B*, **38**(14), 10166.

Anderson, J. C., (1984). *Dielectrics*, Chapman and Hall, London, Chapter: 1.

Azaroff, L. and Brophy, J. J., (1963). *Electronic Processes in Materials*, McGraw-Hill, New York, Chapter: 12.

Blatt, F. J., (1968). *Physics of Electronic Conduction in Solids*, McGraw-Hill, New York, Chapter: 7.

Bloem, J., (1956). *Philips Res. Repts.*, **11**, 273.

Bransky, I. and Tallan, N. L., (1971). "Electrical conduction in low mobility materials," *Physics of Electronic Ceramics*, Part A, Dekker, New York, (L. L. Hench and D. B. Dove, eds.) Chapter: 3.

Charles, R. J., (1965). *J. Am. Cer. Soc.*, **48**, 432.

Epstein, A., et al., (1979). *Sci. Amer.*, **241**, 52.

Friauf, R. J., (1972). "Basic theory of ionic transport processes," *Physics of Electrolytes*, Vol. 1, Academic Press, New York, (J. Hladik, ed.) 210.

Guy, A. G., (1972). *Introduction to Materials Science*, McGraw-Hill, New York, Chapter: 5.

Heckman, R. W., Ringler, J. A., and Williams, E. L., (1967). *Phys. Chem. Glasses*, **8**, 145.

Hench, L. L. and Dove, D. B., (1971). *Physics of Electronic Ceramics*, Part A, Dekker, New York, Chapters: 6–8.

Hummel, R. E., (1985). *Electronic Properties of Materials*, Springer-Verlag, Berlin, Chapter: 9.

Kingery, W. D., Bowen, H. K., and Uhlmann, D. R., (1976). *Introduction to Ceramics*, 2nd ed., Wiley, New York, Chapter: 17.

Kinser, D. L. and Hench, L. L., (1969). *J. Am. Cer. Soc.*, **52**, 638.

Kittel, C., (1986). *Introduction to Solid State Physics*, 6th ed., Wiley, New York, Chapter: 7.

Owen, A. E., (1963). *Progress in Ceramic Science*, J. Burke, ed. Pergamon, N.Y., 77.

Swalin, R. A., (1972). "Thermodynamics of Solids," 2nd Edition, John Wiley and Sons, N.Y.

Tickle, R. E., (1967). *Phys. Chem. Glasses*, **8**, 101.

Van Vlack, L., (1985). *Elements of Materials Science and Engineering*, 5th ed., Addison-Wesley, Reading, Massachusetts, Chapter: 8.

5

LINEAR DIELECTRICS

5.1 IMPORTANCE

Dielectric properties are of special importance when ceramics or glasses are used either as a capacitive element in electronic applications, or as insulation. The dielectric constant, dielectric loss factor, and dielectric strength usually determine the suitability of a particular material for such applications. Variation of dielectric properties with frequency, field strength, and other circuit variables influence performance. Environmental effects such as temperature, humidity, and radiation also influence dielectric applications. Consequently, it is necessary to examine dielectric theory in terms of materials response, circuit response, and environmental response.

Ceramics and glasses as insulating materials have definite advantages over plastics which are major competitors. Lack of flexibility is a problem in most instances where this feature of insulation is required. However, ceramics and glasses possess superior electrical properties, are absent from creep or deformation under stresses at room temperature, and resist environmental changes, particularly at high temperature where plastics oxidize, gassify, or decompose. Ceramics and glasses also can be used to form glass-tight seals with metals and other ceramic components, and thereby become an integral part of an electronic device.

5.2 THEORY

Dielectric properties comprise the nonlong-range conducting electrical characteristics of a material. Dielectric responses result from the *short-range* motion of charge carriers under the influence of an applied electric field. The motion of the charges leads to the storage of electrical energy and the capacitance of the dielectric. Consequently, as a first step in examining dielectric behavior, let us consider the definition of *capacitance*. Capacitance is a measure of the ability of any two conductors in proximity to store a charge Q, when a potential difference V is applied across them.

$$C = Q/V = \frac{\text{coulombs (C)}}{\text{Volt (V)}} = \text{Farad (F)} \tag{5.1}$$

The capacitance of a vacuum capacitor is determined purely by the geometry. It can be shown from elementary electro-statics that the charge density on the plates, Q is proportional to the area A in square meters and the electric intensity applied $\mathbf{E} = V/d$, where d is the distance between the plates (in meters). The proportionality constant is defined as ϵ_0, the permittivity of free space, and is equal to 8.854×10^{-12} C^2/m^2 or F/m. Thus, the capacitance of a parallel-plate capacitor shown in Figure 5.1 will be equal to:

$$Q = qA = \pm\epsilon_0 \mathbf{E} A = \epsilon_0(V/d)A \tag{5.2}$$

$$C_0 = Q/V = \frac{\epsilon_0(V/d)A}{V} = \epsilon_0 A/d \tag{5.3}$$

where q = charge per unit area
d = separation in meters
A = area in square meters

When a material is inserted between the plates the capacitance is increased. The dielectric constant k is defined as the ratio of the capacitance of a condenser, or capacitor, with a dielectric between the plates to that with a vacuum between the plates.

$$k = C/C_0 = \frac{\epsilon A/d}{\epsilon_0 A/d} = \epsilon/\epsilon_0 \tag{5.4}$$

where ϵ is the permittivity of the dielectric material, also in units of C^2/m^2 or F/m.

Thus the dielectric constant of a material is the ratio of the permittivity of the material to the permittivity of free space. Consequently, the dielectric constant k is also called the relative permittivity in some literature. It can be seen in Table 5.1 that typical values of the dielectric constant of most insulating glasses and ceramics are between five and 10.

The charge stored in a typical linear capacitor is on the order of microcoulombs (10^{-6} C) or picocoulombs (10^{-12} C). So, permittivity is

TABLE 5.1 Dielectric Properties of Insulating Glasses and Ceramics at Room Temperature

Glass	Frequency (Hz)	Dielectric constant
SiO_2	10^2–10^{10}	3.78
Al_2O_3 (parallel)	dc	10.50
Al_2O_3 (perpendicular)	dc	8.60
Corning 7040 (Na_2O, K_2O, SiO_2, B_2O_3)	10^2–10^6	4.84–4.73
Pyrex (Na_2O, SiO_2, SiO_2)	10^2–10^6	5.02–4.84
Corning 0080 (Na_2O, CaO, SiO_2)	10^2–10^6	8.30–6.90
Corning 1770 (CaO, Al_2O_3, SiO_2)	10^2–10^6	6.25–6.00
9% Na_2O–91% SiO_2	10^2–10^6	6.4–5.4
20% Na_2O–80% SiO_2	10^2–10^6	10.8–6.6
12.8% Li_2O–87.2% SiO_2	10^2–10^6	9.94–4.95
12.8% Na_2O–87.2% SiO_2	10^2–10^6	8.09–5.66
12.8% K_2O–87.2% SiO_2	10^2–10^6	7.53–6.09
Mg_2SiO_4 (Forsterite)	dc	6.2
Pb-silica glass	dc	19.00
MgO	dc	9.60
NaCl	dc	5.90
α-SiC	dc	9.70
BaO	dc	5.7
Diamond	dc	6.6
LiF	dc	9.0
KBr	dc	4.9

ordinarily in the range of microfarads/meter (10^{-6} F/m) or picofarads/meter (10^{-12} F/m).

Equation (5.4) shows that the presence of a material between the plates of a condenser increases the ability of the plates to store charge. The reason for this is a result of the material containing charged species which can be displaced in response to the field applied across the material. The displaced

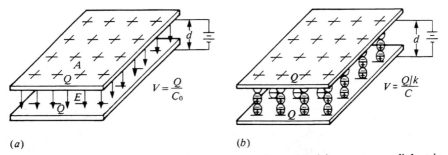

FIGURE 5.1 Charge on a parallel-plane capacitor with (*a*) vacuum as dielectric; (*b*) dielectric material between plates.

charges within the material compose dipoles with a moment $\mu = Q\delta$ (Cm) where δ is the separation distance. The electric dipoles, shown in Figure 5.1b, are oriented with respect to the applied field. The effect of the orientation is to "tie up" charges on the plates of the condenser and thus neutralize part of the applied field. The charge which is not neutralized by dipoles within the material, called *free charge*, equal to Q/k, produces an electric field and voltage towards the outside, $V = (Q/k)/C_0$. Therefore, a smaller external field is required to maintain the same surface charge because some of the charge is held by the polarization in the dielectric.

There are four primary mechanisms of polarization in ceramics and glasses. Each mechanism involves a short-range motion of charge and contributes to the total polarization of the material. The polarization

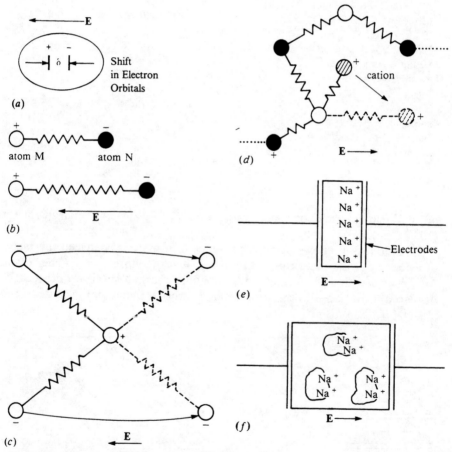

FIGURE 5.2 Schematic of polarization mechanisms in glass and ceramics. (*a*) Electronic. (*b*) Atomic or ionic. (*c*) High-frequency oscillatory dipoles. (*d*) Low frequency cation dipole. (*e*) Interfacial space charge polarization at electrodes. (*f*) Interfacial polarization at heterogeneities.

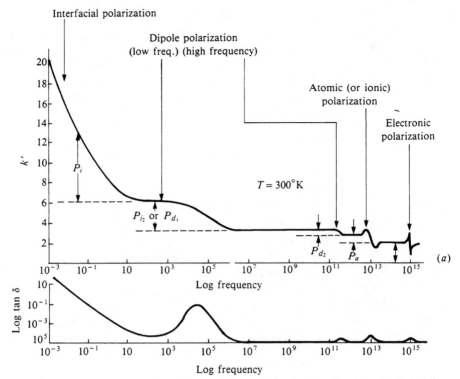

FIGURE 5.3 Frequency dependence of the polarization mechanisms in dielectrics. (a) Contribution to the charging constant (representative values of k'). (b) Contribution to the loss angle (representative values of tan δ).

mechanisms include: electronic polarization (P_e), atomic polarization (P_a), dipole polarization (P_d), and interfacial polarization (P_i). A schematic of the mechanism of operation of each of these major types of polarization as well as two special cases is given in Figure 5.2.

Electronic polarization is due to the shift of the valence electron cloud of the ions within the material with respect to the positive nucleus. This mechanism of polarization occurs at very high frequencies (10^{15} Hz†) which are in the ultraviolet optical range. The mechanism of polarization gives rise to a resonance absorption peak in the optical range, as shown in Figure 5.3. The index of refraction of the material depends on the electronic polarization as discussed in Chapter 8.

At frequencies in the infrared range (10^{12}–10^{13} Hz) (Figure 5.3), atomic or ionic polarization occurs. Atomic polarization is the displacement of positive and negative ions in a material with respect to each other, such as is shown in Figure 5.2b for a Si—O bond. A resonance absorption

† Hz = Hertz = cps = $\sec^{-1}$.

occurs at a frequency characteristic of the bond strength between the ions. If there are several types of ions in an insulator or a distribution in bond strengths such as occurs in glasses, the infrared absorption will be quite broad.

In the subinfrared range of frequencies, dipole polarization contributes to the dielectric properties. Dipolar polarization, also referred to in some texts as orientational polarization, involves the perturbation of the thermal motion of ionic or molecular dipoles, producing a net dipolar orientation in the direction of the applied field. Mechanisms of dipolar polarization can be generally divided into two categories. First, molecules containing a permanent dipole moment may be rotated against an elastic restoring force about an equilibrium position. This effect is especially important for a variety of liquids and gases and polar solids, such as ice and many organic plastic insulators. Two models of this type of polarization in glass are shown in Figure 5.2. Figure 5.2(c) depicts the oscillation of a Si—O—Si bond about an equilibrium position under a sinusoidal ac field. Since such a bond will possess a dipole moment when it is asymmetrical, the oscillation produces a dipolar polarization. The frequency of relaxation of such a mechanism is very high, 10^{11} Hz, at room temperature. A similar mode of dipolar polarization occurs due to the oscillation of an OH$^-$ group about an equilibrium position. The dipole moment is again due to the asymmetry of the Si—OH configurations in a random glass network. Oscillation of these moments also occurs in the range of 10^{11}–10^{12} Hz. Sometimes these mechanisms are termed *Stevels deformation polarization.*

The seond mechanism of dipolar polarization is an especially important contribution to the room temperature dielectric behavior of glasses and ceramics. It involves the rotation of dipoles between two equivalent equilibrium positions. It is the spontaneous alignment of dipoles in one of the equilibrium positions which gives rise to the nonlinear polarization behavior of ferroelectric materials. It is responsible for dielectric constant values of 10^4 or more in such materials. This is discussed in Chapter 6. In linear dielectrics, which we are concerned with in this chapter, orientational polarization occurs largely as a result of motion of charged ions between the interstitial positions within the ionic structure of the insulator. Figure 5.2(d) shows a schematic of dipolar polarization involving the oscillation of a sodium ion between two equivalent positions. Such oscillations occur continuously. The applied field makes the jumps in a direction parallel to the field more probable. Since an appreciable distance is involved in such an ionic transition the polarization occurs at a frequency range of 10^3–10^6 Hz, at room temperature. Because this mechanism involves the same mobile cations that contribute to the dc conductivity it is sometimes referred to as *migration losses.*

The last polarization mechanism, interfacial or space charge polarization, occurs when mobile charge carriers are impeded by a physical barrier that inhibits charge migration. The charges pile up at the barrier producing a localized polarization of the material. When an ac field is of sufficiently low frequency, less than 10^{-3} Hz, a net oscillation of charge can be produced between barriers as far apart as 1 cm (Figure 5.2e), producing a very large capacitance and dielectric constant (Figure 5.3). If the barriers are an internal structural feature (Figure 5.2f), or the density of charges contributing to the interfacial polarization is sufficiently large, the frequency range of sensitivity for interfacial polarization may extend into the kilocycle (10^3 Hz) range (Figure 5.3). In such a case it may be impossible to distinguish the frequency response of a dipole polarization mechanism, such as P_{d_1}, and an interfacial polarization mechanism such as P_{i_2}.

Of the four polarization mechanisms contributing to the dielectric properties of materials, this chapter discusses only dipole (orientational) and interfacial polarization, since they influence the circuit characteristics of insulators from 10^{-3} Hz to 10^9 Hz. Infrared and electronic polarization in ceramics and glasses are discussed as optical properties (Chapter 8).

Now that we have a qualitative understanding of the physical basis of polarization, let us consider in detail: (1) the effects of polarization on the circuit behavior of an insulator, and (2) a quantitative description of dipolar and interfacial polarization models.

5.3 CIRCUIT DESCRIPTION OF A LINEAR DIELECTRIC

If a sinusoidal potential $V = V_0 \exp i\omega t$ is applied to the dielectric, the charge must vary with time, as shown in equation (5.5), which constitutes a charging current, I_c:

$$Q = CV; \quad \text{so} \quad I_c \equiv dQ/dt = C\,dV/dt = i\omega CV = \omega CV_0 \exp\left[i(\omega t + \pi/2)\right] \tag{5.5}$$

The charging current in an ideal dielectric thus leads the applied voltage by $\pi/2$ radians (90°). This relationship is seen vectorially in Figure 5.4.

In addition to the charging current, associated with storage of electric charge by the dipoles, a loss current must also be considered for real dielectrics. The loss current arises from two sources: (1) The long-range migration of charges, e.g., dc Ohmic conduction as discussed in Chapter 4, and (2) the dissipation of energy associated with rotation or oscillation of dipoles. The latter contribution to the dielectric losses is a consequence of the charged particles having a specific mass and, therefore, an inertial resistance to being moved. Electrical energy from the field is lost in the overcoming of this inertia during polarization. The ac conduction from the inertial resistance and the dc conduction both are in phase with the applied voltage. Consequently a loss current in the dielectric can be written as

$$I_l = (G_{dc} + G_{ac})V \tag{5.6}$$

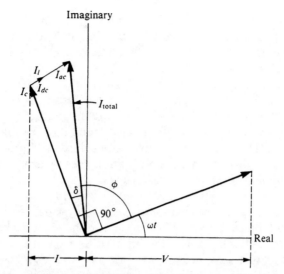

FIGURE 5.4 Vector diagram of charging, loss and total currents in a dielectric.

where G is the conductance in units of Siemens, mho, or ohm^{-1}. The total current I_T for a real dielectric material is thus the sum of equations (5.5) and (5.6):

$$I_T = I_c + I_l = (i\omega C + G_{dc} + G_{ac})V \tag{5.7}$$

As shown in Figure 5.4, the total current in a real dielectric is a complex quantity which leads the voltage by an angle $(90° - \delta)$ where δ is called the *loss angle*.

An alternative way of expressing the concept of a real dielectric possessing both charging and loss processes is to use a complex permittivity to describe the material:

$$\epsilon^* \equiv \epsilon' - j\epsilon'' \quad \text{and} \quad k^* \equiv \epsilon^*/\epsilon_0 = k' - ik'' \tag{5.8}$$

Thus, the total current in the dielectric can now be expressed in terms of the single material parameter k^*, since

$$C \equiv k^* C_0, \quad \text{so} \quad Q = CV = k^* C_0 V \tag{5.9}$$

and

$$i = dQ/dt = C\, dV/dt = k^* C_0 i\omega V = (k' - ik'') C_0 i\omega V_0 \exp i\omega t \tag{5.10}$$

and thus

$$I_T = i\omega k' C_0 V + \omega k'' C_0 V \tag{5.11}$$

The first term on the right-hand side of equation (5.11) describes charge storage in the dielectric and k' is thus called the *charging constant* or often just *dielectric constant* and ϵ'' and k'' are referred to as the *dielectric loss factor* and *relative loss factor*, respectively. The *loss tangent, loss angle,* or *dissipation factor,* tan δ, is defined as

$$\tan \delta \equiv \epsilon''/\epsilon' = k''/k' \tag{5.12}$$

The dissipation factor represents the relative expenditure of energy to obtain a given amount of charge storage. It is the "interest rate," so to speak. The product, k'' or k' tan δ, is sometimes termed the *total loss factor*, and provides the primary criterion for evaluating the usefulness of a dielectric as an insulator. To minimize k'', the losses in the insulator, it is desirable to have a small dielectric constant and most importantly a very small loss angle. The inverse of the loss tangent, $Q = 1/\tan \delta$, is used as a figure of merit in high-frequency insulation applications. In dielectric heating, the critical materials characteristics are the dielectric charging constant k', and the *dielectric conductivity*, $\sigma_T = \omega k''$.

5.4 RELATION OF DIELECTRIC CONSTANT TO POLARIZATION

In order to obtain a quantitative understanding of dielectric properties it is necessary to establish a relationship between the complex dielectric constant k^* and the polarization in the material. This can be done by considering the total electric displacement field **D** in the material. As shown in Figure 5.1b, **D** will be the sum of the electric field established if there was not a dielectric in the condenser, plus the polarization field within the material:

$$\mathbf{D} = \epsilon_0 \mathbf{E} + \mathbf{P} = \epsilon^* \mathbf{E} \tag{5.13}$$

Thus the total electric displacement in the material **D** is related to the external field **E** by the complex permittivity of the material ϵ^*. Consequently, the polarization can be expressed as:

$$\mathbf{P} = \mathbf{E}(\epsilon^* - \epsilon_0) \tag{5.14}$$

since

$$k^* = \epsilon^*/\epsilon_0 \tag{5.15}$$

then

$$\mathbf{P} = \mathbf{E}(\epsilon_0 k^* - \epsilon_0) = \epsilon_0(k^* - 1)\mathbf{E} \tag{5.16}$$

and rearranging:

$$k^* - 1 = \mathbf{P}/\epsilon_0 \mathbf{E} \tag{5.17}$$

or
$$k^* = 1 + \mathbf{P}/\epsilon_0 \mathbf{E} \tag{5.18}$$
and defining:
$$\mathbf{P}/\epsilon_0 \mathbf{E} \equiv \chi, \text{ the electric susceptibility} \tag{5.19}$$
we obtain
$$k^* = 1 + \chi \tag{5.20}$$

Equations (5.18 and 5.20) provide the relationship desired between the dielectric constant and the total polarization in the material. However, it would be even more useful to have a relationship between k^* and the fundamental polarizability of the charge mechanisms contributing to the total polarization $\mathbf{P}$.

Such a relationship can be obtained through the folowing arguments. The polarization $\mathbf{P}$ is equal to the total dipole moment induced in the material by the electric field. Thus

$$\mathbf{P} = N_i \bar{\mu}_i \tag{5.21}$$

where N_i = number of dipoles of type i
$\bar{\mu}_i$ = average dipole moment

Now, the average dipole moment of the charged particles is proportional to the local electric field ($\mathbf{E}'$) which acts on the particles:

$$\bar{\mu}_i = \alpha_i \mathbf{E}' \tag{5.22}$$

where α_i is the polarizability of average dipole moment per unit local field strength and has units of $C^2 \sec^2/Kg$.

Thus the total polarization is

$$\mathbf{P} = N_i \alpha_i \mathbf{E}' \tag{5.23}$$

For gases with little molecular interaction, the locally acting field $\mathbf{E}'$ is the same as the external applied field $\mathbf{E}_a$. However, for insulating dielectric solids, polarization of the surrounding medium substantially effects the magnitude of the local field. Mosotti was the first to derive the local field contribution by the integration of the normal component of the polarized vector over the surface of a spherical cavity in the material (see Kittel, 1986, for the derivation). The result obtained is

$$\mathbf{E}' = \mathbf{E}_a + \mathbf{P}/3\epsilon_0 \tag{5.24}$$

So since $N_i \alpha_i = \mathbf{P}/\mathbf{E}'$ from equation (5.23) then

$$N_i \alpha_i = \frac{\mathbf{P}}{\mathbf{E}_a + \mathbf{P}/3\epsilon_0} \tag{5.25}$$

and by substituting equations (5.14) and (5.15), in that order, we obtain

$$N_i \alpha_i = \frac{1}{\dfrac{1}{(k^* - 1)\epsilon_0} + \dfrac{1}{3\epsilon_0}} \qquad (5.26)$$

Further rearrangement yields

$$N_i \alpha_i = \frac{3\epsilon_0^2(k^* - 1)}{\epsilon_0(k^* + 2)} \qquad (5.27)$$

so

$$\frac{k^* - 1}{k^* + 2} = \frac{1}{3\epsilon_0} N_i \alpha_i \qquad (5.28)$$

this result is the classical Clausius–Mosotti equation which describes the relation between the complex dielectric constant of a material k^* and the number of polarizable species N_i, and the polarizability of the species α_i. As we saw earlier, there are four major classes of polarizable species in ceramics and glasses, α_e, α_a, α_d, and α_i. Thus

$$\frac{k^* - 1}{k^* + 2} = \frac{1}{3\epsilon_0} [N_e \alpha_e + N_a \alpha_a + N_d \alpha_d + N_i \alpha_i] \qquad (5.29)$$

5.5 DIPOLAR POLARIZATION THEORY

As we mentioned earlier, the physical theory of resonance absorption which explains the contribution of α_e, and α_a to k^* in the optical and infrared region is considered in Chapter 8.

It is important to discuss dipolar polarization, however, since it markedly influences the insulation and capacitive applications of ceramics and glass and is strongly affected by composition, structure, and thermal history. For several other theoretical treatments, the reader is referred to Azaroff and Brophy (1963).

Let us consider a bistable dipole model such as shown in Figure 5.2(d). A change in the coordination of the Na^+ ion from the position at left to that at right involves a change in energy as discussed in Chapter 4 for ionic conduction and represented in Figure 4.1. There is a random oscillation of Na^+ ions between these positions at any temperature above 0°K. The probability of a jump, P, between energy wells is exponentially related to the temperature and the energy barrier V (see Figure 2.6)

$$P = Ae^{-V/kT} \quad \text{(no field)} \qquad (5.30)$$

where T = temperature (°K)
$k = 1.38 \times 10^{-23}$ J/°K (Boltzmann constant)

However, with an electric field applied, the potential energy of the two sites will become unequal by an amount

$$\phi_1 - \phi_2 = e(b\mathbf{E}) = eb\mathbf{E}\cos\theta \tag{5.31}$$

where b is distance separating the potential wells and θ is the angle between the field vector and the jump vector. Thus, this model is equivalent to a turn of 180° of a dipole with a dipole moment:

$$\mu_d = \tfrac{1}{2}zeb \tag{5.32}$$

where z = valence of the ion
e = electric charge

Let us assume that there are N bistable dipoles per unit volume, with N being small enough that there will be no dipolar interaction. Also assume that $\cos\theta = 1$ for all dipoles, $P_1 = P_2$ without a field applied, and $V \gg kT$.

The probability of jumps from 1 to 2 can thus be written as

$$P_{12} = A\exp\left[-\frac{V + \mu\mathbf{E}}{kT}\right] \tag{5.33}$$

or

$$P_{12} = A\exp\left[-\left[\frac{V}{kT}\right]\exp\left[-\frac{\mu\mathbf{E}}{kT}\right]\right] \tag{5.34}$$

Since $\mu \approx 10^{-18}$ esu, then as long as $\mathbf{E}$ is less than 10^5 esu where dielectric breakdown begins to occur, $\mu\mathbf{E}/kT$ will be much less than unity. So if the last term in equation (5.34) is expanded using a Taylor series:

$$e^{-x} = 1 - x + \frac{x^2}{2!} + \cdots, \tag{5.35}$$

we obtain

$$\exp\left[-\frac{\mu\mathbf{E}}{kT}\right] = 1 - \frac{\mu\mathbf{E}}{kT} \tag{5.36}$$

which on substitution into equation (5.34) yields

$$P_{12} = \left[A\exp\left(-\frac{V}{kT}\right)\right]\left(1 - \frac{\mu\mathbf{E}}{kT}\right) \tag{5.37}$$

This can be simplified by using the definition for P equation (5.30).

$$P_{12} = P\left(1 - \frac{\mu\mathbf{E}}{kT}\right) \tag{5.38}$$

5.5 DIPOLAR POLARIZATION THEORY

The probability of a jump of the ion in the opposite direction will be

$$P_{21} = A \exp\left[-\frac{V - \mu\mathbf{E}}{kT}\right] \tag{5.39}$$

which likewise yields the following expression:

$$P_{21} = P\left(1 + \frac{\mu\mathbf{E}}{kT}\right) \tag{5.40}$$

Under equilibrium conditions the average population of charges in wells 1 and 2 will not change with time. Consequently, those going into well 1 must come out of well 2 and vice versa. Thus

$$N_1 P_{12} = N_2 P_{21} \tag{5.41}$$

where N_1 and N_2 are the number of charge carriers in each well. Using equations (5.38) and (5.40), we obtain

$$N_1 P\left(1 - \frac{\mu\mathbf{E}}{kT}\right) = N_2 P\left(1 + \frac{\mu\mathbf{E}}{KT}\right) \tag{5.42}$$

which upon rearranging yields

$$N_1 - N_2 = (N_1 + N_2)\frac{\mu\mathbf{E}}{kT} \tag{5.43}$$

and since the total number of wells N occupied per unit volume is constant, the static polarization per unit volume, $\mathbf{P}_s$, is defined as

$$\mathbf{P}_s = (N_1 - N_2)\mu \tag{5.44}$$

By comparing this definition with equation (5.43), we see that

$$\mathbf{P}_s = \frac{N\mu^2 \mathbf{E}}{kT} \tag{5.45}$$

and since

$$\mu = \tfrac{1}{2}zeb \tag{5.46}$$

$$\mathbf{P}_s = \frac{z^2 N e^2}{4}\left(\frac{b^2 \mathbf{E}}{kT}\right) \tag{5.47}$$

Consequently, the dielectric constant of an insulator containing neighboring positions of such bistable dipoles depends on the number of dipoles, oscillation length of the dipole, and temperature. Thus

$$k = 1 + \frac{\mathbf{P}_s}{\epsilon_0 \mathbf{E}} = 1 + \frac{z^2 N e^2 b^2}{4kT} \qquad (5.48)$$

where z = valence
   N = total number of dipoles/unit volume
   e = electron charge
   b = distance between potential wells
   k = Boltzmann constant
   T = temperature

5.6 TIME DEPENDENCE OF DIPOLAR POLARIZATION

In order to describe the ac behavior of the dielectric properties of ceramics and glasses, it is necessary to discuss the time-dependent response of the above bistable dipole model. The change in number of dipoles in site 1 is equal to the outflow to site 2 minus the inflow from site 2; thus

$$\frac{dN_1}{dt} = -N_1 P_{12} + N_2 P_{21} \qquad (5.49)$$

and since $N_1 + N_2 = N$ (constant)

$$\frac{dN_2}{dt} = \frac{dN_1}{dt} \qquad (5.50)$$

Therefore

$$\frac{d(N_1 - N_2)}{dt} = 2\frac{dN}{dt} \qquad (5.51)$$

By using expansions for $\mu \mathbf{E}/kT \ll 1$, as in equation (5.35), we can obtain for the rate of change of occupancy

$$\frac{1}{2}\frac{d(N_1 - N_2)}{dt} = N_1 P_{12} + N_2 P_{21} = -N_1 P\left(\frac{1 - \mu \mathbf{E}}{kT}\right) + N_2 P\left(1 + \frac{\mu \mathbf{E}}{kT}\right) \qquad (5.52)$$

By rearranging, this result is obtained:

$$\frac{1}{2}\frac{d(N_1 - N_2)}{dt} = -P(N_1 - N_2) + P(N_1 + N_2)\frac{\mu \mathbf{E}}{kT} \qquad (5.53)$$

5.6 TIME DEPENDENCE OF DIPOLAR POLARIZATION

which is a differential equation for $N_1 - N_2$. Since $\mathbf{P} = (N_1 - N_2)\mu$, equation (5.53) is also a differential equation describing the time dependence of $\mathbf{P}$:

$$\frac{1}{2}\left(\frac{d\mathbf{P}}{dt}\right)\frac{1}{\mu} = -\frac{P}{\mu}\mathbf{P} + PN\frac{\mu\mathbf{E}}{kT} \tag{5.54}$$

which upon rearranging yields

$$\frac{1}{2P}\left(\frac{d\mathbf{P}}{dt}\right) + \mathbf{P} = \frac{N\mu^2\mathbf{E}}{kT} \tag{5.55}$$

Equation (5.55) is a relaxation equation with a relaxation time, $\tau = 1/2P$, characteristic of the rate of relaxation.
So

$$\tau\frac{d\mathbf{P}_d}{dt} + \mathbf{P}_d = N\alpha_d\mathbf{E} = \mathbf{P} \tag{5.56}$$

where α_d is the dipolar polarizability, μ^2/kT, and the subscript d has been assigned to designate that it is the dipolar polarization that is time dependent. In this sense, $\mathbf{P}_s$ is the static or zero-frequency value of the polarization. Equation (5.56) shows that when $\mathbf{E}$ changes with time, $\mathbf{P}_d$, at a given moment, will generally differ from $\mathbf{P}_s$. As the oscillation time increases there will be a trend towards an equilibrium value of $\mathbf{P}_s$ at a rate of change of $d\mathbf{P}_d/dt$.

Integration of equation (5.56) will lead to a solution of the time or frequency dependence of the dipolar contribution to the dielectric constant. However, several steps of simplification are desirable. For example, since α_e and α_a occur very rapidly (i.e., when $\tau > 10^{-11}$ sec), a high frequency polarization contribution, $\mathbf{P}_\infty$, can be defined as

$$\mathbf{P}_\infty \approx \mathbf{P}_e + \mathbf{P}_a \tag{5.57}$$

and at frequencies of 10^{11} Hz or greater, the dielectric constant will depend only on $\mathbf{P}_\infty$, as

$$k_\infty \approx 1 + \frac{\mathbf{P}_\infty}{\epsilon_0\mathbf{E}} \tag{5.58}$$

Consequently, at frequencies of 10^2 Hz to 10^{11} Hz, the low-frequency or static value of the dielectric constant, k_s, can be expressed as

$$k_s - 1 = \frac{\mathbf{P}_s + \mathbf{P}_\infty}{\epsilon_0\mathbf{E}} \tag{5.59}$$

By inserting equation (5.58) into (5.59) and simplifying, the dipolar polarization can be written in terms of the relaxed static dielectric constant k_s and the unrelaxed high-frequency constant k_∞.

$$k_s - 1 = \frac{\mathbf{P}_s + (k_\infty - 1)\epsilon_0 \mathbf{E}}{\epsilon_0 \mathbf{E}} \tag{5.60}$$

$$(k_s - k_\infty)\epsilon_0 \mathbf{E} = \mathbf{P}_d = \mathbf{P}_s \tag{5.61}$$

Upon inserting equation (5.61) into equation (5.56), one obtains the differential equation describing the change of the dielectric constant from k_s to k_∞ resulting from the time dependence of the dipolar polarization:

$$\tau \frac{d\mathbf{P}_d}{dt} + \mathbf{P}_d = \mathbf{P}_s = (k_s - k_\infty)\epsilon_0 \mathbf{E} \tag{5.62}$$

The general solution of the equation (5.62) is most easily obtained by using complex variables; that is, let $\mathbf{E}^* = \mathbf{E} = \mathbf{E}_0 \exp[i\omega t]$. If we assume that relevant physical laws hold for $\mathbf{E}^*$, as for $\mathbf{E}$, we achieve the time dependence of the dipolar polarization:

$$\mathbf{P}_d^* = \epsilon_0 k^* \mathbf{E}^* + k_s \exp[-\beta t] \tag{5.63}$$

where the coefficient β is to be determined by substitution into the differential equation (5.62).

Then differentiate $\mathbf{P}_d^*$:

$$\frac{d\mathbf{P}_d^*}{dt} = \epsilon_0 k^* i\omega \mathbf{E}^* - \alpha k_s \exp[-\beta t] \tag{5.64}$$

substituting (5.63) and (5.64) back into the differential equation (5.62):

$$\tau \epsilon_0 k^* i\omega \mathbf{E}^* - \tau k_s \beta \exp[-\beta t] + \epsilon_0 k^* \mathbf{E}^* + k_s \exp[-\beta t] = (k_s - k_\infty)\epsilon_0 \mathbf{E}^* \tag{5.65}$$

Combining terms and separating into real and imaginary parts yields

$$(-\tau k_s \beta + k_s) \exp[-\beta t] = (k_s - k_\infty)\epsilon_0 \mathbf{E}^* - \tau \epsilon_0 k^* i\omega \mathbf{E}^* - \epsilon_0 k^* \mathbf{E}^* \tag{5.66}$$

Both sides of the above equation must be equal to the same constant. The only constant possible is zero. Thus, taking first the real part (left-hand side) we have

$$(-\tau k_s \beta + k_s) \exp[-\beta t] = 0 \tag{5.67}$$

Therefore

$$\beta = \frac{1}{\tau} \tag{5.68}$$

5.6 TIME DEPENDENCE OF DIPOLAR POLARIZATION

For the imaginary part (the right-hand side of equation 5.66),

$$0 = (k_s - k_\infty)\epsilon_0 \mathbf{E}^* - \tau\epsilon_0 k^* i\omega \mathbf{E}^* - \epsilon_0 k^* \mathbf{E}^* \tag{5.69}$$

using the imaginary part of $\mathbf{P}_d^*$,

$$\mathbf{P}_d^* = \epsilon_0 k^* \mathbf{E}^* \tag{5.70}$$

equation (5.69) becomes

$$i\omega\tau \mathbf{P}_d^* + \mathbf{P}_d^* = (k_s - k_\infty)\epsilon_0 \mathbf{E}^* \tag{5.71}$$

Therefore

$$(i\omega\tau + 1)\mathbf{P}_d^* = (k_s - k_\infty)\epsilon_0 \mathbf{E}^* \tag{5.72}$$

or solving for $\mathbf{P}_d^*$:

$$\mathbf{P}_d^* = \frac{(k_s - k_\infty)\epsilon_0 \mathbf{E}^*}{(i\omega\tau + 1)} \tag{5.73}$$

Finally recombining the real and imaginary parts, we obtain

$$\mathbf{P}_d^* = k_s \exp\left[-\frac{t}{\tau}\right] + \frac{(k_s - k_\infty)\epsilon_0}{1 + i\omega\tau} \mathbf{E}_0 \exp[i\omega t] \tag{5.74}$$

The first term on the right-hand side of the equation describes the time-dependent decay of the dc charge on the capacitor. The second term describes the ac behavior of the polarization when a field of magnitude $\mathbf{E}_0$ and frequency ω is applied.

Since the electronic and atomic polarization are frequency independent in the range of interest, they can be separated from the time-dependent expression of k^* by the following definition:

$$k^* - k_\infty \equiv \frac{\mathbf{P}_d^*}{\epsilon_0 \mathbf{E}^*} \tag{5.75}$$

Now by substitution of equation (5.73) into the above definition:

$$k^* - k_\infty = \frac{(k_s - k_\infty)\epsilon_0 \mathbf{E}^*}{(1 + i\omega\tau)\epsilon_0 \mathbf{E}^*} \tag{5.76}$$

or

$$k^* = k_\infty + \frac{k_s - k_\infty}{1 + i\omega\tau} \tag{5.77}$$

and, recalling that k^* is defined by

$$k^* \equiv k' - ik'' \tag{5.78}$$

we can separate equation (5.76) into real and imaginary parts as

$$k' = k_\infty + \frac{k_s - k_\infty}{1 + \omega^2 \tau^2} \tag{5.79}$$

and

$$k'' = (k_s - k_\infty)\left(\frac{\omega \tau}{1 + \omega^2 \tau^2}\right) \tag{5.80}$$

and

$$\tan \delta = \frac{k''}{k'} = \frac{(k_s - k_\infty)\omega \tau}{k_s + k_\infty \omega^2 \tau^2} \tag{5.81}$$

These equations (5.79 to 5.81) are the desired frequency-dependent relationships of the charging and loss constants and the loss tangent of the dielectric material. These equations are known as the Debye equations and they yield the graphical relationships shown in Figure 5.5.

At low frequencies, the charging constant is frequency independent at a value which reflects the contribution of $P_e + P_a + P_d$. As the applied

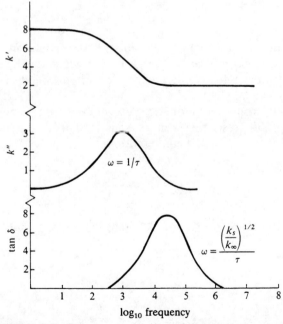

FIGURE 5.5 Frequency variation of dielectric parameters.

frequency, $\omega = 2\pi f$, approaches a value equal to $1/\tau$, k' passes through an inflection and at a higher frequency becomes asymptotic to k_∞ which depends only on $\mathbf{P}_e + \mathbf{P}_a$. When $\omega = 1/\tau$, the oscillating charges are coupled directly with the oscillating field and absorb a maximum in electrical energy and k'' goes through a maxima as a result. The magnitude of the loss peak will be $(k_s - k_\infty)/2$, and thus is directly dependent on the number of oscillating charges and their distance of motion as given in equation (5.48). Since $\mathbf{P}_d$ is inversely proportional to temperature (equation 5.48), the magnitude of the peak height should vary with the reciprocal of the absolute temperature of the dielectric material.

Figure 5.5 and equation (5.81) also show that the loss tangent tan δ goes through a maximum when plotted as a function of log frequency. However, the loss tangent peak is displaced to a higher frequency by the quantity $\sqrt{k_s/k_\infty}$. Consequently, it is often possible to observe a maxima in tan δ where the k'' maxima is at a frequency too low to be measured.

5.7 COLE–COLE DISTRIBUTIONS

In studying bulk relaxation of dielectrics at high frequencies, distributions occur throughout the frequency spectrum. Cole and Cole (1942) modified equation (5.77) to include an exponent α, as follows:

$$k^* - k_\infty = \frac{(k_s - k_\infty)}{(1 + i\omega\tau)} \qquad (5.82)$$

$$\frac{(k^* - k_\infty)}{(k_s - k_\infty)} = \frac{1}{[1 + (i\omega\tau)^{1-\alpha}]} \qquad (5.83)$$

In equation (5.83), the parameter α is used to describe the width of the distribution of relaxation times within the material.

The distribution is obtained by plotting the imaginary part k'' as a function of the real part k', yielding what is termed the Cole–Cole distribution. Figure 5.6 shows schematically the characteristic Cole–Cole plot or Nyquist plot of a RC circuit. The circuit values are

$$\tau = RC_2 \qquad (5.84)$$

$$C_2 = (\epsilon_s - \epsilon_\infty)\epsilon_0 \qquad (5.85)$$

$$C_1 = \epsilon_\infty \epsilon_0 \qquad (5.86)$$

Recalling that

$$k = \frac{\epsilon}{\epsilon_0} \qquad (5.87)$$

and using the Debye equations, the imaginary dielectric constant k''

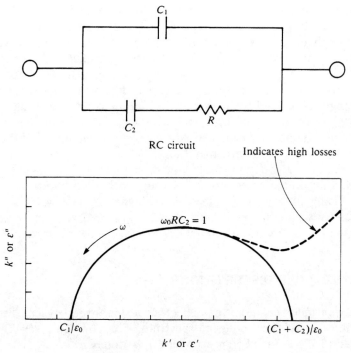

FIGURE 5.6 Cole–cole plot.

becomes

$$k'' = \left(\frac{\epsilon_s - \epsilon_\infty}{\epsilon_0}\right)\left(\frac{\omega R C_2}{1 + \omega^2 R^2 C_2^2}\right) \quad (5.88)$$

or

$$k'' = \frac{C_2}{\epsilon_0^2}\left(\frac{\omega R C_2}{1 + \omega^2 R^2 C_2^2}\right) \quad (5.89)$$

Likewise the real part becomes

$$k' = \frac{\epsilon_\infty}{\epsilon_0} + \frac{\left(\frac{\epsilon_s - \epsilon_\infty}{\epsilon_0}\right)}{(1 + \omega^2 R^2 C_2^2)} \quad (5.90)$$

$$k' = \frac{C_1}{\epsilon_0^2} + \frac{C_2}{\epsilon_0^2 (1 + \omega^2 R^2 C_2^2)} \quad (5.91)$$

Therefore the maximum occurs at

$$\delta R C_2 = 1 \quad (5.92)$$

as shown in Figure 5.6.

The Cole–Cole plot of a material is a measure of the various relaxation times for a specific dielectric material. If a perfect semicircle is produced, then there exists a very narrow distribution of relaxation times. This indicates that only one primary mechanism exists for the polarization within that material. If, on the other hand, there is a tail (or increasing k'' with increasing k') in the distribution, this indicates a large distribution of relaxation times. A large range of relaxation times can indicate multiple polarization mechanisms, but it can also indicate losses due to conduction. As a result, a low loss dielectric would have a Cole–Cole plot that is nearly a semicircle, whereas a poor or high loss dielectric would have a nonbounded increasing k'' with increasing k', (see Figure 5.6).

5.8 TEMPERATURE DEPENDENCE OF DIPOLAR POLARIZATION

The development of the bistable dipole model also provides a basis for understanding the temperature dependence of the dielectric properties of a material. Equation (5.34) expressed the probability of a transition of the oscillating charge in terms of Boltzmann statistics as $P = A \exp[-Q/kT]$. It was also shown in equations (5.55) and (5.56) that P was defined as inversely proportional to the relaxation time of the dipole process, $\tau \equiv 1/2P$. Consequently, the relaxation time is exponentially temperature dependent:

$$\tau = \frac{1}{2A} \exp\left[+\frac{E_a}{kT}\right] \tag{5.93}$$

where E_a is an activation energy for the relaxation process in units of kilocalories per mole, or electron volts. It is also possible to define an intrinsic relaxation time for the relaxation process as

$$\frac{1}{2A} \equiv \tau_0 \tag{5.94}$$

τ_0 will be proportional to a term $\alpha kT/h$, which includes a frequency factor kT/h, describing the average frequency of oscillation of an harmonic oscillator in an ensemble of temperature T. τ_0 also includes the term α (different from the Cole–Cole α) which is an accommodation coefficient that is related to the irreversibility of the oscillations.

By combining equation (5.93) and the Debye equations it is possible to describe the temperature variation in the location of the tan δ loss peak and the ϵ'' loss peak. Since tan δ_{max} occurs when

$$\omega = \left(\frac{k_s}{k_\infty}\right)\tau^{-1} \quad \text{and} \quad \omega = 2\pi f \tag{5.95}$$

then

$$f_{max}(\tan \delta) = \frac{(k_s/k_\infty)^{1/2}}{2\pi\tau_0 \exp[+E_a/kT]} \quad (5.96)$$

or

$$f_m = f_0 \left(\frac{k_s}{k_\infty}\right)^{1/2} \exp\left[-\frac{E_a}{kT}\right] \quad (5.97)$$

and similarly

$$f_m(\epsilon'') = \frac{1}{2\pi\tau} \quad (5.98)$$

so

$$f_m(\epsilon'') = f_0 \exp\left[-\frac{E_a}{kT}\right] \quad (5.99)$$

As shown in Figures 5.7 and 5.8, $\tan \delta_{max}$ will vary with temperature in the manner predicted by equation (5.97). The slope of the $\log f_m$ vs. $1/T°K^{-1}$ plot in Figure 5.8 will be equal to $-E_a/k$ as given in equation (5.97). Thus, an experimental measurement of the dielectric loss peaks as a function of temperature enables calculations of relaxation activation energies to be made.

There are several other important parameters which can also be obtained from dielectric data. τ_0 can be calculated from data such as shown in Figure 5.8. Values of 10^{-13} sec to 10^{-14} sec are indicative that the relaxation

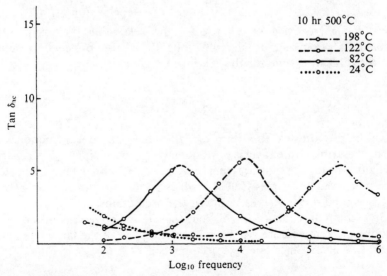

FIGURE 5.7 Shift of the dielectric loss peak in a $Li_2O \cdot 2SiO_2$(L2S) glass due to increasing temperature. After Kinser and Hench (1968).

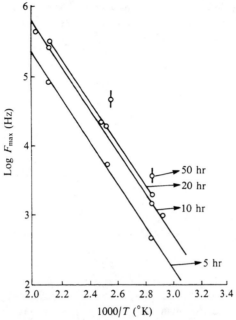

FIGURE 5.8 Logarithmic temperature dependence of the frequency location at the dielectric loss peak for an L2S glass heated at 500°C for various times.

process is ionic in origin. However, very low values of τ_0 are suggestive that the relaxation process is electronic space charge polarization.

Since the quantity $k_s - k_\infty$ is directly related to the dipolar polarization $\mathbf{P}_d$ in the material, the magnitude of $\tan \delta_{max}$ or ϵ''_{max} can be used to calculate the number of oscillating charges and their distance of oscillation by employing equation (5.47). The inverse temperature dependence of $\mathbf{P}_s$ in equation (5.47) also provides something of a test for the presence of dipolar polarization in the material. When the measuring temperature increases, the polarizability of the dipoles decreases and the loss peak height should decrease. However, the rapid increase in σ_{dc} with increasing temperature usually results in the ac loss peak being swamped in magnitude by dc losses at temperatures of one-fourth the softening point or liquidus of the material. Consequently, analysis of a decrease in the height of $\tan \delta$ peaks with temperature is often not possible. If temperature independence can be observed it is suggestive that an interfacial polarization may be operating.

Dielectric loss data in glasses and ceramics can also be employed as a measure of the distribution of ion energies or configurations in the structure. Often, an appreciable broadening in the Debye peaks is observed. The broadening is due to a distribution in relaxation times, which may be a consequence of a distribution of E_a or a distribution in the oscillation distances between equivalent sites, b. There are a variety of

techniques which can be used to analyze distribution of relaxation times. Often a log normal distribution is a satisfactory description:
Let

$$\frac{\epsilon''}{\epsilon'_s - \epsilon'_\infty} = \int_0^\infty \frac{\omega\tau}{1+\omega^2\tau^2} g(\tau)\, d\ln\tau \quad (5.100)$$

where $g(\tau)$ is a weighting function which gives the relative strength of a given relaxation time τ_i. Assuming a log-normal distribution yields:

$$\frac{\epsilon''}{\epsilon'_s - \epsilon'_\infty} = \int_0^\infty \frac{\omega\tau}{1+\omega^2\tau^2} \frac{1}{(2\pi)^{1/2}s} \exp\left[-\frac{1}{2}\frac{\ln(\tau/\tau_0)}{s}\right] d\ln\left(\frac{\tau}{\tau_0}\right) \quad (5.101)$$

where s is the standard deviation of the distribution. This formidable relation comes out to be an error function type of solution for s and has been published in tabular form with instructions as to application by Nowick and Berry (1961).

Analysis of loss peak distribution functions have led to the conclusion that there is a distribution in cation activation energies in various glasses. Several have reached the conclusion that the broadening response is due to variations in ion configurations. Thus, if E_a is represented as $E_a = \Delta F = \Delta H - T\Delta S$, there will be localized variations in the entropy within the material due to the randomness of the structure.

5.9 INTERFACIAL POLARIZATION

Although the atomistic dipolar model may provide a reasonable interpretation of the dielectric behavior of many insulators, unfortunately it does not usually represent a unique interpretation. The frequency response and frequency range of interfacial polarization mechanisms is equivalent to that of the Debye type of model in many cases. Consequently, the possibility of an interfacial interpretation of the dielectric behavior of a material must always be considered as an alternative to the Debye model. Independent evidence that interfaces are or are not involved in the behavior is required to achieve an unambiguous interpretation of dielectric behavior.

There are two principal categories of interfacial polarization and both can be important for glasses and ceramics. The first involves a variation of electrode polarization. The second consists of the dielectric response due to heterogeneities in the material.

If a sharp compositional gradient is presented at one of the electrode–material interfaces due to electrode polarization, heat treatment, or atmospheric attack, a two-layer capacitor results.† As shown in Figure 5.9,

---

† Water films and oxidized electrode layers produce especially strong contributions to interfacial polarization.

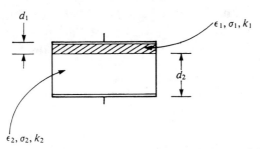

FIGURE 5.9 Schematic two-layer model of a glass dielectric.

such a two layer capacitor involves a region described by a permittivity ϵ_1, a conductivity σ_1, and a thickness d_1, which is in series with the matrix of ϵ_2, σ_2, and d_2. Analysis of this model and numerous other series and parallel RC circuit models of the materials have been described. The frequency dependence of the dielectric parameters of the model in Figure 5.9 are

$$k'(\omega) = \frac{k_s + k_\infty \omega^2 \tau^2}{1 + \omega^2 \tau^2} \tag{5.102}$$

and

$$\sigma(\omega) = \frac{\sigma_s + \sigma_\infty \omega^2 \tau^2}{1 + \omega^2 \tau^2} \tag{5.103}$$

and

$$\tan \delta = \frac{\sigma_\infty}{\omega k_\infty} \left[\frac{\left(\dfrac{k_\infty \sigma_s}{k_s \sigma_\infty}\right) + \omega^2 \tau_\delta^2}{1 + \omega^2 \tau_\delta^2} \right] \tag{5.104}$$

Relaxation times and frequency parameters are defined as

$$\tau \equiv \epsilon_0 \left(\frac{k_s - k_\infty}{\sigma_\infty - \sigma_s}\right) = \epsilon_0 \left(\frac{k_1 d_2 + k_2 d_1}{\sigma_1 d_2 + \sigma_2 d_1}\right) \tag{5.105}$$

$$k \equiv \left[\frac{d_1 + d_2}{\left(\dfrac{d_1}{\sigma_1} + \dfrac{d_2}{\sigma_2}\right)^2} \right] \left(\frac{d_1 k_1}{\sigma_1^2}\right) + \left(\frac{d_2 k_2}{\sigma_2^2}\right) \tag{5.106}$$

and

$$k_\infty \equiv \frac{d_1 + d_2}{\left(\dfrac{d_1}{k_1} + \dfrac{d_2}{k_2}\right)} \tag{5.107}$$

$$\sigma_s \equiv \frac{d_1 + d_2}{\left(\dfrac{d_1}{\sigma_1} + \dfrac{d_2}{\sigma_2}\right)} \tag{5.108}$$

$$\sigma_\infty \equiv \left[\frac{d_1 + d_2}{\left(\dfrac{d_1}{k_1} + \dfrac{d_2}{k_2}\right)^2} \right] \left(\frac{d_1 \sigma_1}{k_1^2} + \frac{d_2 \sigma_2}{k_2^2}\right) \tag{5.109}$$

The frequency response of equations (5.102), (5.103), and 5.104) is equivalent to that shown in Figure 5.5 for the Debye equations. k' goes through a relaxation decay from k_s at low frequencies to k_∞ at high frequencies described by the relaxation time τ. k'' and $\tan \delta$ likewise go through maxima as a function of frequency. Rearranging equation (5.104) to obtain the location of the loss angle maximum on a $\tan \delta$–log frequency plot yields

$$f_m(\tan \delta) = \frac{\sigma_1 d_2 \sigma_2 d_1}{2\pi(k_1 d_2 + k_2 d_1)} \left(\frac{k_s}{k_\infty}\right)^{1/2} \quad (5.110)$$

Equation (5.110) indicates that a glass or ceramic dielectric may be responding as a two-layer capacitor due to a surface film, for example, and exhibit a loss peak in the same frequency range as Debye peaks if the ratios of σ_1/σ_2 and d_1/d_2 are appropriate. Since σ_1 and σ_2 will also be exponentially temperature dependent, the loss peaks of the two-layer capacitor will also shift to higher frequencies with increasing temperature. However, the interpretation of such a shift is not straightforward, since the charge Q_1 and Q_2 for the two conducting layers may be different. Therefore, the ratio of σ_1/σ_2 may change with temperature as well which would produce a frequency shift proportional to $\Delta(\sigma_1/\sigma_2)$.

Heterogeneities are another source of polarization in polycrystalline ceramics and glasses. Wagner (1914) was the first to extend Maxwell's multilayer dielectric analysis to include dispersed spherical phases. Sillars (1937) later generalized heterogeneous dielectrics to include dispersed ellipsoidal phases. An important consequent of the final Maxwell–Wagner–Sillars (M–W–S) theory is that the dielectric parameters of an insulator may be interpreted in terms of a distribution of ellipsoidal particles of a given volume fraction and size, instead of atomistic dipoles as Debye theory requires. Since the frequency location of loss peaks and the frequency and temperature dependence of k' and k'' can be equivalent in both the M–W–S and Debye theories independent confirmation of a model for the dielectric behavior of a particular material must be obtained.

A schematic of the M–W–S heterogeneous dielectric model is shown in Figure 5.10. The diagram shows isolated ellipsoidal particles of conductivity σ_2 in a matrix of conductivity σ_1. When σ_2 is greater than σ_1, the phase boundary between the particles and the matrix will serve as a barrier to charge migration, and ac charge oscillation and interfacial polarization will occur (Figure 5.2). Sillars (1937) has shown that for a volume fraction of spheroids, q, of conductivity σ_2 and permittivity ϵ_2 and axial ratio of a/b, imbedded in a perfect dielectric matrix of conductivity σ_1 and permittivity ϵ_1, the dielectric parameters of the composite material are

$$\epsilon' = \epsilon_\infty + \frac{\epsilon_1' N}{1 + \omega^2 \tau^2} \quad (5.111)$$

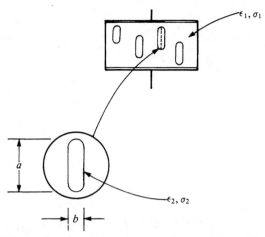

FIGURE 5.10 Schematic of a dielectric containing high conductivity ellipsoidal heterogeneities of axial ratio $= a/b$, with $\sigma_2 \gg \sigma_1$.

$$\epsilon'' = \frac{\epsilon_1' N \omega \tau}{1 + \omega^2 \tau^2} \tag{5.112}$$

$$\tan \delta = \frac{\epsilon_1' N \omega \tau}{\epsilon_\infty (1 + \omega^2 \tau^2) + \epsilon_1' N} \tag{5.113}$$

where

$$\tau = \frac{(\epsilon_1'(\lambda - 1) + \epsilon_2')\epsilon_0}{\sigma_2} \tag{5.114}$$

and

$$N = q \frac{\lambda^2 \epsilon_1'}{\epsilon_1'(\lambda - 1) + \epsilon_2'} \tag{5.115}$$

and

$$\epsilon = \epsilon'\left(1 + q\frac{\lambda(\epsilon_2' - \epsilon_1')}{\epsilon_1'(\lambda - 1) + \epsilon_2'}\right) \tag{5.116}$$

The relationship between λ and a/b is shown in Figure 5.11.

The similarity of the form of the above equation with the Debye equations is readily apparent. However, the location of the M–W–S dielectric relaxation is primarily a function of the axial ratio of the ellipsoidal particles and the conductivity of the dispersed phase. For example,

$$f_{m(\epsilon'')} = \frac{1}{2\pi\tau} = \frac{\sigma_2}{2\pi(k_1'(\lambda - 1) + k_2')} \tag{5.117}$$

If the axial ratio of particles changes during a growth process from spherical

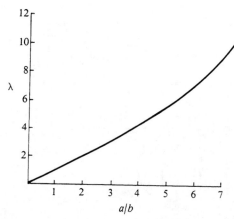

FIGURE 5.11 Variation of Sillars' dimensional parameter λ.

to dendritic with an equivalent ellipsoidal ratio of $a/b = 7$, then the loss peak location would shift downward by an entire order of magnitude. If the conductivity of the dispersed phase were decreasing as a result of growth, the downward shift would even be more marked. However, elongated growth accompanied by a change in σ_2 would tend to be self-compensating.

The M–W–S model also shows that the magnitude of both ϵ' and ϵ'' are dependent on the volume fraction of the dispersed particles and their axial ratio. Consequently, a change in magnitude of ϵ' or ϵ'' cannot be unambiguously interpreted unless other evidence is available to estimate independently the axial ratios or the volume fraction. With either information, though, it is possible to obtain the other structural features from the above equations.

5.10 COMPOSITION AND STRUCTURAL EFFECTS OF GLASSES

The next objective is to attempt to interpret the dielectric behavior of some selected glass and ceramic compositions in terms of the various theories presented above. Special emphasis will be placed on the relation of structural models to the behavior observed.

A number of dielectric studies of vitreous SiO_2 have demonstrated the presence of two dielectric loss peaks at cryogenic temperatures. Results have shown that the magnitude of the loss peak at 13°K can be related to the H_2O content of the glass (Figure 5.12). The second cryogenic fused silica loss peak occurs in the temperature range of 28–60°K at audio frequencies and has an activation energy of 1.2 Kcal/mole. Consequently, the peak shifts to a frequency range of 10^{11}–10^{12} Hz at room temperature, beyond the range of concern for audio or radio circuit applications of vitreous silica, but certainly within the range of interest for microwave devices.

5.10 COMPOSITION AND STRUCTURAL EFFECTS OF GLASSES

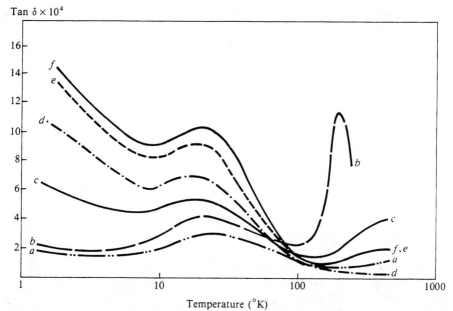

FIGURE 5.12 Dielectric loss behavior of a variety of vitreous silica glasses as a function of temperature. (a) Spectrosil WF, 20 ppm OH. (b) Infrasil, 30 ppm OH. (c) Vitreosil OG, 430 ppm OH. (d) Corning 7940, 940 ppm OH. (e) Silica, 1160 ppm OH. (f) Spectrosil B, 1190 ppm OH. After Mable and McCammon (1969).

The water-based loss peak has been analyzed as an oscillating permanent dipole and can be described by the P_{d2} mechanism shown on the right of Figure 5.2. The higher temperature peak is associated with what Stevels terms *deformational losses*. The dipolar mechanism P_{d2} that has been suggested for the peak involves the oscillation of the bridging oxygens in nearly colinear Si–O–Si bonds. The model suggests that if the Si–O–Si bond is nearly 180°, it will be possible for the oxygen to oscillate between two equivalent positions, as Figure 5.2, with only a small energy barrier. No spatial rearrangements or bond breaking would be involved. This interpretation would seem to be in reasonable agreement with the structural interpretation of vitreous silica advanced by Mozzi and Warren (1969) where the bond angles could be best described by a skewed Gaussian distribution about 144°. The half breadth of the distribution is nearly 35° and about 10% of the bond angles lie between 175°–180°. It is presumably the oscillation of bonds within the tail of this distribution that produces the "deformational losses."

Perhaps the most detailed atomistic analysis of a dipole relaxation mechanism in glass has been made by Charles (1963), who attempts a quantitative correlation of dielectric relaxation and dc conductivity in an alkali silicate glass. There are two basic postulates, (1) that near a given nonbridging oxygen ion in a glass there are several equivalent sites for an alkali metal ion, and (2) that the alkali ion will respond to the field in two major types of motion. The alkali ion may oscillate between equivalent sites about the same nonbridging oxygen which leads to ac polarization but not dc conduction. Or, the alkali ion may exchange nonbridging oxygen ions which will usually involve both ac polarization and dc conduction. Charles analyzed these two types of alkali ion motion in detail using a modification of the ionic conductivity and bistable dipole models presented herein. He obtained reasonable agreement between theory and observation.

Charles also examined the influence of microstructure on the dielectric behavior of several alkali silicate glasses. In this classic experiment he found that a rapid quench of 7.5 mole% and 15 mole% Li_2O-SiO_2 glasses changed a dispersed Li_2O-rich glassy phase to a tubular-interconnected phase. The dielectric losses of the rapidly quenched sample were increased by an order of magnitude and the location of the loss peak also shifted by a factor of 10 to higher frequencies. Binary silicate glasses containing higher percentages of Li_2O, up to 40 mole%, did not show appreciable changes in dielectric loss upon changes in quench rate and the microstructure also remained that of a dispersed lithia-rich phase in an SiO_2-rich matrix. His interpretation of the change in loss behavior with thermal history of the various glasses is based on M–W–S interfacial polarization. The conclusion reached was that the continuously distributed lithia-rich phase associated with spinodal decomposition gives rise to a large ac loss. It was due to a very high a/b axial ratio in the M–W–S heterogeneous loss model. The nearly spherical dispersed phase associated with a nucleation and growth phase separation exhibits a much lower loss due to the low axial ratio in the interfacial polarization model.

An interfacial M–W–S polarization model has also been used by Kinser (1971) to interpret the changes in dielectric properties associated with the thermal treatment of several Li_2O-SiO_2 and Na_2O-SiO_2 glasses. It was found, as shown in Figure 5.7, that appreciable losses occur in 30 mole% Li_2O-SiO_2 glasses after a heat treatment of 10 hours at 500°C (Figure 5.7). These results also show, Figure 5.13, that loss peaks are not present for the same samples before they are heat treated, in contrast to Charles's data. The source of the dielectric losses in the heat treated glasses was attributed to the presence of a high conductivity metastable crystal phase of lithium metasilicate. Guinier X-ray diffraction, hot-stage electron microscopy, and X-ray small-angle scattering were also used to establish the presence of the metastable phase. Analysis of the change in frequency of the dielectric loss peak with thermal treatment employing the M–W–S theory led to the conclusion that the axial ratio of the metasilicate precipitate was decreasing from 100/1 to 5/1 during the nucleation of the equilibrium lithium disilicate phase. A summary of the changes in dielectric loss with thermal treatment is given in Figure 5.14. Similar results have also been found in 26.4% Li_2O-SiO_2 glasses as well as 33 mole% Li_2O-SiO_2 glasses, which are not

5.10 COMPOSITION AND STRUCTURAL EFFECTS OF GLASSES

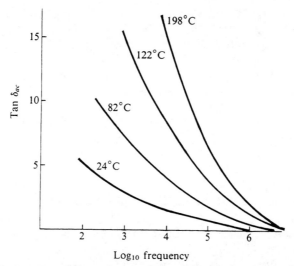

FIGURE 5.13 Frequency dependence of the dielectric loss angle of a 30 mole% Li_2O–SiO_2 glass as quenched.

phase separated, and 33 mole% Na_2O–SiO_2 glasses. The activation energy for the loss process (Figure 5.8), was calculated to be 15 Kcal/mole, equivalent to the activation energy for dc conduction. Extension of the temperature range has shown that the appearance and disappearance of the dielectric loss peaks in the Li_2O–SiO_2 glasses occurs after as little as 24 hr at

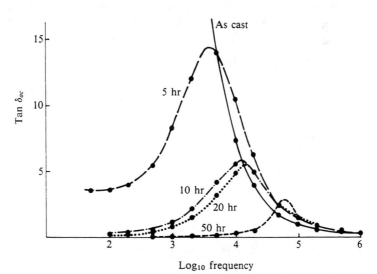

FIGURE 5.14 Variation in loss angle of 30 mole% Li_2O–SiO_2 glass as treated at 500°C. After Kinser and Hench (1968).

475°C and as high as 525°C. Consequently, the presence of dielectric loss peaks in some glass systems may be attributable to annealing treatments that result in precipitation of metastable crystallites, which give rise to M–W–S interfacial losses.

Thus, based on the case histories cited, the detailed analysis of dielectric behavior of glasses with well-characterized water contents, complete amorphous X-ray structure analysis, transmission electron micrographic evaluation of phase separation, and controlled thermal histories are all needed in order to develop firmly a generalized analysis of the dielectric properties of a particular glass.

5.11 SOL-GEL SILICA DIELECTRIC SPECTRA

The polarization mechanisms of low-frequency dielectric relaxations have been studied by Wallace and Hench (1988). Their observations were made on sol-gel-derived porous silica monoliths, i.e., type VI silica, (Hench, et. al., 1988). Their analysis is reviewed as an example of the complexity of dielectric phenomena in multiphase materials. The dielectric properties were measured as a function of adsorbed water in the ultraporous silica. The dielectric constant increased linearly with water content when the measurements were taken at a frequency of 10^7 Hz (Figure 5.15).

Pure, dry, dehydrated silica gel is a porous form of amorphous silica which exhibits no dielectric relaxation between 1 Hz and 1 THz. Liquid water has one dipolar relaxation at 23.4 GHz. However, the dielectric relaxation spectra, between 1 Hz and 1 THz, of pure water adsorbed in the pores of particulate silica gel can exhibit up to three relaxations ($R1$, $R2$, and $R3$), depending on the water content and texture of the gel.

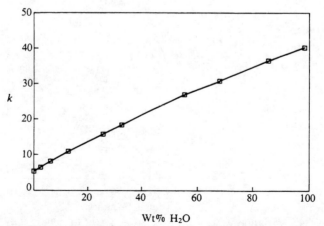

FIGURE 5.15 Dielectric Constant for gel silica at 1×10^7 Hz.

A low frequency relaxation (R1) is observed at frequencies below 10 MHz and is generally attributed to Maxwell–Wagner-type interfacial polarization, that is, conducting inclusions in an insulating matrix. This polarization is related to the "bulk water" adsorbed in the granules of the particulate gels used in previous investigations, and is caused by the conduction of protons dissociated from surface silanols. The absolute magnitude and frequency of R1 for a specific gel sample depends upon a complex interdependence of gel structure and adsorption variables. These variables include: gel particle size, average pore size $\bar{r}$, pore volume V, surface area S, silanol concentration [SiOH], impurity ion concentration, and adsorbed water structure. A relaxation can be characterized by the frequency $F\delta$ of the maximum of its dielectric loss factor spectra $E''(f)$, where f = frequency. The characteristic frequency $F\delta$ of R1, called $F\delta 1$, shows a logarithmic-type dependence on water content, W = gm H$_2$O/gm gel, and an Arrhenius-type dependence on temperature T, e.g.,

$$F\delta 1 = CW^n \exp\left[-\frac{E_a}{kT}\right] \qquad (5.118)$$

where n varies between three and six, depending on the sample and E_a is the activation energy. The magnitude of the dielectric loss factor at $F\delta 1$, called $|\delta 1|$, which is roughly constant for all values of W, also varies from sample to sample:

$$3 < |\delta 1| < 8 \qquad (5.119)$$

For a gel containing no adsorbed water, e.g., $W = 0$ g/g, only R2, due to dipolar orientation of the surface silanols, SiOH, around the Si—O bond axis is seen. $|\delta 2|$ is proportional to [SiOH] which depends on thermal history of the sample. The absolute value of $F\delta 2$ varies from 1 to 10 GHz depending on [SiOH]. When water molecules are adsorbed into pores, they form hydrogen bonds with the silanols. This causes $|\delta 2|$ to increase as W increases. At $T < 0°$C, $F\delta 2$ decreases as W increases. As T increases to $0°$C, for all values of W, $F\delta 2$ increases to a constant value equal to that for $W = 0$ g/g, i.e., 10 GHz in this case. For $W > 0$ g/g, R2 is attributed both to relative lifetimes of hydrogen bonds formed by clusters of water molecules H-bonded to the surface SiOH groups and to the rate of transfer of protons between adjacent H-bonded water molecules adsorbed on surface SiOH groups.

Zhilenkov and Nekrasova (1980) observed a third relaxation at $F\delta 3 = 14$ GHz at $0°$C, which appears for $W > 0.2$ g/g for their sample. $F\delta 3$ and $|\delta 3|$ increase as W and T increase. R3 appears after the formation of a second phase in the adsorbed water. The second phase forms upon completion of the "bound" water layer at a critical water content $W_c =$

0.2 g/g. This phase exhibits no melting transition. Two phases of adsorbed water have also been observed using proton NMR spectroscopy.

When the temperature is lowered far enough for the second phase to freeze, $F\delta 3$ disappears, whereas $F\delta 2$ does not disappear but keeps decreasing as T decreases. Thus $R2$ must be related only to the bound phase, while $R3$ must involve the second "free" water phase.

The dielectric relaxation of water adsorbed in monolithic silica gels is similar to $R1$ in its frequency dependence on W, but the value of $F\delta 1$ which was observed was much lower and $|\delta 1|$ was much larger than the values seen for particulate gels with the same W. The Debye-type shape of $R1$ could be attributed to either dipolar or interfacial polarization but it is generally attributed to the interfacial polarization. The unique geometry, compared with particles, of monolithic silica gel, shows mechanisms of relaxation due to water adsorption, using a digital ac impedance bridge.

The admittance spectra, $Y(f)$, from which the dielectric relaxation spectra can be calculated, of sol-gel-derived cylindrical silica gel samples were measured between 5 Hz and 13 MHz using a Hewlett Packard HP4192A ac impedance analyzer. All gels were made from acid-catalyzed tetramethoxysilane sols, and have a pore radius of about 14 Å. Their texture and structure depends on thermal history. The impedance analyzer allows the amplitude of the ac signal to be varied and a dc bias to be applied. Silver paint was used as the measuring electrode material. The values of the texture of the sample are: $S = 752 \, m^2/g$, $V = 0.553 \, cc/g$, and $R = 14.7 \, Å$, as measured by isothermal gas adsorption, with a bulk density of 1.04 g/cc.

The dielectric relaxation spectra of water adsorbed in monolithic silica gel samples consists of an $R1$ relaxation superimposed on another relaxation occurring at lower frequencies (called RS). In an equilibrated gel, $R1$ is stable and reproducible. If water is allowed to evaporate from the surface of a saturated gel the low frequency relaxation RS decreases in magnitude. This implies that RS is due to conduction on the external surface of the gel. The $F\delta$ frequency of RS, called $F\delta S$, is far too low to be measured, but the intensity of the relaxation is large enough for the tail of the relaxation to obscure the loss factor spectra of $R1$ and make accurate measurement of $F\delta 1$ difficult. The dielectric loss tangent spectra, $\tan \delta(f)$, can be used to separate these two relaxations. The tail of RS is just visible on the low frequency side of the loss tangent peak of $R1$ for samples measured using Ag paint electrodes (Figure 5.16, curve a). The frequency of the maximum of the loss tangent spectra, $F\delta 1$, can be accurately measured and is directly proportional to $F\delta 1$. Thus the loss tangent spectra and $F\delta 1$ can be used to characterize $R1$, and also be related to previous investigations.

$F\delta 1$ of $R1$ of a cylindrical gel is an exponential function of the sample thickness D, (i.e. the electrode separation). Figure 5.17 shows this as a log–log plot for a saturated gel with $W = 0.154 \, g/g$ that is

$$F\delta 1 = 471 \, D^{-1.929} \tag{5.120}$$

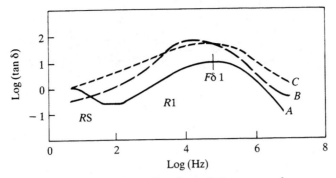

FIGURE 5.16 Loss tangent as a function of frequency for porous silica gel monoliths with different electrodes.

The same exponential relationship is seen with other samples, but the exponent varies between −0.6 and −2.0.

The material used as measurement electrodes influences the intensity, $|\delta 1|$, and frequency, $F\delta 1$, of the tan $\delta(f)$ spectra of $R1$. The overall shape of the relaxation is the same for both silver (curve a) and carbon paint (curve b), and vapor deposited Pt electrodes (curve c) (Figure 5.16).

Figure 5.17 shows the dependence of $F\delta 1$ on the thickness of a silica gel sample. This data are replotted in Figures 5.18 and 5.19 as log–log and linear plots respectively, showing a transition from an exponential to a linear dependence on water content, W, at $W_c = 0.275$ g/g. The equations best fitting these curves are:

$$F\delta 1 = 6.88 \times 10^7 \, W^{5.353} \quad \text{for} \quad 0.06 < W < 0.275 \text{ g/g} \quad (5.121)$$

$$F\delta 1 = 2.34 \times 10^7 \, W - 305{,}000 \quad \text{for} \quad 0.275 < W < 0.50 \text{ g/g} \quad (5.122)$$

The loss tangent spectra of $R1$ is independent of the magnitude of the ac voltage signal from 0.01 V to 1.0 V, whereas RS is strongly affected. Applying a dc offset voltage to the ac signal also effects RS, whereas $R1$ is unaffected.

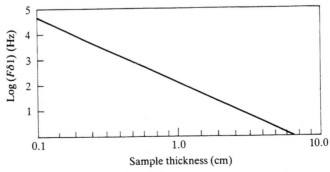

FIGURE 5.17 Variation of loss tangent with sample thickness for porous silica gel monoliths.

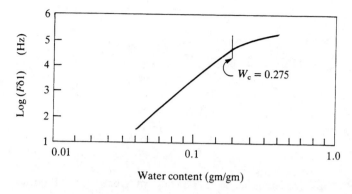

FIGURE 5.18 Exponential dependence of $F\delta 1$ to the water content of porous silica gel mololiths.

The properties of $R1$ due to water adsorption in silica gel monoliths are explained by an interfacial polarization mechanism due to charge build up at the electrodes, (i.e., electrode polarization). The charge carriers are protons which conduct via a hopping mechanism.

The evidence for the electrode polarization mechanism includes:

1. The frequency $F\delta 1$ of $R1$ observed in monolithic silica gel samples varies with sample thickness (Figure 5.17) and electrode material (Figure 5.16). RS shows a similar behavior.
2. The loss factor magnitude, $|\delta 1|$, is 1000 times larger, and $F\delta 1$ is much lower, in monolithic gels than in particulate gels for the same W.
3. An intrinsic dipolar polarization is modelled by a parallel RC circuit, and an extrinsic electrode polarization is modelled by a series RC circuit. $R1$ is modelled by a series RC circuit.

In pure water, conductivity is due to the hopping of protons created by

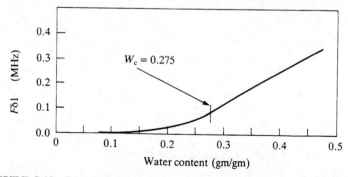

FIGURE 5.19 Linear dependence of loss tan frequency with water content.

an autodissociation mechanism. Proton hopping occurs via the tunnelling of protons between the electron clouds of hydronium ions and correctly oriented adjacent water molecules. The hopping of protons dissociated from acidic surface silanols is also believed to be the conduction mechanism of water adsorbed in silica gel. The specific hopping mechanism depends on the statistical thickness of the adsorbed water layer, i.e., $W/S(gm\ H_2O/1000\ m^2)$. When only the bound nonfreezing water layer is adsorbed, i.e., $W < W_c$, the conductivity G of water adsorbed in silica gel is proportional to W raised to the power n, where n varies between three and seven, depending on the sample. For $W > W_c$, G is directly proportional to W.

The *nuclear correlation relaxation time*, τ_n, of protons in water adsorbed in silica gel is calculated from the proton NMR spectra. τ_n is associated with the movement of protons in the two phases, the bound and free water, present in the adsorbed system. The bound water is the phase showing no melting or freezing transition and is the water adsorbed up to $W = W_c$, and exhibits a suppressed melting point. τ_n exhibits the same type of dependence on W as does G, that is, exponential for $W < W_c$, and linear for $W > W_c$.

The charge carriers for $R1$ could be protons or impurity ions, such as sodium. Wallace concludes that the charge carriers are protons, since:

1. $F\delta 1$ of $R1$ of the silica gel monoliths shows the same type of dependence on W as reported for G and τ_n (Figures 5.18 and 5.19). The transition from exponential to linear dependence on W occurs at the same statistical thickness of adsorbed water, i.e., W/S, of about $0.36\ g/1000\ m^2$ for $F\delta 1$, G and τ_n.
2. ICP analysis of a saturated gel detected no impurity ion concentrations greater than 1 ppm.
3. The diffusion of a 0.1 M NaCl solution into a gel does not affect $R1$, but increases the magnitude of RS.

Proton conduction could occur via a bound hopping or an ionic diffusion mechanism. Neither increasing the amplitude of the ac signal, applying a dc bias, or diffusing a 0.1 M NaCl solution into the gel affects $R1$, whereas RS is strongly affected. This means that $R1$ has a different conduction mechanism, e.g., proton hopping, than RS, e.g., ionic diffusion.

$R1$ is due to electrode polarization resulting from proton conduction in the adsorbed water layer via a hopping mechanism. Therefore the interfacial polarization attributed to $R1$ in particulate gels is correct for monolithic gels as well. The similarity of the dependence of $F\delta 1$, G and τ_n on W and the change from an exponential to linear dependence at the same statistical thickness ($W/S \simeq 0.36\ g\ H_2O/1000\ m^2$) suggest that the proton hopping mechanism depends on the structure of the adsorbed water which also depends on the statistical thickness. In contrast, the surface conduction relaxation RS is due to the diffusion of impurity ions.

5.12 DIELECTRIC BREAKDOWN

When ceramics are used in the electrical industry for insulation, capacitors, and encapsulation, the material is exposed to a voltage gradient which it must withstand for the operating life of the system. Failure occurs when an electrical short circuit develops across the material. Such a failure is called dielectric breakdown. The voltage gradient, expressed in volts/centimeter, sufficient to cause the short is termed the breakdown strength of the material.

Breakdown strengths of ceramics vary widely due to many factors. Some of the most important include: thickness, temperature, ambient atmosphere, electrode shape and composition, surface finish, field frequency and wave form, porosity, crystalline anisotropy, amorphous structure, and composition. A general comparison of some of these variables can be made from the data accumulated in Table 5.2 [Walther and Hench (1971)].

Strengths as large as several million volts/centimeter are reported for micron-thick thin films. Because they are so thin, the voltage insulation of such films is small, however. Thus, bulk ceramics of many centimeters in thickness are required to insulate large electrical power voltages. Table 5.2 shows that the breakdown strength unfortunately decreases drastically to levels of only several thousand volts/centimeter for bulk ceramics. The changes in strength occurring with increased thickness are due to changes in breakdown mechanisms. The temperature effects on breakdown shown in Table 5.2 are due to the strong influence of thermal energy on breakdown mechanisms.

Dielectric breakdown occurs when the temperature, i.e., thermal energy, of the lattice or its electrons reaches a value during the application of an electric field such that the conductivity increases rapidly resulting in permanent damage to the material. There are three basic types of breakdown: intrinsic, thermal, and avalanche (O'Dwyer 1964). There are also three pseudo-types of breakdown called discharge, electro-chemical and mechanical breakdown. Pseudo-types can be considered to be produced by one or more of the three basic mechanisms.

Dielectric discharge is associated with a gaseous breakdown in the pores or at the surface of a solid material. Electro-chemical breakdown is a result of a gradual deterioration of insulating properties through chemical reactions until breakdown is due to cracks, defects, and other stress raisers distorting the applied field and thus precipitating failure.

Experimentally, intrinsic breakdown is found to be primarily field dependent in that the applied field determines when the electron temperature reaches the critical level for breakdown. Observations that intrinsic

TABLE 5.2 Breakdown Voltage For Glass And Ceramics[a]

Material	Form	Thickness	Temperature	Breakdown strength 10^6 V/cm (dc)
Al_2O_3	Anodized film	300 Å	25°C	7.0
Al_2O_3	Anodized film	6000 Å	25°C	1.5
Al_2O_3	Anodized film	1000 Å	100°C	16.0
99.5% Al_2O_3	Polycrystalline ceramic	0.63 cm	25°C	0.18
94.0% Al_2O_3	Polycrystalline ceramic	0.63 cm	25°C	0.26
Alumina porcelain		0.63 cm	25°C	0.15
High Voltage porcelain		0.63 cm	25°C	0.15
Steatite porcelain		0.63 cm	25°C	0.10
Forsterite porcelain		0.63 cm	25°C	0.15
Low voltage porcelain		0.63 cm	25°C	0.03
Lead glass		0.02 cm	25°C	0.25
Lead glass		0.02 cm	200°C	0.05
Lime glass		0.004 cm	25°C	2.5
Borosilicate glass (BSI)		0.003 cm	20°C	5.8
Borosilicate glass (BSI)		0.003 cm	100°C	2.5
Borosilicate glass (BSI)		0.0005 cm	20°C	6.5
Quartz crystal		0.005 cm	20°C	6.0
Quartz crystal		0.005 cm	−60°C	4.1
Quartz, fused		0.005 cm	20°C	6.6
Quartz, fused		0.005 cm	−60°C	7.0
NaCl	[100] Single crystal	0.002 cm	25°C	2.5
NaCl	[111]	0.002 cm	25°C	2.2
NaCl	[110]	0.002 cm	25°C	2.0
NaCl	Single crystal	0.014 cm	25°C	1.26
KCl	Single crystal	0.014 cm	25°C	1.11
KBr	Single crystal	0.014 cm	25°C	0.82
TiO_2	Rutile (∥ opt. axis)	0.01	25°C	0.02
TiO_2	(⊥ opt. axis)	0.01	25°C	0.12
$BaTiO_3$	Single crystal	0.02	0°C	0.040
$BaTiO_3$	Single crystal	0.02	150°C	0.010
$BaTiO_3$	Polycrystal	0.02	25°C	0.117
$SrTiO_3$	Single crystal	0.046	25°C	0.414
Mica	(muscovite crystal)	0.002	20°C	10.1
Mica	(muscovite crystal)	0.006	20°C	9.7
$PbZrO_3$	Polycrystal (0% porosity)	0.016 cm	20°C	0.079
$PbZrO_3$	(10% porosity)	0.016 cm	20°C	0.033
$PbZrO_3$	(22% porosity)	0.016 cm	20°C	0.020

[a] After Walther and Hench (1971).

breakdown occurs at or below room temperature and occurs in very short time intervals, approximately one microsecond or less, is strong evidence that it is electronic in nature. The name "intrinsic" is used because breakdown by this mechanism is independent of the sample or electrode geometry used (provided no field distortion occurs) or of the waveform applied. Hence the value of applied field required to cause intrinsic breakdown at a given temperature is a property solely dependent on the material.

Breakdown observed between room temperature and approximately 300°C is described as thermal breakdown, because reaching the critical thermal energy for breakdown is primarily influenced by the ambient temperature, and not the electric field. Also, thermal breakdown is dependent on the rate of application of the field. Slow increases in field cause breakdown in milliseconds to minutes, with the value of breakdown being influenced by sample geometry. For faster field pulses, breakdown is independent of geometry, and breakdown strength increases with shorter pulse times.

Avalanche breakdown is related to intrinsic breakdown, in that it occurs at relatively low temperatures and short times. However, thermal properties of the material are used to describe the breakdown behavior, so it is properly a combination of thermal and intrinsic mechanisms. Thin samples, such as dielectric films, undergo avalanche destruction. There is a statistical variation in breakdown strength with short pulses for this mechanism of breakdown. Prebreakdown noise is found with slower pulses, which indicates a sequential type of lattice destruction is occurring.

5.13 INTRINSIC BREAKDOWN MECHANISMS

The basic theoretical approach used to describe dielectric breakdown in solids is based upon an energy balance equation of the form

$$A(T_0, \mathbf{E}, \alpha) = B(t_0, \alpha) \tag{5.123}$$

where $A(T_0, \mathbf{E}, \alpha)$ = energy gained by the material from the applied field
$B(T_0, \alpha)$ = energy dissipated by the material
T_0 = lattice temperature
$\mathbf{E}$ = applied field
α = energy distribution parameter, which depends on the model proposed

Thus

$$A = B \tag{5.124}$$

is the limiting condition for breakdown.

Theories of intrinsic breakdown can be classified as those dealing primarily with the electron–lattice energy transfer, and those considering changes in the electron energy distribution in the material.

In electron–lattice interaction models the behavior of the material is approximated by considering a single electron of average E, i.e., $\alpha = E$. Consequently, there is a low electron density, a small probability of electron–electron interaction, and only electron–lattice energy transfer is allowed. The mechanisms contributing to this transfer are:

1. Lattice vibrations in a dipolar field,
2. Electron shell distortion accompanying the dipolar field-lattice vibrations, and
3. Short-range electron shell distortion in a nonpolar field.

The mechanism chosen to describe the electron–lattice energy transfer depends on the model of the material used. The problem is to determine the value of E required for the energy balance and then to calculate $\mathbf{E}_c$, the critical field strength, that causes the critical temperature for breakdown, T_c, to be reached.

One model, called the *Fröhlich high energy criterion*, (Fröhlich, 1937) assumes that multiplying the number of conduction electrons to a very large value will destroy the lattice. Thus, $E = I$, the ionization energy or gap between the valence and conduction band. Breakdown occurs when the ionization rate exceeds the recombination rate, causing the conduction electron density to increase irreversibly. Figure 5.20 shows the energy balance equation schematically. The value of E at the intersection of the curves A and B indicates the energy an electron will have to be accelerated by the appropriate field, $\mathbf{E}$. The high energy electron resulting from a recombination collision will have $E \leq I$, so that only those fields $\mathbf{E} \geq \mathbf{E}_c$ can cause the electron to be accelerated, resulting in another ionization and thus increasing the density of conduction electrons.

Another model is the *low energy criterion*, due to von Hippel (1935) and Callen (1949). They assumed the limiting condition for breakdown to occur when $B(E', T_0)$ is a maximum, with E' being the maximum average electron energy required for steady state conditions. This means that $\mathbf{E}'_c$ is the field needed to accelerate all the conduction electrons against the lattice influence and is also shown on Figure 5.20.

In general, the actual breakdown field, $\mathbf{E}_B$, must be greater than $\mathbf{E}_c$, the value necessary to cause an irreversible imbalance in the energy transfer equation.

Intrinsic theories based on changes in the electron energy distribution of the solid have also been advanced. These theories are based on five

CHAPTER 5: LINEAR DIELECTRICS

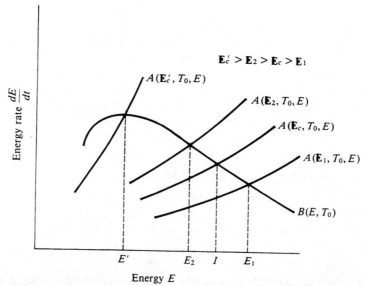

FIGURE 5.20 Schematic representation of the energy balance relation for the Fröhlich high energy criterion $A(E_c, T_0 E) = B(E, T_0)$, and the von Hippel–low energy criterion, $A(\mathbf{E}'_c, T_0, E) = B(E', T_0)$.

contributing factors:

1. Acceleration of electrons due to the applied field,
2. Collisions between conduction electrons,
3. Collisions between conduction electrons and the lattice,
4. Ionization or recombination of electrons to and from the valence band or traps, and
5. Diffusion due to a field gradient.

This large number of theoretical variables can be reduced by ignoring ionization–recombination processes. This is possible because it can be assumed that most electrons in dielectrics occupy energy states intermediate between those low enough to allow recombination or high enough to permit ionization. The diffusion mechanism can also be dismissed by not considering the influence of the field source.

The energy distribution in the solid can be classified under assumptions of either a low or high density of conduction electrons. Assuming a low electron density permits ignoring electron–electron collisions. For low energy elections, a Boltzmann distribution is obtained. However, for high energy electrons, Fröhlich (1947) showed that a valid distribution function could not be described. The distribution approaches infinity with increasing

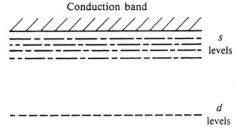

FIGURE 5.21 Trap level representation for the Fröhlich amorphous material model.

energy unless ionization–recombination properly makes the theory an avalanche type.

For high electron density theories it is assumed that electron–electron collisions are so important that they determine the electron energy distribution. Electron–lattice interactions determine the critical value of electron temperature T_c, usually greater than the lattice temperature T_0. Critical conduction electron densities of $n_c = 10^7 \, \text{cm}^{-3}$ and $n_c = 10^{14} \, \text{cm}^{-3}$ have been calculated for polar and nonpolar materials, respectively. Thus for these models, $(\alpha - T_c)$ and the breakdown strength are calculated by determining the field, $\mathbf{E}_i$, necessary to make T increase to infinity.

An electron density distribution model for a pure crystalline solid has been proposed by Fröhlich and Paranjape (1956). With few or no defects there would be few traps. Consequently, in the energy dissipation function B, ionizing collisions were also assumed to be of minor significance and therefore is one point of criticism. Another less than satisfactory aspect of the model is that for $n_c = 10^{17} \, \text{cm}^{-3}$ and $\mathbf{E} < \mathbf{E}_c$, the energy gain function A is a very large value of 600μ eV per ion volume, where μ is the mobility. For normal values of μ, this magnitude of energy absorption would destroy the crystal at less than critical fields.

Fröhlich (1947) has also proposed a model for amorphous materials in which he assumes a trap distribution as shown in Figure 5.21. These are shallow s traps just below the conduction band and deep d traps within the gap. The density of electrons in the traps is $n_c < n_s \ll n_d$ and $n_c = 10^{17} \, \text{cm}^{-3}$. n_c plus n_s thus determines the conduction electron energy distribution and the electron temperature T. The main mechanism of energy transfer in this model is emission of phonons associated with the change of electrons between s levels and the conduction band. O'Dywer (1964) has extended this theory to include isolated defect levels in crystals.

5.14 THERMAL BREAKDOWN MECHANISMS

Thermal breakdown theory also is based on an energy balance relation, but it is the balance between heat dissipated by the sample and the heat generated due to Joule heating, dielectric losses, and discharges in the ambient. Hence, it is the lattice temperature and not an electron temperature that must reach a critical level for breakdown to occur. The influence of the applied field is only indirectly felt as it influences the heat generating mechanisms and does not play the determining role evident in the intrinsic theories based on the electron temperature in the solid. Because of the relatively weak dependence between field and temperature, the value T_c is not too important and the actual lattice temperature at breakdown, T_0', is usually somewhat greater.

The basic relation for thermal breakdown is

$$C_v \frac{dT}{dt} = \text{div}\,(K\,\text{grad}\,T) = \sigma(\mathbf{E}, T_0)\mathbf{E}^2 \qquad (5.125)$$

where C_v = heat capacity of the material
 div $(K\,\text{grad}\,T)$ = heat conduction of a volume element
 $\sigma(\mathbf{E}, T_0)\mathbf{E}^2$ = heat generation term

There is no charge accumulation, so the current is continuous.

For slow field applications, a steady state is assumed and the $C_v\,dT/dt$ term can be ignored. Calculations employing this model show that breakdown strength is inversely proportional to the square root of the thickness. This square root dependence agrees well for thin samples, where a uniform temperature can be realized, but for thicker samples the breakdown strength is experimentally observed to be inversely proportional to the thickness itself. The criterion for determining if a sample is thick or thin depends on several materials constants, namely the thermal conductivity as well as the pre-exponential σ_0, and the activation energy E_a, for the conductivity–temperature dependence of the solid:

$$\sigma = \sigma_0 \exp\left[\frac{E_a}{kT}\right] \qquad (5.126)$$

When the field pulse is fast enough, negligible heat transfer can take place. Consequently, the heat conduction term can be ignored and the electrodes only influence the field distribution and not the heat flow, as in the previous case. By integration, it is possible to calculate t_c, the time for breakdown to occur after reaching the critical temperature T_c:

$$t_c = \int_{T_c}^{T_0'} \frac{C_v\,dT}{\sigma(\mathbf{E}_c, T_0)\mathbf{E}_c^2} \qquad (5.127)$$

The strong dependence of t_c on $\mathbf{E}_c$ is easily seen in the above expression.

Similarly, it can be shown that

$$\mathbf{E}_c = \frac{KT_0}{t_c^{1/2}} \exp\left[\frac{E_a}{2kT}\right] \qquad (5.128)$$

thus, the critical breakdown field is essentially independent of T_c.

Numerical solutions to thermal breakdown models must be used when the field frequency is intermediate and either of the above simplifications can not be made. Each case becomes a different problem with different boundary conditions. Numerical techniques can also be extended to cover multidimensioonal samples or different waveforms with some success.

5.15 AVALANCHE BREAKDOWN MECHANISMS

While thermal breakdown theory may be more practical for the application temperatures of many ceramic materials, if the material geometry approaches that of a thin film, *avalanche theory* may prove more useful. Avalanche breakdown theory is an attempt to combine features of intrinsic and thermal theories, since an electron distribution instability will have thermal consequences. The electron behavior is described with an intrinsic theory and the breakdown criterion is based on thermal properties. Avalanche theory considers the gradual or sequential buildup of charge rather than the sudden change in conductivity, even though the charge buildup may occur in a very short time.

Avalanche theories can be easily classed as to their initiation mechanisms which are either field emission or ionization collision. Field emission assumes that the conduction electron density increases by tunneling from the valence band to traps or to the conduction band. Considering that the probability for emission is

$$P = a\mathbf{E} \exp\left[-\frac{bI^2}{\mathbf{E}}\right] \qquad (5.129)$$

this shows that P is small until $\mathbf{E}$ is quite high (a and b are constants). Using this development gives an order of magnitude calculation of $\mathbf{E} = 10^7$ V/cm, with the impulse thermal criteria $T = T_c$ being the critical parameter.

The simpler single electron approach to an ionization collision theory seems to be preferred to the more complex many electron or avalanche multiplication view. The single electron model assumes that at least 10^{12} conduction electrons per cubic centimeter are needed to disrupt the lattice. When one starting electron-ionizing collision liberates two electrons, which in turn liberate four and so on, it will take about 40 such collisions to achieve breakdown. This simple approach gives a critical field $\mathbf{E}_c$ that is dependent on thickness. The sample must be thick enough to have a least

40 mean free path lengths. However, the sample can not be so thick that the conduction electron density becomes so large that electron–electron collisions occur, thus limiting the electron energy to that below the ionization level.

The many electron avalanche theory gives a relation between the electron energy distribution function and the ionization–recombination rates. There are two views of this approach to dielectric breakdown. One, due to Heller (1951), assumes that the ionization rate is so large that recombination cannot keep up. It is assumed in the other model by Franz (1956) and Veelken (1954), that recombination can be dismissed, since ionization increases the number of conduction electrons so rapidly that no recombination mechanism can quench the process. The actual case is likely to be a combination of the two. The electron source is ignored, because the density of the source electrons has no relation to E_c.

O'Dywer (1964) developed a space charge modified field emission model for avalanche breakdown. His criticism of the single electron or 40 generation model is based on its lack of a continuity of current. Assuming that the freed electrons and holes occurring after the fortieth collision from a parallel plate capacitor, calculations show that a field of 10^{11} V/cm would be needed to maintain this charge separation. Since such a large breakdown strength is not observed, he suggests a continuous current model involving cold cathode field emission with collision ionization occurring after emission, for short breakdown times. Only electrons of energy kT, i.e., room temperature, are assumed to be accelerated by the field to the ionization level, although intermediate electron energies could cause ionization, or some electrons with sufficient energy might not. Calculation of a field vs. distance relation indicates that a space charge of low mobility holes is

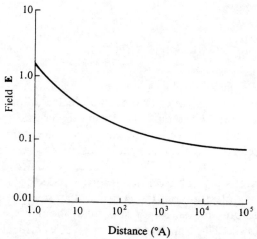

FIGURE 5.22 Representation of the field gradient calculation of O'Dwyer.

influencing the field distribution and hence breakdown strengths, as seen by the drop in field near the cathode, shown in Figure 5.22. Comparison of the theoretical results with Al_2O_3 and NaCl thin film breakdown data gives good agreement.

5.16 EXPERIMENTAL VARIABLES

Sample geometry influences field distribution and heat dissipation considerably. Figure 5.23 shows the effect of thickness of the breakdown path on the breakdown strength for thin films, and indicates that the breakdown strength decreases sharply with distance at some critical value. The same general pattern is observed for thicker specimens, although some doubt exists with regard to amorphous materials.

The temperature dependence of many materials is shown in Figure 5.24. In the low temperature region, the breakdown strength increases with increasing temperature and then declines. The location of the maxima depends on the material, and large deviations from this relation can occur, including temperature independent regions.

The breakdown strength generally increases when the surrounding environment, such as special oils, has a high dielectric strength. The applied field magnitude, frequency, duration, waveform, and distribution are also important variables. These can be very important, especially for intrinsic-type breakdown, which is directly field dependent. In general, dc fields produce lower breakdown strengths than ac fields for thin specimens. The reverse is true for thick specimens, due to dielectric heating and poor heat

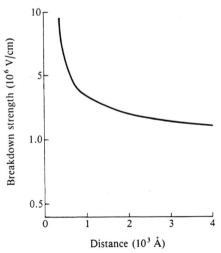

FIGURE 5.23 Thickness dependence of breakdown strength typical of thin films.

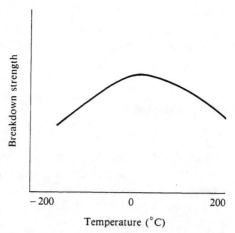

FIGURE 5.24 Nominal temperature dependence of breakdown strength for many materials.

dissipation. The greatest discrepancies to this generalization occur at very high and very low frequencies. Increased breakdown values are obtained with shorter square wave pulses and faster rise times. Prestressing has received too little attention, especially since the influence of space charge on breakdown strength is being considered more favorably. Experiments on glass indicate a very definite influence of prestressing or applying a field less than breakdown for some time. Also, there is a threshold field below which breakdown does not occur in any time interval.

Electrodes are also variable, with vacuum depositions of aluminum or gold being the most reliable. Even polarity can make a difference if electrode geometry does not provide a uniform field intensity.

In glasses, breakdown strength follows the change in resistivity with additions of alkali ions. The influence of composition on ceramics can be great. For example, in $Ba_{1-x}La_xZrO_3$, for $x = 0.06$, the breakdown strength increased 175%, apparently due to electron–hole recombinations reducing the conductivity. A similar effect is observed for additions of Nb in place of Zr.

Microstructure also affects breakdown strength, including material texture, particle alignment, stress distribution and porosity. Other material properties that influence dielectric breakdown are surface condition, grain boundary effects, point defects, dislocations, and dielectric permittivity. There is evidence, for example, that dislocations increase the free charge in dielectrics, and consequently grinding, polishing, surface treatments, or mechanical history can influence breakdown behavior as well.

PROBLEMS

Given the following dielectric data for a research ceramic material based on a parallel plate capacitor configuration where the electrode diameter = 1.0 in. and the sample thickness = 0.2 in.

Measuring frequency f (Hz)	Conductance G (μOhm^{-1})	Capacitance C (pF)
72°C		
20K	0.885	4.76
10K	0.792	5.44
5K	0.748	6.12
2K	0.723	7.11
1K	0.706	8.91
500K	0.676	14.50
340K	0.627	22.53
260K	0.603	32.4
200K	0.598	47.08
100K	0.574	139.0
90°C		
20K	3.764	5.72
10K	3.672	6.47
5K	3.636	7.90
2K	3.605	15.21
1K	3.572	36.12
700K	3.548	60.82
500K	3.514	104.7
200K	3.292	455.4
112°C		
20K	6.827	6.36
15K	6.781	6.83
10K	6.732	8.04
5K	6.678	12.85
3K	6.628	21.8
2K	6.557	39.4
1.5K	6.534	62.2
1.2K	6.480	91.1
1K	6.405	118.7
500K	6.058	353.4

Conduct the following analysis:

5.1 Plot k' vs. $\log \omega$ (all three temperatures on same plot).

5.2 Plot k'' vs. $\log \omega$ (all three temperatures on same plot).

5.3 Plot tan δ vs. log ω (all three temperatures on same plot). *Hint*: tan $\delta = G/2\pi fC$.

5.4 Plot log ω_{max} vs. $1/T$ (°K^{-1})

5.5 Calculate the activation energy E_a for the relaxation process.

5.6 Based on the value of E_a obtained discuss possible mechanisms for the relaxation.

5.7 Determine τ_0 and discuss the significance of the result.

5.8 Plot log σ_{dc} vs. $1/T$ (°K^{-1}) remembering the definition for σ_{dc} and employing extrapolations when required.

5.9 Calculate $E_{a(dc)}$ and compare with 5.5 and 5.6.

5.10 Is there evidence for electrode polarization? Discuss.

5.11 Discuss the physical basis for the type of curves found in 5.1–5.4.

5.12 Discuss the temperature-dependent changes in the curves.

5.13 Obtain your best evaluation of k_s and k_∞ from the data (use extrapolations when required).

5.14 Discuss the magnitudes obtained and origin of difference between the two values.

5.15 Using the values obtained calculate the theoretical Debye relaxation curves.

5.16 Show on a plot tan δ vs. log ω measured (vs. tan δ vs. log ω Debye).

5.17 Discuss any differences observed.

5.18 Use one of the methods for obtaining distribution of relaxation times to evaluate the distribution in the dielectric material.

5.19 Discuss possible causes for distribution in τ.

5.20 If the ceramic capacitor was made up of two parallel layers of dielectrics, what parameters of the two layers would be required to result in the value of τ observed? (see Chapter 30, Halliday and Resnick 1965).

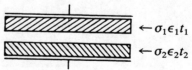

5.21 If the ceramic were a heterogeneous material containing a second phase (2) with $\sigma_2 \gg \sigma_1$ and of the shape of ellipsoids, what combination of volume fraction and axial ratio (a to b) of second-phase particles, would have to be present to result in the observed properties?

5.22 Would this material make a good insulator? Discuss.

5.23 Would this material make a good capacitor? Discuss.

5.24 Discuss what type of real-life ceramic could give rise to the behavior observed in the above sample.

5.25 Predict the dielectric breakdown strength of this material.

READING LIST

Anderson, J. C., (1964). *Dielectrics*, Chapman and Hall Ltd, London, Chapters: 2, 3, 4, 5, 6, 7, 8, 10.

Azaroff, L. and Brophy, J. J., (1963). *Electronic Processes in Materials*, McGraw-Hill, New York, Chapter: 12.

Blatt, F. J., (1968). *Physics of Electronic Conduction in Solids*, McGraw-Hill, New York, Chapter: 7.

Callen, H. B., (1949). *Physics Review*, **76**, 1394.

Charles, R. J. *J. Am. Ceram. Soc.* **46**, 235.

Cole, K. S. and Cole, R. H., (1942). *Chem. Phys.* **10**, 98.

Franz, W., (1956). *Handbuch der Physik* (S. Flugge, ed.), Vol. 1. Springer-Verlag, Berlin.

Frölich, H. (1947). *Proc. Phys. Soc. London, Sect. A* **188**, 521.

Fröhlich, H., and Paranjape, B. V. (1956). *Proc. Phys. Soc. London, Sect. B* **69**, 21.

Frohlich, H., (1937). *Proc. Roy. Soc. London, Ser. A*, **160**, 230.

Halliday, D. and Resnick, R. (1965). *Physics*, Combined Edition, Wiley, New York, Chapters: 26, 27, 28, 29, 30, 31.

Heller, W. R. (1951). *Phys. Rev.* **84**, 1130.

Hench, L. L. and Dove, D. B., (1971). *Physics of Electronic Ceramics*, Part A, Dekker, New York, Chapters: 17–20.

Hench, L. L., Wang, S. H., and Nogues, J. L., (1988). "Gel-silica optics." *Proc. SPIE Int. Soc. Opt. Eng.* **878**, 76–85.

von Hippel, A. (1935). *Ergeb. Exakten Naturwiss.* **14**, 79.

Horwick, A. S., and Berry, B. S., (1961). *IBM J. Res. Dev.*, **5**, 297 and 312.

Kingery, W. D., Bowen, H. K. and Uhlmann, D. R., (1976). *Introduction to Ceramics*, 2nd ed., Wiley, New York, Chapter: 18.

Kinser, D. L. and Hench, L. L., (1968). *J. Amer. Ceram. Soc.*, **52**, 445.

Kinser, D. L., (1971). "Electrical conduction in glass and glass ceramics," in *Physics of Electronic Ceramics*, (L. L. Hench and D. B. Dove, eds.) Dekker, New York, pg. 523.

Kittel, C., (1988). *Introduction to Solid State Physics*, 6th ed., Wiley, New York.

Mable, S. H. and McCammon, R. D., (1969). *Phys. Chem. Classes*, **10**, 222.

Mozzi, R. L., and Warren, B. E., (1969). *Appl. Cryst.*, **2**, 164.

Nowick, A. S., and Berry, B. S., (1961). *IBM J. Research*, **5**, 297–312.

Rosenberg, H. M., (1978). *The Solid State,* 2nd ed. Oxford Univ. Press, London and New York, Chapter: 13.

Sillars, R. W., (1937). *Proc. Inst. Elect. Engrs.,* (London), 80, 378.

Veelken, R., (1954). *Z. Phys.* **142,** 476.

Wagner, C. W., (1914). *Archiv fur Elektrotechnik,* **2,** 371.

Wallace, S., and Hench, L. L., (1988). *Materials Research Society Symp.,* **121,** 589.

Walther, G. C., and Hench, L. L. (1971). "Dielectric breakdown of ceramics," in *Physics of Electronic Ceramics,* (L. L. Hench and D. B. Dove, eds.,) Part A. Dekker, New York.

Zhilenkov, I. V. and Nekrasova, E. G. (1980). *Russ. J. Phys. Chem. (Engl. Transl.)* **54,** 1503.

6

NONLINEAR DIELECTRICS

6.1 INTRODUCTION

Nonlinear dielectrics are an important class of crystalline ceramics which can exhibit very large dielectric constants ($k_s > 1000$), due to spontaneous alignment or polarization of electric dipoles. The spontaneous alignment of electric dipoles results in a crystallographic phase transformation below a critical temperature T_c. The electric dipoles are ordered parallel to each other within the crystal in regions called *domains*. When an electric field is applied, the domains can switch from one direction of spontaneous alignment to another. This gives rise to very large changes in polarization and dielectric constant. Hence the name *nonlinear dielectrics*. This nonlinear electric polarization is analogous to the nonlinear magnetic behavior of ferromagnetic materials. Consequently, certain nonlinear dielectrics are also called ferroelectrics even though they do not contain iron. Because of a strong electro-mechanical coupling, these materials have widespread use as pressure transducers, ultrasonic cleaners, loudspeakers, gas ignitors, relays, and in a variety of electro-optical applications (see Chapter 10).

6.2 CRYSTALLOGRAPHIC CONSIDERATIONS

Of the 32 crystal classes or point groups (see Table 6.1) 11 are centrosymmetric and consequently cannot possess polar properties or spontaneous polarization. One of the remaining 21 noncentrosymmetric point groups (cubic 432) has symmetry elements which prevent polar characteristics.

TABLE 6.1 The 32 Crystal Point Groups

Optical axes	Crystal class	Centrosymmetric point groups		Noncentrosymmetric point groups			
				Polar		Nonpolar	
Biaxial	Triclinic	$\bar{1}$		1		none	
	Monoclinic	2 or m		2	m	none	
	Orthorhombic	mmm		$mm2$		222	
Uniaxial	Tetragonal	4 or m	4 or mmm	4	$4mm$	$\bar{4}$	$\bar{4}2m$	22
	Trigonal	$\bar{3}$	$\bar{3}m$	3	$3m$		32
	Hexagonal	6 or m	6 or mmm	6	$6mm$	$\bar{6}$	$\bar{6}m2$	622
Optically isotropic	Cubic	$m3$	$m3m$	none		432	$\bar{3}m$	23
Total number		11 groups		10 groups		11 groups	

However, the other 20 point groups have one or more polar axes (Table 6.1) and thus can exhibit various polar effects such as *piezoelectricity*, *pyroelectricity* and *ferroelectricity*.

The crystal point group notation is straightforward. The count of the number of operations that will keep the crystal structure the same is a measure of the symmetry. The count must be unique; you cannot count an orientation twice in the same structure. Table 6.2 indicates the notation or count along with the operation.

All crystals can be subdivided into six crystal classes depending upon their minimum symmetry requirements (Table 6.3). The relation of crystallographic axes and angles of the unit cells to their symmetry is shown in Table 6.3.

Let us consider an example. One cubic crystal point group can be represented by the point group notation for x-axis, y-axis, and z-axis as

$$4\ 3\ 2$$

This means there are four 90° rotations that can be made while holding the

TABLE 6.2

Notation	Symmetry operation
$1, 2, 3, \ldots, n$	Axis having n rotations of symmetry
m	Axis representing a mirror plane
$\bar{1}, \bar{2}, \bar{3}$, etc.	Axis operation of a rotation followed by an axis inversion

TABLE 6.3

System	Minimum symmetry	Unit cell a, b, c = dimensions	Unit cell α, β, γ = angles
Triclinic	1 or ($\bar{1}$)	$a \neq b \neq c$	$\alpha \neq \beta \neq \gamma$
Monoclinic	2 (or m)	$a \neq b \neq c$	$\alpha = \beta = 90° \neq \gamma$
Orthorhombic	222 (or mmm) or (2)	$a \neq b \neq c$	$\alpha = \beta = \gamma = 90°$
Tetragonal	4 (or $\bar{4}$)	$a = b \neq c$	$\alpha = \beta = \gamma = 90°$
Hexagonal	6 (or 3)	$a = b \neq c$	$\alpha = \beta = 90°$ and $\gamma = 120°$
Isometric (cubic)	333 (or $\bar{3}$)	$a = b = c$	$\alpha = \beta = \gamma = 90°$

x-axis. Each of these four rotations look identical to the original orientation. If the y-axis is now held, only three rotations give unique positions. The fourth position was counted while the x-axis was being held. Finally, when the z-axis is held there is only two positions that are unique. The other two were counted earlier. The symmetry is so great for this cubic crystal that no dipole moment can be maintained about any axis.

There are a total of 44 different point groups for molecules in general but only 32 are possible for crystal formation.

The other two operations include a mirror image and a rotation followed by an inversion. The mirror image operation places a mirror perpendicular to the axis being held. If the molecule or structure does not change in appearance then this is a valid symmetry operation for that axis.

The third operation is the dual process of holding an axis, rotating it and then inverting the axis being held. Thus, if the x-axis is being held, it is inverted by moving it to the negative x-axis position. After inversion the negative x-axis is being held where the positive x-axis was previously.

Piezoelectricity is the property of a crystal to exhibit electric polarity when subjected to a stress; that is, when a compressive stress is applied a charge will flow in one direction in a measuring circuit. A tensile stress causes charge to flow in the opposite direction. Conversely, if an electric field is applied, a piezoelectric crystal will stretch or compress depending on the orientation of the field with the polarization in the crystal (see Figure 6.1).

Of the 20 piezoelectric classes of crystals, 10 have a unique polar axis; an axis which shows properties at one end different than the other. Crystals in these 10 classes are called polar crystals because they are spontaneously polarized. The spontaneous polarization is usually compensated through external or internal conductivity or twinning (domain formation).

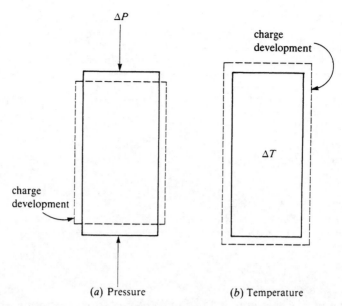

FIGURE 6.1 Applied stress causes: (a) Piezoelectric effect; (b) Pyroelectric effect.

The magnitude of the spontaneous polarization depends upon temperature. Consequently, if a change of temperature is imposed on the crystal an electric charge is induced on the crystal faces perpendicular to the polar axis. This is called the *pyroelectric* effect. Each of the 10 classes of polar crystals are pyroelectric.

Ferroelectric crystals are also pyroelectric. However, ferroelectric crystals are only those crystals for which the spontaneous polarization can be reversed by applying an electric field. *Thus, a ferroelectric is a spontaneously polarized material with reversible polarization.*

6.3 FERROELECTRIC THEORY

Ferroelectricity is the reversible spontaneous alignment of electric dipoles by their mutual interaction. Ferroelectricity occurs due to the local field $\mathbf{E}'$ increasing in proportion to the polarization which is increased by the aligning of dipoles in a parallel array with the field. The alignment is spontaneous at a temperature T_c, where the randomizing effect of thermal energy kT is overcome.

The defining equation for the onset of ferroelectricity follows from the definition of electric polarization (see Chapter 5):

$$\mathbf{P} = (k' - 1)\epsilon_0 \mathbf{E} = N\alpha \mathbf{E}' \qquad (6.1)$$

where

$$E' = E + \frac{P}{3\epsilon_0} \quad \text{and} \quad k' = \text{relative dielectric constant} \qquad (6.2)$$

thus

$$P = N\alpha\left(E + \frac{P}{3\epsilon_0}\right) = N\alpha E + \frac{N\alpha P}{3\epsilon_0} \qquad (6.3)$$

Rearranging yields

$$P - \frac{N\alpha P}{3\epsilon_0} = N\alpha E \qquad (6.4)$$

and

$$P\left(1 - \frac{N\alpha P}{3\epsilon_0}\right) = N\alpha E \qquad (6.5)$$

so

$$P = \frac{N\alpha E}{\left(1 - \frac{N\alpha}{3\epsilon_0}\right)} \qquad (6.6)$$

Since the electric susceptibility, χ, is defined as

$$\chi = k' - 1 = \frac{P}{\epsilon_0 E} \qquad (6.7)$$

then substituting equation (6.6) into equation (6.7) yields

$$\chi = k' - 1 = \frac{N\alpha/\epsilon_0}{\left(1 - \frac{N\alpha}{3\epsilon_0}\right)} \qquad (6.8)$$

Recalling that $k' = \epsilon/\epsilon_0$, then equation (6.8) is called the Clausius–Mosotti equation when rearranged as

$$\frac{N\alpha}{\epsilon_0} = \frac{\epsilon - 1}{\epsilon + 2} \qquad (6.9)$$

We can see from equation (6.8) that when

$$\frac{N\alpha}{3\epsilon_0} \to 1 \qquad (6.10)$$

then P, χ, and k' must go to infinity.

Recall from Chapter 5 that the orientation of a dipole is inversely proportional to temperature:

$$\alpha_0 = C/kT \qquad (6.11)$$

where C is the Curie constant of a material. If we consider materials where $\alpha_0 \gg \alpha_e + \alpha_a + \alpha_i$, then a critical temperature T_c will be reached, where

$$\frac{N\alpha_0}{3\epsilon_0} = \frac{N}{3\epsilon_0}\left(\frac{C}{kT_c}\right) = 1 \qquad (6.12)$$

Consequently, T_c occurs when the following condition is met:

$$T_c = \frac{NC}{k3\epsilon_0} \qquad (6.13)$$

Below this critical temperature, spontaneous polarization sets in and all the elementary dipoles have the same orientation.

Combining the above equations yields the *Curie–Weiss law* which defines the temperature onset of ferroelectricity:

$$\chi = k' - 1 = \frac{N\alpha_0 T}{\left(1 - \dfrac{N\alpha}{3\epsilon_0}\right)} \qquad (6.14)$$

When equation (6.14), the Curie–Weiss law, is combined with the defining equation for the critical temperature, for example,

$$\frac{T_c}{T} = \frac{N\alpha}{3\epsilon_0} \qquad (6.15)$$

the following relation is obtained which describes the temperature dependence of the electric susceptibility of a ferroelectric and the onset of ferroelectric behavior at T_c.

$$\chi = \frac{3N\alpha/3\epsilon_0}{\left(1 - \dfrac{N\alpha}{3\epsilon_0}\right)} = \frac{3T_c/T}{\left(1 - \dfrac{T_c}{T}\right)} = \frac{3T_c/T}{(T - T_c)/T} = \frac{3T_c}{T - T_c} \qquad (6.16)$$

Figure 6.2 shows the linear Curie–Weiss dependence of χ^{-1} on temperature above the onset of ferroelectricity for $BaSrTiO_2$. At the Curie point there is a spontaneous alignment of the dipoles leading to a discontinuity in the temperature dependence.

Values of the Curie constant T_c, spontaneous polarization, and dielectric constants for various ferroelectric materials are given in Table 6.4.

6.3 FERROELECTRIC THEORY

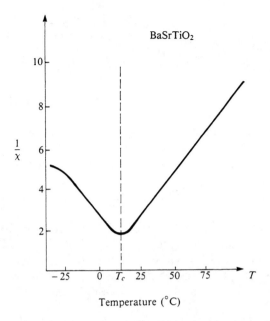

FIGURE 6.2 Curie–Weiss law.

TABLE 6.4 Ferroelectric Properties of Several Materials

Materials	Curie constant (°C)	Curie temperature (°C)	Spontaneous polarization ($\mu C/cm^2$) at [T°C]	Dielectric constant (at T_c)
	Ferroelectric: complex crystal structure			
Rochelle salt ($NaKC_4H_4O_6 \cdot 4H_2O$)	2.2×10^2	+24°	0.25 [23]	5000
KH_2PO_4	3.3×10^3	−150°	4.7	10^5 (c-axis) 70 (a-axis)
	Ferroelectric: perovskite-type crystal structure			
$BaTiO_3$	1.7×10^5	+120°, 5, 90	26 [23]	1600
$PbTiO_3$	1.1×10^5	+490°	750 [23]	—
$KNbO_3$	2.4×10^5	+415, 225, −10	30	4200, 2000, 900
	Antiferroelectric: perovskite type crystal structure			
$PbZrO_3$	1.6×10^5	+230°	—	3500

There are Curie temperatures up to 490°C, spontaneous polarizations up to 750 $\mu C/cm^2$ and dielectric constants up to 5000 (see Table 6.5 for other ferroelectrics).

6.4 STRUCTURAL ORIGIN OF THE FERROELECTRIC STATE

The spontaneous alignment of dipoles which occurs at the onset of ferroelectricity is often associated with a crystallographic phase change from a centrosymmetric, nonpolar lattice to a noncentrosymmetric polar lattice. Barium titanate ($BaTiO_3$) is an excellent example to illustrate the structural changes that occur when a crystal changes from a nonferroelectric (*paraelectric*) to a ferroelectric state.

The Ti ions of $BaTiO_3$ are surrounded by six oxygen ions in an octahedral configuration (Figure 6.3). The octahedral coordination is expected from the radius ratio of 0.468 (see Chapter 1, Figure 1.4). All crystals possessing the TiO_6 configuration have a high dielectric constant, as a result of a large dispersion stemming from infrared vibrations (see Chapter 8).

Since a regular TiO_6 octahedron has a center of symmetry, the six Ti—O dipole moments cancel in antiparallel pairs. A *net* permanent moment of the octahedron can result only by a unilateral displacement of the positively

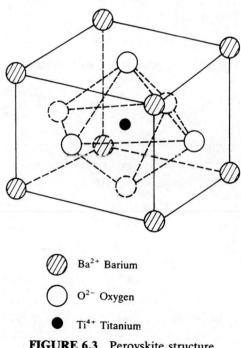

Ba^{2+} Barium

O^{2-} Oxygen

Ti^{4+} Titanium

FIGURE 6.3 Perovskite structure.

6.4 STRUCTURAL ORIGIN OF THE FERROELECTRIC STATE

charged Ti^{4+} ion against its negatively charged O^{2-} surroundings. Ferroelectricity requires the coupling of such displacements and the dipole moments associated with the displacements.

For TiO_2, each oxygen ion has to be coupled to three Ti ions if each Ti is surrounded by six oxygens. In rutile, brookite, and anatase (three crystal modifications of TiO_2) the TiO_6 octahedra are grouped in various compensating arrays by sharing two, three, and four edges respectively with their neighbors. Consequently, all the Ti—O dipole moments cancel and none of the TiO_2 crystal forms are ferroelectric.

However, in the ABO_3 or $BaTiO_3$ (perovskite-like) structure, named after the $CaTiO_3$ perovskite mineral, each oxygen has to be coupled to only two Ti ions. Consequently, the TiO_6 octahedra in $BaTiO_3$ can be placed in identical orientations, joined at their corners, and fixed in position by Ba ions. This gives the opportunity for an effective additive coupling of the net dipolar moment of each unit cell.

Thus, in $BaTiO_3$ the Ba and O ions form a fcc lattice with Ti ions fitting into octahedral interstices (see Figures 6.3). The characteristic feature of the Ba, Pb, and Sr titanates is that the large size of Ba, Pb, and Sr ions increases the size of the cell of the fcc BaO_3 structure so that the Ti atom is at the lower edge of stability in the octahedral interstices. There are consequently minimum energy positions for the Ti atom which are off-center and can therefore give rise to permanent electric dipoles.

At a high temperature $T > T_c$, the thermal energy is sufficient to allow the Ti atoms to move randomly from one position to another, so there is no fixed asymmetry. The open octahedral site allows the Ti atom to develop a large dipole moment in an applied field, but there is no spontaneous alignment of the dipoles. In this symmetric configuration the material is paraelectric, (i.e., no net dipole moment when $\mathbf{E} = 0$).

When the temperature is lowered to below T_c, the position of the Ti ion and the octahedral structure changes from cubic to tetragonal symmetry with the Ti ion in an off-center position corresponding to a permanent dipole. These dipoles are ordered, giving a domain structure with a net spontaneous polarization within the domains.

The crystallographic dimensions of the $BaTiO_3$ lattice change with temperature, as shown in Figure 6.4a, due to distortion of the TiO_6 octahedra as the temperature is lowered from the high temperature cubic form. Because the distorted octahedra are coupled together, there is a very large spontaneous polarization, giving rise to a large dielectric constant and large temperature dependence of the dielectric constant as shown in Figure 6.4b. The anisotropy of the ferroelectric crystal phases below T_c are shown in Figure 6.5. The spontaneous polarization is considerably stronger in the c-direction which results in the larger dielectric constant in this orientation (see Figure 6.4b).

Let us examine how the ferroelectric phase transformation occurs at T_c. As shown in Figure 6.5, above T_c $BaTiO_3$ is isotropic. The Ti atoms

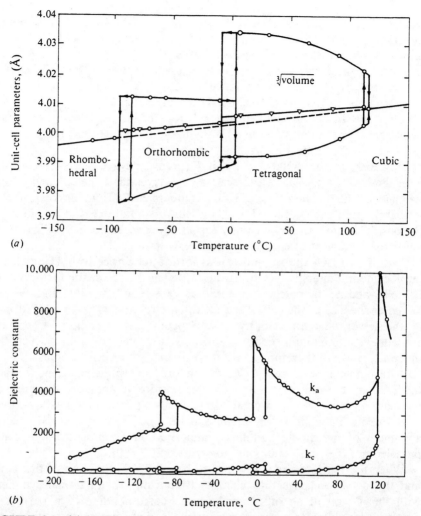

FIGURE 6.4 (a) Lattice parameters of $BaTiO_3$ as a function of temperature (after Merz). (b) Dielectric constants of $BaTiO_3$ as a function of temperature (after Kay and Vousdan, 1949).

are all in equilibrium positions in the center of their octahedra. Thermal agitation produces strong fluctuations around the equilibrium position. An external field will make the net moment nonzero by displacing Ti atoms unilaterally. However, in the absence of an external field the isotropic crystals are nonpolar. As T_c is approached, there is an increase in probability that one of the TiO_6 octahedra will be permanently polarized with a Ti^{4+} ion in an off-center position. How this can occur is illustrated in Figure 6.6.

If Ti_A moves toward O_I the dipole moment $O_I \rightarrow A$ becomes stronger and the $O_{II} \rightarrow A$ moment becomes weaker. Consequently, O_I moves toward A

6.4 STRUCTURAL ORIGIN OF THE FERROELECTRIC STATE

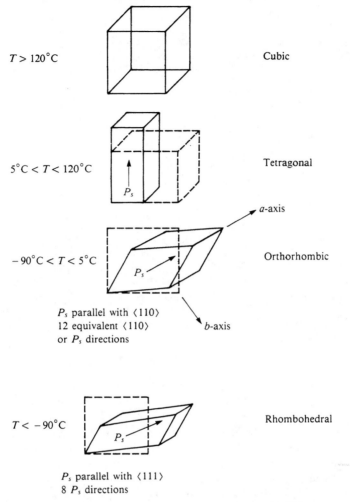

FIGURE 6.5 Crystallographic changes of BaTiO$_3$.

and away from O$_{II}$. B and C follow the motion of A and D. E tends to follow suit because of the coupling between ions O$_{III}$ and O$_{IV}$. They tend to move downward, repelled by O$_I$. The magnitude of displacement of the Ti in its oxygen coordination octahedron is 0.12 Å and the oxygen displacements are ~0.03 Å (Figure 6.7). Thus, a net dipole moment has been produced in the octahedron by the permanent displacement of the Ti ion against its surrounding oxygen ions. The coupling between neighboring octahedra increases the displacements and increases the internal field. Thermodynamic, statistical mechanical and quantum-mechanical treatments of this structural transformation are reviewed in Burfoot (1967).

A tetragonal phase is the first low temperature phase which corresponds

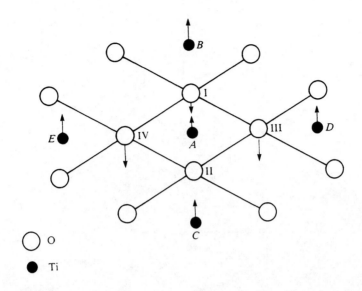

FIGURE 6.6 Atomic displacements as BaTiO$_3$ approaches T_c.

to the Ti ions being located in a nonsymmetric position in the TiO$_6$ octahedra (Figure 6.5). In the tetragonal phase spontaneous polarization can occur along any of the three previously cubic [100] axes. Consequently, domains can be formed 90° to each other as well as antiparallel (180°). However, as shown in Figures 6.4 and 6.5 two additional low-temperature phase changes also occur for BaTiO$_3$. At 5°C, the tetragonal phase distorts even further along the c-axis resulting in an orthorhombic phase. There are 12 equivalent $\langle 110 \rangle$ directions in the mm orthorhombic crystal lattice and therefore there are 12 equivalent $\mathbf{P}_s$ directions in the BaTiO$_3$ crystal between 5°C and −90°C (see Figure 6.5).

At −90°C, a third phase transformation occurs where a rhombohedral phase is formed. In this crystal with a $3m$ space group, the polarization is parallel to the $\langle 111 \rangle$ direction, leading to eight equivalent $\mathbf{P}_s$ modes of polarization.

The change from the cubic to tetragonal phase is a second-order phase transition. The continuous second-order phase change mean that the energy, volume, and structure changes more or less continuously. The temperature derivatives of these quantities have singularities. At the Curie temperature, the first derivatives of energy and the specific heat increases as described by λ-type curves. This behavior appears where the dipoles and the saturation polarization, $\mathbf{P}_s$, are oriented parallel to the cubic $\langle 110 \rangle$ direction (Figure 6.8).

6.4 STRUCTURAL ORIGIN OF THE FERROELECTRIC STATE 249

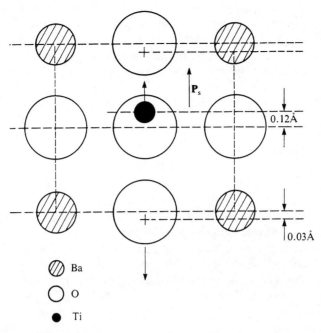

FIGURE 6.7 TiO$_6$ octahedra displacements in the ferroelectric transition of BaTiO$_3$ (after Azaroff and Brophy).

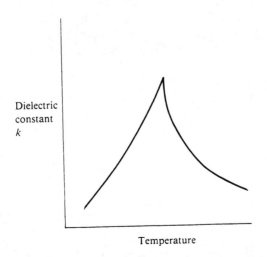

FIGURE 6.8 Singularities for λ-type functions.

The second-order change is a consequence of the necessity to nucleate polarized domains. The domain growth proceeds as the temperature approaches the Curie point until they divide the crystal into an array of domains determined by prehistory and energy balance. In the absence of a strong electric field, a multidomain structure results to neutralize the formation of an external field due to dipoles ending on the crystal surface.

6.5 HYSTERESIS

The result of the spontaneous polarization of a ferroelectric at T_c is the appearance of very high k' and a hysteresis loop for polarization. The hysteresis loop is due to the presence of crystallographic domains within which there is complete alignment of electric dipoles.

At low field strengths in unpolarized (also called virgin) material, the polarization **P** is initially reversible and is nearly linear with the applied field. The slope gives k'_i, the initial dielectric constant, as indicated in Figure 6.9a and equations 6.17 and 6.18. The value of k_i will be similar to k' of the cubic phase.

$$\tan \alpha = \frac{\mathbf{P}}{\mathbf{E}} \qquad (6.17)$$

$$\tan \alpha = (k'_i - 1)\epsilon_0 \qquad (6.18)$$

where k' is the initial dielectric constant.

At higher field strengths, the polarization increases considerably as a result of the switching of the ferroelectric domains. The polarization switches so as to align with the applied field by means of domain boundaries moving through the crystal. (Figure 6.9b and c).

Figure 6.9a shows that at high field strengths, the change in polarization is small due to polarization saturation; that is, all the domains of like orientation are aligned with the field. Extrapolation of the high field **E** curve back to $\mathbf{E} = 0$ gives $\mathbf{P}_s$, the saturation polarization, corresponding to the spontaneous polarization with all the dipoles aligned in parallel. Tables 6.4 and 6.5 compares values of $\mathbf{P}_s$ for a variety of ferroelectric materials. For many ferroelectric ceramics, the values range between 25 and 50 $\mu C/cm^2$.

When the applied field continues to be applied at values greater than required to achieve $\mathbf{P}_s$, the polarization continues to increase, but only proportional to k'_i. This is because all of the domains are oriented parallel to each other. However the individual TiO_6 polarizable units can continue to be distorted increasing the unit polarization. This is an important contrast to ferromagnetic and ferrimagnetic materials where application of a magnetic field greater than required for $\mathbf{M}_s$ does *not* increase the net magnetic moment of the material (Chapter 7).

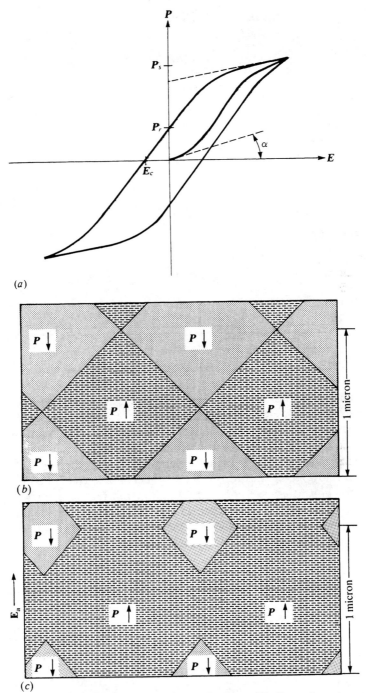

FIGURE 6.9 (a) Hysteresis loop for polarization. (b) Domain microstructure without an applied field. (c) Domain growth in direction of an applied field.

TABLE 6.5 Additional Ferroelectric Oxides

Formula	T_c(°C)	P_s(μC/cm^2)	at T(°C)
Li NbO$_3$	+1210°	71	+23°
Na NbO$_3$	−200°	12.0	−200°
K NbO$_3$	+435°	30.3	+250°
Pb(0.5Sc 0.5Nb)O$_3$	+90°	3.6	+18°
Pb(0.33Mg 0.67Nb)O$_3$	−8°	24.0	−170°
Pb(0.33Zn 0.67Nb)O$_3$	+140°	24.0	+125°
Li TaO$_3$	665°	50.0	+25°
Pb Ta$_2$O$_6$	260°	10.0	+25°
Pb(0.5Fe 0.5eTa)O$_3$	−40°	28.0	−170°
Sr Bi$_2$Ta$_2$O$_9$	335°	5.8	+25°
Sm(MoO$_4$)$_3$	197°	0.24	+50°
Eu$_2$(MoO$_4$)$_3$	180°	0.14	+25°
Pb$_5$GeO$_{11}$	178°	4.6	+25°
SrTeO$_3$	485°	3.7	312°

When **E** is cut off, **P** does not go to zero but remains at a finite value, called the *remanent polarization*, **P**$_r$. This is due to the oriented domains being unable to return to their random state without an additional energy input by an oppositely directed field (Figure 6.9a). The strength of **E** required to return **P** to zero is the coercive field, **E**$_c$. Values of **E**$_c$ for several ferroelectric ceramics are given in Table 6.6.

There is a substantial effect of temperature on the shape of the hysteresis loop. At low temperature, the loops become fatter and **E**$_c$ increases corresponding to a large energy required to reorient domain walls; that is, the domain configuration is frozen in. As the temperature is increased, **E**$_c$ decreases until at T_c no hysteresis remain and k' is single valued at a value characteristic of the paraelectric phase (see Figure 6.10).

6.6 DESCRIPTION OF FERROELECTRICITY BASED ON LOCAL FIELDS

A ferroelectric crystal can be divided into units which each possess dipole moments, e.g., Ti—O units. The interaction of these units with one another is the basis of a local field theory of ferroelectricity (see Burfoot, 1967, or Jona and Shirane, 1962 for additional details).

Assume that all units surrounding a reference unit can be replaced by a continuum with polarization **P**. Then the continuum will contribute to the

6.6 DESCRIPTION OF FERROELECTRICITY BASED ON LOCAL FIELDS

TABLE 6.6 PBO·La$_2$O$_3$·ZrO$_2$·TiO$_2$(PLZT) Ferroelectrics and Antiferroelectrics[a]

Composition (PbZrO$_3$/PbTiO$_3$) + %La$_2$O$_3$	Curie temperature T_c(°C)	Remanent polarization, P_R (μC/cm^2)	Dielectric constant k'	Coercive field, E_c (k C/cm)	Electromechanical coupling factor, k_p (planar, thin disk) no units	Strain coefficient, d_{33} ($\times 10^{12}$ m/V)	Voltage coefficient, g_{33} ($\times 10^{-3}$ Vm/N)
(70/30) +8	30°	26	4050	2	—	—	—
(65/35) +2	320°	40	650	14	0.45	150	23
(65/35) +6	240°	35	1210	9	—	—	—
(65/35) +7	150°	34	1850	5	0.62	400	22
(65/35) +8	65°	30	3400	4	0.65	682	20
(62/38) +8	105°	33	3550	4	—	—	—
(55/45) +2	340°	43	1830	13	—	—	—
(55/45) +4	285°	38	2290	13	0.67	490	22
(55/45) +5	250°	38	3270	10	0.70	578	18
(40/60) +8	240°	28	980	18	0.34	—	—
(40/60) +12	140°	25	1300	13	0.47	235	12
(40/60) +15	40°	18	5270	8	—	—	—
(10/90) +8	355°	29	355	38	—	—	—
BaTiO$_3$	135°	25	1600	0.5–2	0.36	190	13

[a] After Haertling (1986).

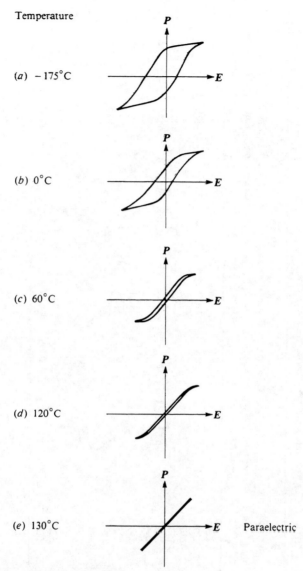

FIGURE 6.10 Effect of temperature on BaTiO$_3$ hysteresis: $T_c \approx 125°C$. (a)–(d) ferroelectric state. (e) Paraelectric state.

local field acting on the reference unit, that is, $\mathbf{E}' = \mathbf{E} + \beta \mathbf{P}$. $\mathbf{E}'$ can vary throughout the crystal even if $\mathbf{E}$ is zero, because $\beta \mathbf{P}$ can have different values at different points in the crystal cell although it will average out to zero. Above T_c:

$$\beta \mathbf{P} = 0 \tag{6.19}$$

6.6 DESCRIPTION OF FERROELECTRICITY BASED ON LOCAL FIELDS

The coefficient, β, represents the electric cooperative interaction between the units. Cooperation occurs when the interaction energy is large compared with the energy of the assembly, that is, the state of each unit is influenced by the states of the others. Cooperation occurs when the alignment of any two dipoles depends upon the alignment of the whole.

Let us examine the physics of the cooperative situation. We first assume a lattice of identical polarizable units. As we saw earlier in equation (6.1)

$$\mathbf{P} = \frac{\alpha \mathbf{E}'}{v} \quad \text{or} \quad \mathbf{P}v = \alpha \mathbf{E}' \tag{6.20}$$

where v is the volume of the unit. When

$$\mathbf{E}' = v\mathbf{P} \tag{6.21}$$

is plotted, curve N is generated as shown in Figure 6.11. When

$$\mathbf{E}' = \mathbf{E} + \beta \mathbf{P} \tag{6.22}$$

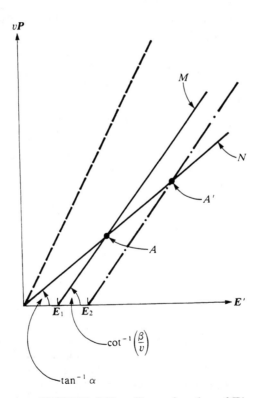

FIGURE 6.11 $v\mathbf{P}$ as a function of $\mathbf{E}'$.

is plotted, then curve M is created. Point A is a stable solution where

$$v\mathbf{P} = \alpha(\mathbf{E} + \beta\mathbf{P}) \tag{6.23}$$

$$v\mathbf{P} = \alpha\mathbf{E} + \alpha\beta\mathbf{P} \tag{6.24}$$

$$v\mathbf{P} - \alpha\beta\mathbf{P} = \alpha\mathbf{E} \tag{6.25}$$

$$\mathbf{P}(v - \alpha\beta) = \alpha\mathbf{E} \tag{6.26}$$

$$v\mathbf{P}\left(1 - \frac{\alpha\beta}{v}\right) = \alpha\mathbf{E} \tag{6.27}$$

$$v\mathbf{P} = \frac{\alpha\mathbf{E}}{\left(1 - \frac{\alpha\beta}{v}\right)} \tag{6.28}$$

This means that for any particular applied $\mathbf{E}$ there will be a magnitude of $\mathbf{P}$ established by the value of α and β. This is a stable situation because any change in $\mathbf{P}$ will be opposed by a change in $\mathbf{E}'$; that is, only at point A will the $\mathbf{P}$ and $\mathbf{E}'$ both be maximum for the given value of $\mathbf{E}$.

A zero field, $\mathbf{E} = 0$, gives no polarization and a linear dielectric results. The extent of movement of A to A' is X. This leads to the Clausius–Mosotti equation obtained by defining:

$$\mathbf{D} = \epsilon_0 \mathbf{E} + \mathbf{P} = \epsilon \mathbf{E} \tag{6.29}$$

Then substituting,

$$v\mathbf{P} = \frac{\alpha\mathbf{E}}{\left(1 - \alpha\frac{\beta}{v}\right)} \tag{6.30}$$

and using the Lorentz β:

$$\beta_L = \frac{1}{3\epsilon_0} \tag{6.31}$$

where

$$\epsilon - \epsilon_0 \quad \text{is small} \tag{6.32}$$

Note that the above linear case holds only as long as $\alpha < v/\beta$. When α becomes large then the unstable condition for solution A arises (see Figure 6.12). As solution A becomes unstable, it yields a negative value for χ.

However, if there is a nonlinearity or anharmonicity in $\mathbf{P}$ vs. $\mathbf{E}'$, a stable solution can occur at point B. In other words, microscopic nonlinearity in α produces a macroscopic nonlinearity of the motion of point B as a function of $\mathbf{E}$, as shown by the dotted lines.

6.6 DESCRIPTION OF FERROELECTRICITY BASED ON LOCAL FIELDS

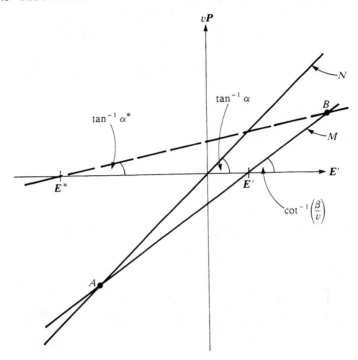

FIGURE 6.12 Unstable solution spontaneous polarization.

What can cause the nonlinearity? Consider the origin of the dipoles as perturbations of electron clouds. A harmonic oscillator of spherical symmetry of an electron cloud in and out of an applied electric field appears as shown in Figure 6.13 (see Chapter 2). As the displacement of the ions d due to the local field becomes larger, the **PE** vs. d curve becomes progressively more anharmonic. This makes $d\mathbf{P}/d\mathbf{E}'$ become less and less. The lower values of α is therefore ultimately related to the onset of the exclusion principle, as described in Chapter 2.

Lets consider the implications of the nonlinerity of α. The **P–E** equation now must be

$$v\mathbf{P} = \frac{\alpha^*(\mathbf{E} + \mathbf{E}^*)}{\left(1 - \alpha^* \dfrac{\beta}{v}\right)} \qquad (6.33)$$

where $\mathbf{E}^*$ is the magnitude of $\mathbf{E}$ defined by the tangent to B, and indicated in Figure 6.12. The value of **P** is now much larger and is finite even at a zero applied field (i.e., $\mathbf{P}_s$). Therefore, $\mathbf{P}_s$ arises because of the large α_o and the nonlinearity. The very large values of $\mathbf{P}_s$ occur when β is large, then α_o can

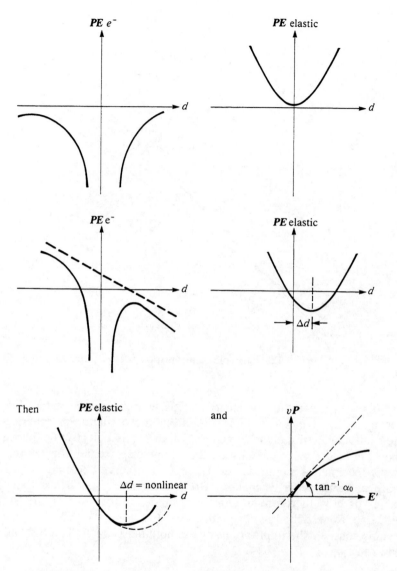

FIGURE 6.13 Nonlinear solution to dipolar polarization.

become very large before nonlinearity sets in. Cooperation also causes homogeneous switching, as illustrated in Figure 6.14.

As long as $\mathbf{E}$ lies between $\pm \mathbf{E}_c$ (i.e., line V), B is a lower energy for $+\mathbf{E}$ and $\mathbf{E}'$ is lower for $-\mathbf{E}$. Cooperation holds the value B as $\mathbf{E}$ is decreased until $|\mathbf{E}| > -\mathbf{E}_c$. Then the $\mathbf{P}$ jumps in value. The extent of the hysteresis is determined by $\alpha^*(\mathbf{E}) = v/\beta$.

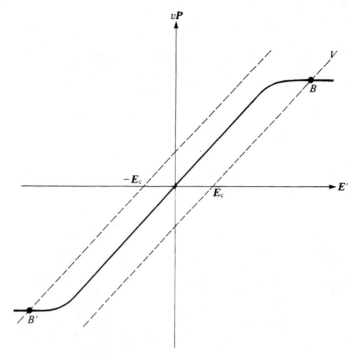

FIGURE 6.14 Homogeneous switching.

6.7 EFFECT OF TEMPERATURE ON POLARIZABILITY

Thermal oscillations increase as temperature increases and results in a decrease in polarizability because of the sharpness of the anharmonic terms (Figure 6.15). The state of spontaneous polarization is unstable when $\alpha^*(0)$ is less than v/β. Conversely on cooling, the state of zero polarization is unstable when α^* has increased sufficiently to be greater than v/β. The order of the transition will depend on the type of anharmonicity. As shown in Figure 6.16, a second-order transition occurs because α_0 and $\alpha^*(0)$ become equal to v/β simultaneously, i.e., point S. However, under conditions shown in Figure 6.17, a first-order transition takes place with α_0 equal at lower temperature than α^*. Between these temperatures, curve R, neither zero polarization or $\mathbf{P}_s$ is an unstable solution. Therefore, the transition becomes first order with a thermal hysteresis (again, for additional details, see either Burfoot, 1967, or Jona and Shirane, 1962).

6.8 FERROELECTRIC DOMAINS

A domain is a region in a crystal where the polarization $\mathbf{P}_s$ is homogeneous, i.e., in the same direction, separated by a domain wall from a neighboring

260 CHAPTER 6: NONLINEAR DIELECTRICS

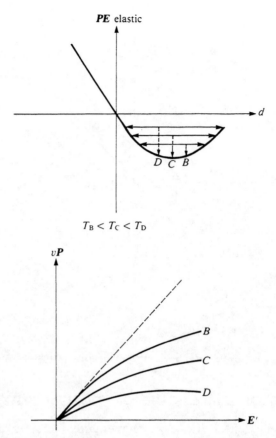

FIGURE 6.15 Polarizability as a function of increasing temperature.

region of different direction of $\mathbf{P}_s$. $\mathbf{P}_s$ may be slightly different in neighboring domains because of electro-strictive effects, as discussed below. *Electrostriction* is the change in physical dimensions and shape of a material due to application of an electric field.

In order to determine domain configurations (see Figure 6.18) it is necessary to characterize the bulk of the domain and the conditions controlling wall thickness. This is a very difficult calculation which can be simplified by considering the wall to have a single unit energy g.† The energy expression to be minimized in calculating the width of the domain is

$$U = U_e + U_P + U_X + U_w + U_d \tag{6.34}$$

† See Burfoot (1967) or Jona and Shirane (1962) for additional details on domain calculations. For predictions of domain wall orientations see Cross (1971).

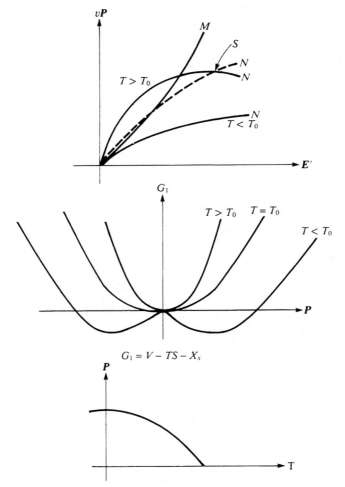

FIGURE 6.16 Second-order anharmonicity as a function of increasing temperature.

where

$$U_e = \text{effect of applied field on domain energy} \tag{6.35}$$

$$U_P \propto \mathbf{P}^2; \quad U_X \propto X^2, \quad \text{the bulk electric and elastic energies} \tag{6.36}$$

$$U_d = \text{depolarization energy} \tag{6.37}$$

U_d is the energy related to the internal field set up in the crystal by the polarization and not compensated. The internal field opposes the applied field $\mathbf{E}_a$, and is therefore called the depolarizing field $\mathbf{E}_d$ (see Figure 6.19). The

262 CHAPTER 6: NONLINEAR DIELECTRICS

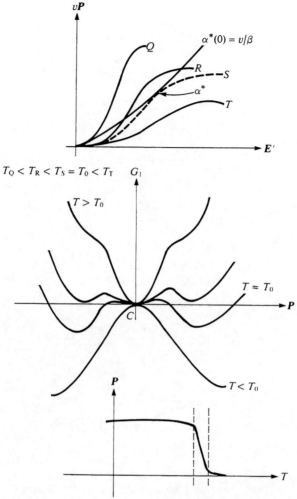

FIGURE 6.17 First-order transition as a function of increasing temperature.

terms U_P and U_X are the same in each domain, irrespective of wall position, and therefore are independent of D the domain thickness.

Under a surface area of A there will be a volume AL (see Figure 6.18), and amount of wall area, such that the total wall energy of

$$U_w = g\frac{AL}{D} \tag{6.38}$$

and g is the surface energy of the wall. Therefore the problem is a minimization of

$$U_w + U_d = \frac{gAL}{D} + QA \tag{6.39}$$

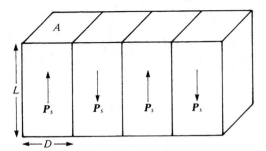

FIGURE 6.18 Bulk domain geometry.

with respect to D; where Q is the depolarization energy per unit area and $Q \propto D$.

Q falls off as $1/y$ where y is distance from the end of domain (see Figure 6.20). Calculations show that equilibrium value of D is

$$D = [gL(i + \sqrt{\mathbf{E_a E_d}})]^{1/2}/\mathbf{P}_s\sqrt{3,4}; \left[\text{i.e., } D \propto \left(\frac{[gL]}{\mathbf{P}_s}\right)^{1/2}\right] \quad (6.40)$$

Therefore thin crystals will have small domains because the depolarization end effects will override bulk effects. If wall energy is made large by segregation of ions, for example, domains will be large. Thus, solid solutions that produce large values of $\mathbf{P}_s$ will produce small domains.

The measured values of g are in the range of 2–4 erg/cm² for 90° walls and 7–10 erg/cm² for 180° walls; equilibrium values of D are in the range of 10 μm.

What is the wall structure? The 180° switching walls of BaTiO₃ are believed to be one lattice spacing thick. This is appreciably less than the walls separating magnetic domains. The reason is as follows. Equilibrium wall thickness is determined by two opposing tendencies.

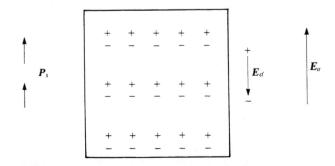

FIGURE 6.19 Depolarizing field within polarizable material.

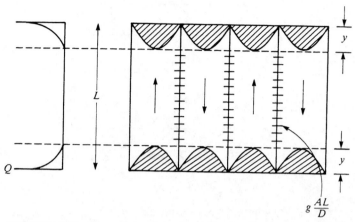

FIGURE 6.20 Depolarization of the domain ends to a depth y.

1. Dipoles want to align with others in the bulk of the domain which tends to force domain walls to be thin.
2. Neighboring dipoles want to align parallel to each other which tends to force the walls to be thick.

So, let the dipoles which are not in one of the two bulk orientations contribute a positive energy $(+A)$. Also, let the the tendency for dipoles to be mutually parallel be described as an energy $(+B\delta^2)$, where δ is a small angle of departure from the mutually parallel condition. Then, in magnetic materials B is very large, since it represents the exchange energy (see Chapter 7). However, for the 180° ferroelectric wall, where n is the wall thickness in number of dipole spacings

$$\delta = \frac{\pi}{n} \tag{6.41}$$

so

$$U = An + Bn\left(\frac{\pi}{n}\right)^2 \tag{6.42}$$

This energy is a minimum with respect to n when

$$n = \pi\left(\frac{B}{A}\right)^{1/2} \tag{6.43}$$

Therefore, because of the smaller values of B in ferroelectric materials the walls are much thinner than in magnetic materials, as illustrated in Figure 6.21.

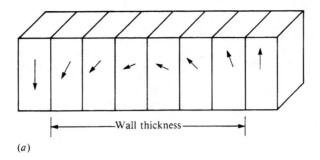

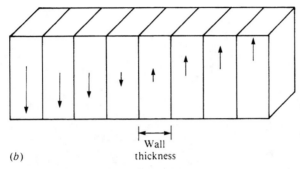

FIGURE 6.21 Wall thickness for (a) ferromagnetic and ferrimagnetic material, (b) ferroelectric material.

The significance of the thin wall is that the wall energy U_w will be highly localized, so that it can greatly exceed kT and will be difficult to move under thermal energy.

The difference in domain walls also produces differences in the hysteresis loop of ferroelectric and magnetic materials. For a magnetic material there is reversible wall motion at a low applied magnetic field strength **H**, and irreversible wall motion at higher **H**. At a high **H**, domain reorientation occurs into an easy direction nearest the field, and at highest fields, magnetic vectors not parallel to the field are being pulled by it out of easy directions.

In contrast, for a ferroelectric material, because of the very high anisotropies, only easy directions are polarized into domains. Thus, domain growth is not impeded appreciably and a major part of polarization occurs over a very small range of field. Therefore, the sides of the hysteresis loop are nearly vertical and the top is nearly flat. This is true because the dielectric polarization adds only one part in 1000 to the spontaneous value of polarization. These differences can be noted by comparing the hysteresis curves for a $BaTiO_3$ ferroelectric and a zinc ferrite in Figure 6.22.

The coercive field in some ferroelectrics also appears to be nearly nonexistent. The typical switching loop occurs only when the rate of change of

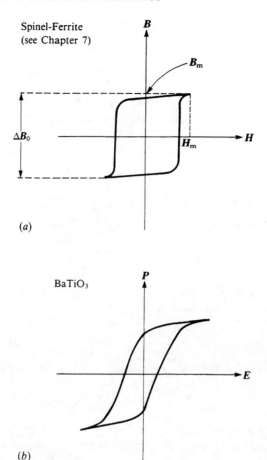

FIGURE 6.22 Hysteresis difference between (a) a Spinel Ferrite and (b) Barium Titanate.

field lies within certain limits. At very low frequencies, any field will cause switching, at very high frequencies the walls will not have time to switch.

6.9 EFFECT OF ENVIRONMENT ON SWITCHING AND TRANSITIONS

Small changes in external conditions (field, stress, temperature) may result in large changes in polarization **P** for ferroelectrics. For example, Figure 6.23 is an equilibrium diagram for Rochelle salt for which large changes in polarization **P** occur with changes in temperature and pressure. The "phases" in Figure 6.23 represents changes in polarization behavior. When the temperature T is above the ferroelectric transition temperature T_c, it has no spontaneous polarization when $T < T_c$; $\mathbf{P} \neq 0$. To demonstrate this requires measuring **P**

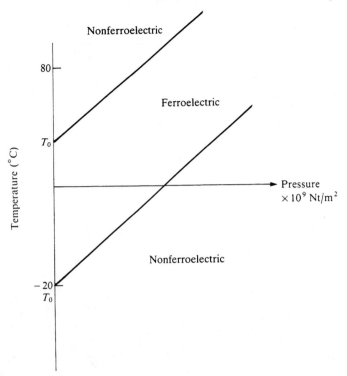

FIGURE 6.23 Phase changes of Rochelle salt as a function of environment.

but this is difficult because charges in crystal faces are quickly compensated, either through the crystal or by the external circuit. The measurement of **P** is possible by measuring the change in **P** induced by changes in E, X, or T. Thus, $d\mathbf{P}$ exists as a current in a circuit connected to electrodes on the crystal, $d\mathbf{P}/dt = i$. Ferroelectrics have a distinguishing response to a changing field (see Figure 6.24).

6.10 EFFECT OF AN ELECTRIC FIELD ON T_c

The magnitude of applied field **E** has a large effect on the switching of domains and also on the temperature of the onset of ferroelectricity, T_c. As shown in Figure 6.25, step A–B desigantes the switching transition, i.e., reversing domain orientation, due to changing **E** at a constant temperature. Step C–D is a transition to the nonferroelectric state which is due to thermal disordering of the dipoles in the domains.

Step A–H is a transition from ferroelectric to nonferroelectric state under a field bias. As indicated, the transition temperature T_c is higher. Step F–G is double loop switching. These field effects on polarization switching are

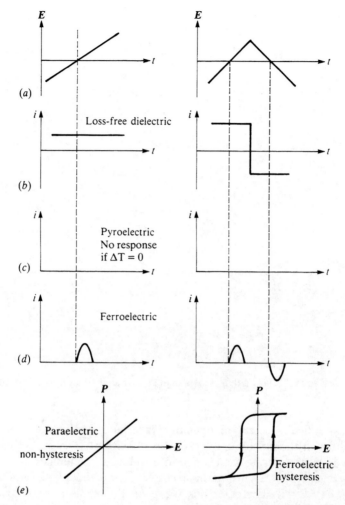

FIGURE 6.24 (a) Applied field, (b) response of a loss-free dielectric, (c) a pyroelectric, (d) a ferroelectric, and (e) the hysteresis of a paraelectric and a ferroelectric.

shown in Figure 6.26. When the bias does not exceed 6 KV/cm, the bias causes finite polarization even at temperatures above T_c; a linear dielectric response of the material results. When the bias is sufficiently large the spontaneous polarization $\mathbf{P}_s$ and the dielectric polarization merge and T_c loses meaning (Figure 6.27).

Physical properties are tensor relations between internal conditions such as $\mathbf{P}$, S, shape and volume (strain) and external conditions such as field, $\mathbf{E}$, T, and stress X. Thus, piezoelectricity must be represented

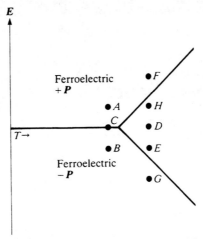

FIGURE 6.25 Ferroelectric transitions.

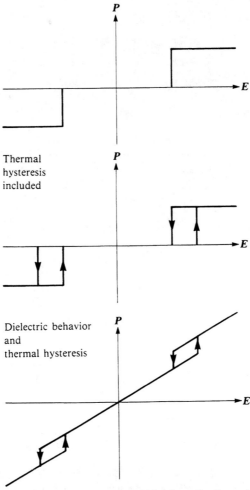

FIGURE 6.26 Field effects on polarization of a ferroelectric.

269

270 CHAPTER 6: NONLINEAR DIELECTRICS

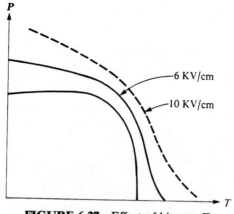

FIGURE 6.27 Effect of bias on T_c.

by four relations

$$\frac{d\mathbf{P}}{dX}, \quad \frac{d\mathbf{P}}{dx}, \quad \frac{d\mathbf{E}}{dx}, \quad \frac{d\mathbf{E}}{dX} \tag{6.44}$$

since it is a coupled property.

6.11 COMPOSITIONAL FACTORS

There are some limitations associated with ferroelectric materials. Although they make it possible to obtain high dielectric constants k' the measured

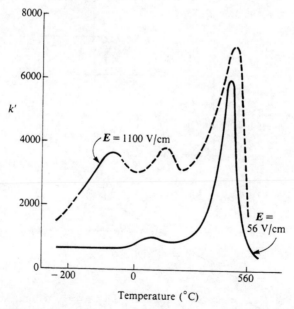

FIGURE 6.28 Field strength dependence of the dielectric constant of $BaTiO_3$.

6.11 COMPOSITIONAL FACTORS

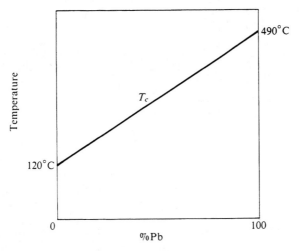

FIGURE 6.29 The effect of solution of $PbTiO_3$ in $BaTiO_3$ on T_c.

value of k' depends on the applied field, as shown in Figure 6.28. For high field strengths, the domains are more effectively oriented, and a higher k' results. There is also a strong temperature dependence on k' so that circuit characteristics change over a moderate temperature range.

The temperature dependence of T_c, and k' can be modified by forming solid solutions over a wide range of compositions. Table 6.5 compares the properties of several ferroelectric solid-solution series. For example, Pb^{2+}, Sr^{2+}, Ca^{2+}, Cd^{2+} can be substituted for Ba^{2+} in the titanate lattice (Figure 6.29). Also, Sr^{4+}, Hf^{4+}, Zr^{4+}, Ce^{4+}, Th^{4+} can be substituted for Ti^{4+}. The zirconates ($XZrO_3$), niobates ($XNbO_3$), tantalates ($XTaO_3$), tungstates (XWO_3) and molybdates (XMO_3) also form ferroelectrics. The possibility of forming solid solution in all of these crystals offers a wide range of values for k', P_s, $\Delta k'/\Delta T$, and T_c.

For example, substitution of Sr^{2+} for Ba^{2+} lowers the Curie point. As shown in Figure 6.29, substitution Pb^{2+} for Ba^{2+} increases T_c to a maximum of 490°. The effect is due to the lower M—O bonding for Pb < Ba < Sr which offers less resistance to the motion of the Ti ions in their six fold oxygen coordination field.

The strain energy introduced by electro-strictive effects on cooling through and below the T_c, when some domains change their orientation in relation to others causes a time dependence for the dielectric constant known as *aging*. The rate of change increases as the initial value of k' increases. Both composition and heat treatment affect aging and also k'. In a polycrystalline ceramic, domain reorientation is affected by grain size, impurities, and pores which prevent domain movement due to stresses imposed by surrounding grains.

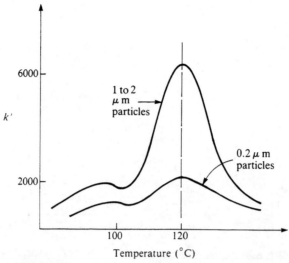

FIGURE 6.30 Ferroelectric behavior of ultrafine particle size $BaTiO_3$.

6.12 EFFECT OF GRAIN SIZE ON FERROELECTRIC BEHAVIOR

Figure 6.30 shows the ferroelectric transition of $BaTiO_3$ at 120°C for ultrafine particles. When this is compared with single crystalline $BaTiO_3$ in Figure 6.4 we can see some markedly different behavior. For single crystalline material the transition is extremely sharp. In fact the derivative goes to infinity. In the case of fine particles (1 to 2 μm) the transition is gradual. This indicates that there is a relationship between the size of the crystalline structure and the equilibrium positions of titanium ions in the polarized state. In ultrafine powders (0.2 μm) there exists little or no orientational relationship. Likewise, the increase in dielectric constant is much less for ultrafine particles. This again shows the interrelationship of the microstructure and the ferroelectric domains. The domain orientation in an ultrafine powder is random. This randomization tends to broaden the ferroelectric transition, as seen in Figure 6.30.

6.13 ANTIFERROELECTRIC–FERROELECTRIC TRANSITIONS

Antiferroelectric crystals contain spontaneously polarized ions, or dipoles, arranged in antiparallel directions. The antiferroelectric structure of $PbZrO_3$ is illustrated in Figure 6.31 where the arrows represent the direction of shift of Pb ions with respect to the oxygen lattice. The result is an orthorhombic cell with two alternating sublattices of dipoles of equivalent but opposite polarizations, i.e., $\mathbf{P}_2 = -\mathbf{P}_1$. Thus, the net polarization of the crystal is zero. In each sublattice, $\mathbf{P}$ goes to zero above T_c. Because of the sublattice polarization antiferroelectric crystals are birefringent (see Chapter 10), even though the overall polarization is zero.

6.13 ANTIFERROELECTRIC–FERROELECTRIC TRANSITIONS

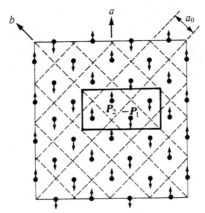

FIGURE 6.31 Antiferroelectric structure of $PbZrO_3$. Arrows represent direction of shifts of Pb atoms, the solid line outlines an orthorhombic unit cell.

An antiferroelectric configuration of dipoles is energetically very close to a ferroelectric configuration with a free energy difference of only a few calories/mole. Thus, an antiferroelectric can be switched to a ferroelectric state by slightly varying the composition or by applying a strong electric field or pressure. Figure 6.32 illustrates the effect on T_c and the stable crystal phases when Ti is substituted for Zr in $Pb(ZrTi)O_3$, also called PZT. Only 7% of solid solution of Ti in the PZT system is required for the dominant phase to be a rhombohedral ferroelectric phase F rather than antifer-

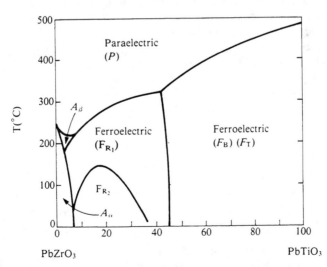

FIGURE 6.32 Phase diagram of the $PbZrO_3$–$PbTiO_3$ system. A_α and A_β are antiferroelectric phases. (After F. Sawaguchi and Jaffee (1953) *J. Phys. Soc.*, Japan, **8**, 615.)

roelectric A. The Curie temperature increases from 230°C for the antiferroelectric to paraelectric transition $A \rightleftarrows P$ to 490°C for the $F \rightleftarrows P$ transition.

The F phase can be stabilized by substituting Ba for Pb or Nb for Zr in the PZT‡ crystals. The A phase is stabilized when La is substituted for Pb or Sn for Zr due to differences in relative electronic bond strengths.

The additional energy associated with applied pressure favors formation of antiferroelectric phases. Consequently, when PZT‡ compositions are close to a phase boundary such as 7% in Figure 6.32, hydrostatic pressure can induce a transition between phases.

These pressure induced transitions are technologically important since they can be used in explosively driven power supplies which operate by generating a shock wave that releases the bound surface charge on a polarized (*poled*)† ferroelectric material. Since the F phase has a larger volume than the A phase, a suitable compressive stress produces a $F \rightarrow A$ phase transition. When the ferroelectric has been poled (by cooling through the $P \rightarrow F$ transition under an electric field) then virtually all the bound charge is released during the pressure induced $F \rightarrow A$ phase transition. When the transition occurs very rapidly, as under an explosive shock wave, megawatts of power are generated for a very short period of time and a substantial voltage appears across the ferroelectric if the load impedance is high.

The effect of pressure on the phase relations for a PZT‡ composition, with 5% Ti and 2% Nb replacing 7% Pb, is shown in Figure 6.33. At zero applied pressure, the sequence of phases with decreasing temperature is $P \rightarrow F_{R2} \rightarrow F_{R1}$, where F_{R2} is the rhombohedral ferroelectric phase and F_{R1} is related to F_{R1} by a phase transformation involving a doubling of the unit cell. At applied pressures greater than 0.2GPa, antiferroelectric phases become stable, with the range of stability depending on the temperature. The phase sequence with decreasing temperature at an applied pressure of 0.6GPa is $P \rightarrow A_T \rightarrow A_0$, where A_0 is the orthorhombic antiferroelectric phase and A_T is the tetragonal antiferroelectric phase. Applying 0.3GPa pressure at ambient temperature will induce the $A_0 \rightarrow F_{R1}$ phase transition used in an explosively driven power supply. For discussion of the effects of other PZT compositions on the pressure-temperature phase diagrams see Fritz and Keck (1978).

The anisotropic crystal structure of ferroelectrics and antiferroelectrics gives rise to anisotropic optical properties, as discussed in Chapter 10. Consequently, phase transitions from $P \rightleftarrows A$, $P \rightleftarrows F$, or $F \rightleftarrows A$ can produce changes in optical properties accompanying the changes in electrical and mechanical properties. Figure 6.34 shows the intensity of reflection in the

---

† Poling is the term used to describe the alignment of domains so that there is net electric moment of a polycrystalline ferroelectric material. Poling involves cooling the material through T_c in an electric field. The field enhances nucleation of domains with dipoles oriented in the same direction.

‡ Registered Trademark of Vernitron, Inc.

6.13 ANTIFERROELECTRIC–FERROELECTRIC TRANSITIONS

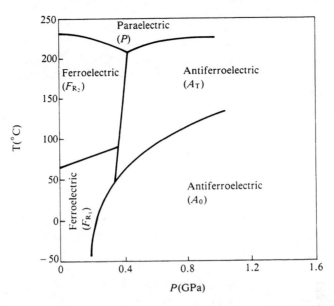

FIGURE 6.33 Pressure-temperature phase diagram for PZT 95/5–2Nb with increasing pressure. (After Fritz and Keck, (1978) *J. Phys. Chem. Solids*, **39**, 1163–1167.)

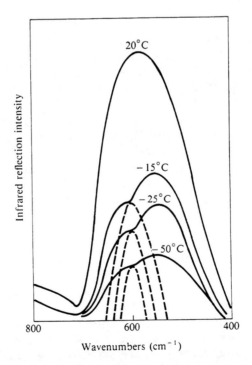

FIGURE 6.34 Change in infrared reflection peaks of a PZT–Nb ferroelectric due to a low temperature phase change. (Courtesy of G. LaTorre and L. Storz.)

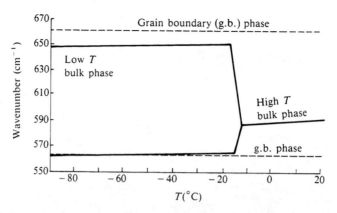

FIGURE 6.35 Effect of temperature on the location of infrared reflection peaks of bulk and grain boundary phases of a PZT–Nb ferroelectric (courtesy of G. LaTorre and L. Storz).

infrared region of the optical spectrum for a PZT–Nb composition similar to that of Figure 6.33. At ambient temperature there is a single molecular vibrational mode (see Chapter 8) giving rise to a single peak centered at 590 cm^{-1}. When the temperature is lowered below $-12°C$ the IR reflection peak divides into two peaks, indicative of formation of a phase of lower symmetry.

Figure 6.35 shows the temperature dependence of the IR modes for the bulk ferroelectric material compared with a fracture surface of the same sample. The optical spectrum of the fracture surface does not show evidence of the phase change. This is due to the fracture being integranular which exposes a grain boundary phase with a composition different from the bulk. As shown in Figure 6.32 a substitution of only a few percent Ti for Zr can change a PZT from antiferroelectric to ferroelectric. The strain associated with a grain boundary phase of different crystal anisotropy may account for some of the complex mechanical behavior, aging, and reliability of these materials. There is also the potential for transformation toughening. The dielectric behavior of these complex materials is inextricably linked to mechanical performance. See Cook et al. (1983), Dungan and Storz (1985) and problem 6.6 for additional treatment of these relationships.

6.14 PLZT† CERAMICS

Substitutional solid solutions of lanthanum oxide (La_2O_3) for lead oxide (PbO) in the completely miscible lead zirconate ($PbZrO_3$) and lead titanate ($PbTiO_3$) system yields ceramics with important piezoelectric and electro-optic characteristics. The La^{3+} ions replace Pb^{2+} ions in the A site of the ABO_3 perovskite lattice, as illustrated in Figure 6.36 for a composition with a ratio

---

† La modified PZT® Registered Trademark of Vernitron, Inc.

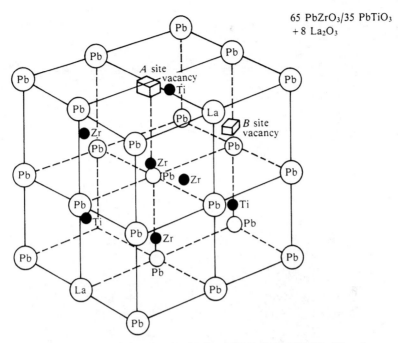

FIGURE 6.36 Crystal unit cell for $65PbZrO_3/35PbTiO_3/8La_2O_3$.

of 65% $PbZrO_3$ to 35% $PbTiO_3$ and approximately 8% La_2O_3. (Note: The oxygen ions are not shown in their face centered positions in Figure 6.36 for clarity.) Thus, the general formula for PLZT ceramics is:

$$Pb_{1-x}La_x(Zr_y Ti_{1-y})_{1-x/4}O_3 \qquad (6.45)$$

Because La^{3+} replaces Pb^{2+}, electrical neutrality is maintained by lattice site vacancies (see Chapter 4). Vacancies can exist on either the A sites or B sites, as shown in Figure 6.36, although equation (6.45) assumes the vacancies are all on B sites. (See Haertling, 1986 in the Reading List for discussion of the effect of vacancies and excess PbO content on processing of PLZT ceramics.)

Addition of La tends to stabilize the paraelectric cubic and the antiferroelectric phases of the PZT system (Figure 6.32) at the expense of the ferroelectric phases. Consequently, the Curie temperature of PLZT compositions decreases as the percentage of La increases (see Table 6.6). For example, at a ratio of $PbZrO_3$ to $PbTiO_3$ of 65/35 only 9% La_2O_3 is sufficient to reduce the region of stability of the rhombohedral ferroelectric phase to below room temperature. Thus, the virgin or unpoled 9/65/35 ceramic is cubic. The phase boundary between cubic PE and ferroelectric phases FE is especially important in electro-optic applications (see Chapter 10).

TABLE 6.7 Additional Antiferroelectric Oxides

Formula	T_c(°C)
Na NbO$_3$	354°
Pb$_2$Nb$_2$O$_7$	−258
Pb$_2$GaNbO$_6$	100
Pb$_2$BiNbO$_6$	−235
Pb(0.33Li, 0.33Nb, 0.33W)O$_3$	110°
Pb(0.33Yb, 0.33Nb)O$_3$	302
Pb HfO$_3$	163°
Cd HfO$_3$	605°
Bi FeO$_3$	850°
Cd TiO$_3$	960°
(0.5K, 0.5Bi)TiO$_3$	380°, 410°
(0.5Na, 0.5Bi)TiO$_3$	220
Pb(0.5Mg, 0.5W)O$_3$	38
Pb(0.5Mn, 0.5W)O$_3$	150
Pb(0.67Mn, 0.33W)O$_3$	200

Table 6.6 shows the effects of PbZrO$_3$/PbTiO$_3$ ratio and % La$_2$O$_3$ on various physical properties of PLZT ceramics. For a constant PZ/PT ratio, as the La content increases many properties are improved, that is, the dielectric constant increases, the hysterisis loop is more square with a smaller coercive field, and the electromechanical coupling coefficients increase. However, these advantages are at the expense of a lower T_c and slightly lower spontaneous and remanent polarization. Another major advantage of the La additions is that these compositions are much easier to sinter to optical transparency, a requirement for the electro-optical applications described in Chapter 10.

Both piezoelectric and electro-optical applications use PLZT compositions near the phase boundary between the FE rhombohedral and FE tetragonal phases where most properties are optimized (Figure 6.32 and problem 6.3). La contents <5% are usually used for piezoelectrics and >6% for electro-optics. Compositions that yield a tetragonal FE phase usually have a high coercive field and are termed "hard" whereas the rhombohedral FE phase has a low coercivity and are considered "soft" materials. (See Haertling, 1986 for discussion of various PLZT compositions, applications, and additional data.) For analysis of several applications of these materials see problems 6.5 and 6.6. Additional antiferroelectric oxides and their Curie temperatures are listed in Table 6.7.

PROBLEMS

6.1 Calculate the susceptibility, χ, for $0\,\text{K} < T < 1000\,\text{K}$ and compare the following graphically:
   (a) Rochelle salt.
   (b) KH_2PO_4.
   (c) $BaTiO_3$.
   (d) $PbTiO_3$.
   (e) PLZT (65/35/2La).

6.2 Determine the crystal group notation for the following:
   (a) diatomic molecule SiC.
   (b) Cs_4I_4 orthorhombic cluster.
   (c) Are there any mirror plane, m, operations for either the above SiC or CsI? If so what are they?

6.3 Fritz and Keck (1978) show that decreasing the Pb content in a PZT–Nb ferroelectric from 95% to 86% reduces the $F_{R2} \rightleftharpoons F_{R1}$ phase transition from 60°C, to 15°C, at $P = 0$. Assume that the difference between the grain boundary and bulk phases in Figure 6.35 is due to Pb content:
   (a) Would an excess or deficiency in Pb in the grain boundary phase explain the results of Figure 6.35.
   (b) How much difference is required?
   (c) Predict the effects of applied pressure on the interfacial strains between the bulk and grain boundary phases.
   (d) Assume that the thickness of the grain boundary phase is 10 microns. Estimate the interfacial stresses as the material goes through the -12°C transition in Figure 6.35.

6.4 Use the data in Table 6.6 and Figure 6.32 to construct a phase equilibrium diagram for the $PbZrO_3$–$PbTiO_3$–La_2O_3 system.

6.5 Barium Titanate is used extensively in the ceramic capacitor industry. (See Halliday and Resnick, 1965.)
   (a) Calculate the dc capacitance for a ceramic capacitor that is $1\,\text{cm}^2$ in area with a 1 micron thick piece of $BaTiO_3$ separating the plates and operating near T_c.
   (b) The breakdown voltage for a particular lot of these capacitors was lower than usual. The average was determined to be 144 V. If this was due to contamination during the processing of the ceramic dielectric, what is the contamination concentration (in ppb) necessary to produce this effect? (Hint: see Chapter 10 if necessary.)

(c) The reduction in breakdown voltage could also be due to a second phase produced by improper processing. Assume that the second phase has a dielectric constant of $k = 100$, and can be modeled by series capacitors. The capacitance contributed by each phase is proportional to its volume fraction. What is this volume fraction in the defective capacitor if it has a capacitance of only 8.8 nanofarads. Plot the capacitance for $0.1 <$ vol frac < 0.5.

(d) Show why a parallel capacitor model would not account for the reduced capacitance in the defective dielectric material.

(e) What is a likely second phase responsible for the reduced capacitance?

6.6 A small, portable high voltage power supply can be made using a poled piezoelectric ceramic disk. The voltage coefficient, g_{33}, is defined as the ratio of the open circuit electric field to the applied stress. Remembering that the electric field is the voltage over the thickness of the ceramic, calculate the voltage that can be generated for a 5000 psi stress onto a 1/2 inch thick ceramic disk of:

(a) $BaTiO_3$.
(b) PLZT (65/35/7).
(c) What gap can be jumped by the above voltages for:
Dry air?
Fused quartz?
Titania?
Teflon?
Paper?
Pure $BaTiO_3$?

6.7 Draw a set of polarization/electric field hysteresis curves for the list of PLZT compositions in Table 6.6.

6.8 High performance ceramic microphones and speakers are also based on piezoelectric materials. A piezoelectric tweeter is lighter, more efficient, more reliable, has better transient response, and is cost competitive with magnetic designs. Assume that the voltage sensitivity is proportional to the square of the electromechanical coupling factor, k_p.

(a) Arrange the materials in Table 6.6 in order of their voltage sensitivity.
(b) The strain induced in the ceramic disks is directly proportional to the strain coefficient, d_{33}. Calculate the strain for the materials for a constant voltage of 5 V.
(c) Plot the strain as a function of the voltage sensitivity.
(d) Which PLZT composition appears to yield the most efficient electro-mechanical response? Explain?

READING LIST

Anderson, J. C., (1964). *Dielectrics*, Chapman and Hall, London, Chapter: 9.

Azaroff, L. and Brophy, J. J., (1963). *Electronic Processes in Materials*, McGraw-Hill, New York, Chapter: 12.

Bunget, I. and Popescu, M., (1984). *Physics of Solid Dielectrics*, Elsevier, New York.

Burfoot, J. C., (1967). *Ferroelectrics, an Introduction to the Physical Principles*, Nostrand, London.

Cady, W., (1946). *Piezoelectricity*, McGraw-Hill, New York.

Cook, R. F., Freiman, S. W., Lawn, B. R., and Pohanka, R. C., (1983). *Ferroelectrics*, **50**, 267–272.

Cross, L. E. (1971). *Physics of Electronic Ceramics* (L. L. Hench and D. B. Dove, eds.), Part B, Chapter 22. Dekker, New York.

Dungan, R. H. and Storz, L. J., (1985). *J. Amer. Ceram. Soc.*, **68**(10), 530–533.

Fritz, I. J. and Keck, J. D., (1978). *J. Phys. Chem. Solids*, **39**, 1163–1167.

Haertling, G. H., (1986). *Ceramic Materials for Electronics*, (R. C. Buchanan, ed.), Dekker, New York, Chapter 3.

Haliday, D. and Resnik, R., (1965). *Physics for Students of Science and Engineering*, combined edition, Wiley, New York.

Hench, L. L. and Dove, D. B., eds. (1971). *Physics of Electronic Ceramics*, Part A. Dekker, New York.

Hench, L. L. and Dove, D. B., (1971). *Physics of Electronic Ceramics*, Part B. Dekker, New York, Chapters: 21, 22.

Jona, F. and Shirane, G., (1962). *Ferroelectric Crystals*, Pergamon Press, New York.

Kay, H. F. and Vousdan, P., (1949). *Phil. Mag.*, **7**, 40, 1019.

Kingery, W. D., Bowen, H. K., and Uhlmann, D. R., (1976). *Introduction to Ceramics*, 2nd ed., Wiley, New York, Chapter: 18.

Kittel, C., (1986). *Introduction to Solid State Physics*, 6th ed., Wiley, New York, Chapter: 8.

Landolt-Bornstein New Series Group III, (1975). Volume 9, *Ferroelectric and Antiferroelectric Substances*, Springer-Verlag, Berlin.

Lines, M. E. and Glass, A. M., (1977). *Principles and Applications of Ferroelectrics and Related Materials*, Clarendon Press, Oxford, England.

Mason, W., (1950). *Piezoelectric Crystals and Their Application to Ultrasonics*, Nostrand, New York.

Merz, W. J., (1949), *Phys. Rev.*, **76**, 1221.

Pepinsky, R., (1972). *Physics of Electronic Ceramics*, Part B, (L. L. Hench and D. B. Dove, eds.), Dekker, New York, Chapter: 21.

7

MAGNETIC CERAMICS

7.1 INTRODUCTION

Magnetic ceramics have developed as an important commercial field since the 1940s because they have strong magnetic coupling, high electrical resistivity, and low loss characteristics. A wide range of substitutional solid solubility also makes it possible to "tailor" the magnetic properties of polycrystalline magnetic ceramics over a very wide range of Curie points, remanent polarization, saturation polarization, and coercive fields. Thus, ceramic magnets can be molecularly designed and processed for specific electronic applications. These applications include: audio and video recording tapes, motors, generators, videocassette recorders, and computer disks.

7.2 BASIC THEORY

The magnetic properties of materials are a consequence of the net magnetic moments of the electrons in the atoms composing the material. The electrons have a magnetic moment as a result of their motion. Both orbital and spin motions of the electron yield a magnetic moment. The orbital magnetic moment due to the electron moving around the nucleus can be likened to a current in a loop of wire of zero resistance.

According to Ampere's law a moving current (in emu) encircling an area of loop $a = \pi r^2$ results in a magnetic field at right angles to the plane of the current. Thus, if e is the charge on the electron in esu and c the velocity of light, then e/c is the charge in emu. The current, which is charge per unit time, therefore is

$$i = (e/c)t = (e/c)(v/2\pi r) \tag{7.1}$$

where r is the orbital radius and v is the electron velocity. Therefore,

$$\mu_{(orbit)} = A \cdot i = (\pi r^2)\left(\frac{ev}{2\pi rc}\right) = \frac{evr}{2c} \tag{7.2}$$

Remember from Chapter 2 that the Bohr theory also postulates that the angular momentum of the electron must be an integral multiple of $h/2\pi$, where h is Planck's constant. Consequently,

$$mvr = nh/2\pi \tag{7.3}$$

where m is the mass of the electron.

Combining equations (7.2) and (7.3) yields:

$$\mu_{(orbit)} = n\left(\frac{eh}{4\pi mc}\right) \tag{7.4}$$

for the magnetic moment in the nth Bohr orbit of the electron.

As discussed in Chapter 2, the electron behaves as if it were spinning about its own axis as well as moving about the nucleus. This spin motion results in the *magnetic spin quantum number* and a *spin magnetic moment*, $\mu_{(spin)}$ equal to

$$\mu_{(spin)} = \frac{eh}{4\pi mc} = 0.927 \times 10^{-20} \frac{\text{erg}}{\text{Oe}} \tag{7.5}$$

By comparing equations (7.4) and (7.5), we see that the magnetic moment due to spin is equal to that due to orbital motion for electrons in the first Bohr orbital where $n = 1$. Consequently, this amount of magnetic moment is called a *Bohr magneton*, with the symbol μ_B.

Thus,

$$\mu_B = \frac{eh}{4\pi mc} = \frac{e\hbar}{2mc} = 0.927 \times 10^{-20} \frac{\text{erg}}{\text{Oe}} \tag{7.6}$$

The magnetic moment of an atom is the vector sum of all its electronic spin and orbital moments. Two alternatives exist for any given atom: (1) The sum of $\mu_{(orbit)}$ and $\mu_{(spin)}$ of all the electrons cancel, which yields

$$\sum \mu_{(orbit)} + \mu_{(spin)} = 0 \quad \text{(Diamagnetic)} \tag{7.7}$$

or (2) the cancellation of the magnetic moments is only partial so the atom is left with a net magnetic moment, that is, a "magnetization."

$$\sum \mu_{(orbit)} + \mu_{(spin)} > 0 \quad \begin{pmatrix} \text{Paramagnetic} \\ \text{Ferromagnetic} \\ \text{Antiferromagnetic} \\ \text{Ferrimagnetic} \end{pmatrix} \quad (7.8)$$

Materials that satisfy equation (7.7) are "nonmagnetic" and are called *diamagnetic*. Materials that include atoms that satisfy equation (7.8) can

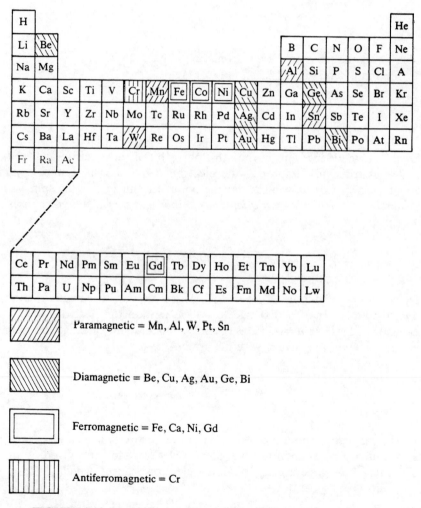

FIGURE 7.1 Magnetic state of the elements at room temperature.

exhibit a variety of magnetic behaviors depending upon the ordering of the magnetic atoms in the material. The alternative magnetic states, termed *paramagnetic, ferromagnetic, antiferromagnetic,* and *ferrimagnetic,* will be described in subsequent sections.

The magnetic states of the elements at room temperature and their number of Bohr magnetons are shown in Figure 7.1. (See Cullity, 1972, for the magnetic properties of rare earth elements.) The magnetic moment, or number of Bohr magnetons of a metal atom, is based upon Hund's rule, derived from optical spectroscopy measurements, which states, "The spins in a partly filled electron shell are arranged so as to produce the maximum spin unbalance consistent with the Pauli exclusion principle." The orbital contribution is unimportant in ferrites, the ferrimagnetic ceramics of concern in this chapter; consequently Hund's rule is simplified to only spin contributions for ferrites.

Consider Figure 7.2 showing the spin alignments of ions in the first transition metal series. The outermost $3d$ electron shell can contain five electrons with spin moments oriented up, $+\frac{1}{2}$, and five electrons with spin moments oriented down, $-\frac{1}{2}$. Hund's rule requires that the first five electrons occupy energy states with *all* spins either up or down in order to maximize the moment. This results in $\mu_B = 5$ for the transition metal ions of Mn^{2+}, Fe^{3+}, and Co^{4+}, as illustrated in Figure 7.2.

The sixth d-electron, such as for Fe^{2+}, Co^{3+}, or Ni^{4+}, must have its spin oriented down due to the exclusion principle (Figure 7.2). Consequently, ions with $3d^6$ electrons have a spin-only moment of $5 - 1 = 4\mu_B$. The number of spin-only moments of the transition metal ions range from zero, for both the $3d^0$ ions (Sc^{3+} and Ti^{4+}) and the $3d^{10}$ ions (Cu^+ and Zn^{2+}) to a maximum of $\mu_B = 5$ for the $3d^5$ ions, Mn^{2+}, Fe^{3+}, and Co^{4+}. As we see later for ferrimagnetic ceramics, this wide range of μ_B values and the variable oxidation states (cationic charge) of transition metal cations in an oxide crystal make it possible to "tailor" the net magnetic moment of magnetic ceramics.

The number of unpaired electron spins greatly influences the interaction of a material with an applied magnetic field. The relationship between the magnetic flux density, **B**, and the magnetic field strength, **H**, as defined by Maxwell, is

$$\mathbf{B} = \mu \mathbf{H} \tag{7.9}$$

where μ is the permeability of the material.† In free space μ_0 is

$$\mu = \mu_0 = 1 \tag{7.10}$$

† Note that the symbol μ is often used for both magnetic moment and magnetic permeability. In order to avoid confusion we generally apply a subscript to magnetic moments, such as μ_m, μ_B, $\mu_{(spin)}$ or $\mu_{(orbit)}$.

Transition metal ions	Number of 3d electrons	Electron spin alignments		Spin-only moment in μ_B
Sc^{3+}, Ti^{4+}	$3d^0$	☐☐☐☐☐	$4s^2$ ↑↓	0
Ti^{3+}, V^{4+}	$3d^1$	↑☐☐☐☐	$4s^2$ ↑↓	1
Ti^{2+}, V^{3+}, Cr^{4+}	$3d^2$	↑↑☐☐☐	$4s^2$ ↑↓	2
V^{2+}, Cr^{3+}, Mn^{4+}	$3d^3$	↑↑↑☐☐	$4s^2$ ↑↓	3
Cr^{2+}, Mn^{3+}, Fe^{4+}	$3d^4$	↑↑↑↑☐	$4s^2$ ↑↓	4
Mn^{2+}, Fe^{3+}, Co^{4+}	$3d^5$	↑↑↑↑↑	$4s^2$ ↑↓	5
Fe^{2+}, Co^{3+}, Ni^{4+}	$3d^6$	↑↓↑↑↑↑	$4s^2$ ↑↓	4
Co^{2+}, Ni^{3+}	$3d^7$	↑↓↑↓↑↑↑	$4s^2$ ↑↓	3
Ni^{2+}	$3d^8$	↑↓↑↓↑↓↑↑	$4s^2$ ↑↓	2
Cu^{2+}	$3d^9$	↑↓↑↓↑↓↑↓↑	$4s^2$ ↑↓	1
Cu^+, Zn^{2+}	$3d^{10}$	↑↓↑↓↑↓↑↓↑↓	$4s^2$ ↑↓	0

FIGURE 7.2 Spin alignments and spin-only magnetic moments of transition metal ions.

Thus, **B** and **H** are identical. However, for real materials and especially magnetic materials this is not the case.

Magnetic materials, that is materials with net unpaired magnetic spins, will respond to an external magnetic field. A force will be measurable. The force is proportional to the gradient of the magnetic field strength:

$$F_x = V\chi \mathbf{H} \frac{d\mathbf{H}}{dx} \tag{7.11}$$

where V = volume of the sample
$\mathbf{H}$ ≡ magnetic field strength and a function of x
x = direction of x-axis
F_x = force in x-direction
χ ≡ susceptibility of the material

Therefore, the permeability of the material and the susceptibility are related by the following definition:

$$\frac{\mu}{\mu_0} \equiv 1 + 4\pi\chi \tag{7.12}$$

expressed in electromagnetic units or "emu." If we express equation (7.12) in MKS, cgs, or SI units, the 4π factor is omitted:

$$\frac{\mu}{\mu_0} = 1 + \chi \tag{7.13}$$

In this case χ is called the relative susceptibility of the material.

We can substitute equation (7.13) into equation (7.9) where we obtain

$$\mathbf{B} = \mu_0(1 + \chi)\mathbf{H} \tag{7.14}$$

or

$$\mathbf{B} = \mu_0\mathbf{H} + \mu_0\chi\mathbf{H} \tag{7.15}$$

This can also be rewritten in the form

$$\mathbf{B} = \mu_0(\mathbf{H} + \mathbf{M}) \tag{7.16}$$

where $\mathbf{M}$ ≡ magnetization or magnetic moment per unit volume, or

$$\mathbf{M} = \frac{\mu_m}{V} \tag{7.17}$$

where μ_m = magnetic moment
V = volume

$$\mathbf{M} = \chi\mathbf{H} \tag{7.18}$$

The permeability μ of a material can be used to classify the various magnetic properties of materials as indicated in Table 7.1. There are five main classes of magnetic behavior: diamagnetic, where the permeability is less than unity, paramagnetic and antiferromagnetic where the permeability is greater than unity but small, and ferromagnetic and ferrimagnetic, where the permeability is large due to spontaneous alignment of spins.

TABLE 7.1 Characterization of Magnetic Materials

Magnetic property	Permeability μ	Susceptibility χ
Diamagnetic	$\mu < 1$	$-M/H \approx -10^{-6}$
Paramagnetic and Antiferromagnetic	$\mu > 1$	$+M/H \approx 10^{-6}$
Ferromagnetic and Ferrimagnetic	$10^2 > \mu > 10^6$	Spontaneous

In order to understand more fully the concept of magnetization, and the different classes of magnetic materials, consider a simple example. In Figure 7.3, a coil of electrically conducting wire is shown. Referring to equation (7.16):

$$\mathbf{B} = \mu_0(\mathbf{H} + \mathbf{M}) \qquad (7.19)$$

Assume for a moment that there is no material within the coil, then

$$\mathbf{B} = \mu_0\mathbf{H}, \quad \text{in} \quad \frac{\text{webers}}{\text{m}^2} \qquad (7.20)$$

$$\mathbf{B} = \mu_0(nI) \qquad (7.21)$$

where n = number of turns per meter
I = current in coil in amperes
μ_0 = permeability in webers per ampere meter = 1

If we now insert a magnetic material into the source coil, the magnetic flux density **B** will become

$$\mathbf{B} = \mu_0 nI + \mu_0\mathbf{M} \qquad (7.22)$$

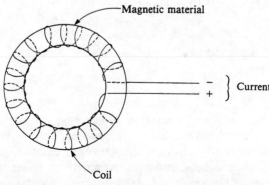

FIGURE 7.3 Coil on core of magnetic material.

where **M** is the magnetic moment per unit volume or the magnetization in Amperes per meter.

By comparing equations (7.20) and (7.21) we see that the magnetic field strength **H** is in units of ampere-turns. This is sometimes referred to as the *magnetizing force*.

If we compare the magnetic flux density with and without a magnetic core material we find

$$\frac{\mathbf{B}_{core}}{\mathbf{B}_{no\,core}} = \frac{\mu_0(\mathbf{H}+\mathbf{M})}{\mu_0 \mathbf{H}} \quad (7.23)$$

$$\frac{\mathbf{B}_{core}}{\mathbf{B}_{no\,core}} = \frac{(\mathbf{H}+\mathbf{M})}{\mathbf{H}} \quad (7.24)$$

Thus, the permeability of the core is

$$\mu = \frac{\mathbf{B}_{core}}{\mathbf{B}_{no\,core}} \quad (7.25)$$

In summary, an external magnetic field induces a magnetic field in the material which interacts with the external field. When a material placed in an inhomogeneous magnetic field opposes the field, it is *diamagnetic*. When a material placed in an inhomogeneous magnetic field aids or reinforces the field, it is *paramagnetic*. The total magnetization induced in the material is proportional to the applied-field strength. For the *ferromagnetic* or *ferrimagnetic* effect, the magnetism M of the material is spontaneous and requires no external field.

7.3 DIAMAGNETISM

Atoms with closed or filled outer electron orbitals are diamagnetic. In this case the electron spin moments cancel because each electron is paired. The electron orbitals precess about an axis parallel to the external magnetic field. Precession of electron orbitals is very similar to the precession of a spinning top. The precession creates a current and a corresponding magnetic moment. In this case, the magnetic moment of a diamagnetic material opposes the externally applied magnetic field. This effect is very small. Examples include noble gases, alkali halides, silica, and covalent crystals, such as diamond because each orbital contains two electrons of paired opposite spin.

The theory of diamagnetism, developed by Langevin in 1905, is based upon *Lenz's law* that states "The magnetic field produced by an induced current opposes the change in magnetic field which produces it." As discussed above, the motion of an electron in orbit around a nucleus is equivalent to a current in a loop. As soon as the applied magnetic field **H** increases from zero, the change in flux $d\phi$ through the loop (the electron's orbit) induces an emf ϵ in the loop, according to *Faraday's law*:

$$\epsilon = -10^{-8}\frac{d\phi}{dt} = -10^{-8}\frac{d(\mathbf{H}A)}{dt}\,\mathrm{V} \qquad (7.26)$$

where A is the area of the loop (the electron orbital). The minus sign indicates that the induced emf acts in such a way as to oppose the change in flux, as required by Lenz's law. The electron velocity slightly decreases, due to the back emf, resulting in a decrease in the magnetic moment of the loop. Since the electron orbital behaves like a current loop with zero resistance the decrease in moment is not momentary, it lasts as long as the field H is applied. Thus, $\mu_{(\text{orbit})}$ is changed by $\Delta\mu$ in the negative direction when H is applied. Consequently, the orbital magnetic moments of all the electrons in a diamagnetic atom can be oriented such that they cancel in a zero field, their lowest energy state, but all react to an applied field by producing a moment opposite to that field.

The change in moment of an electron caused by the field can be shown (see Cullity, 1972) to be

$$\Delta\mu = -\frac{e^2 r_{xy}^2 \mathbf{H}}{6mc^2} \qquad (7.27)$$

where r_{xy} is the radius of an orbit that can take on all possible orientations to the applied field in the z direction.

An atom with Z electrons will have a field induced change in moment of

$$\Delta\mu_{(\text{atom})} = -\frac{e^3 \mathbf{H}}{6mc^2}(Zr_{xy}^2) \qquad (7.28)$$

Where r_{xy}^2 for an atom is the average of the squares of the various orbital radii.

On a volumetric basis, since the number of atoms per unit volume is

$$\frac{\text{number of atoms}}{\text{cm}^3} = \frac{N\rho}{A} \qquad (7.29)$$

where N = Avogadro's number
ρ = density
A = atomic weight

we can write

$$\Delta\mu(\text{per cm}^3) = -\left(\frac{N\rho}{A}\right)\left(\frac{e^2 Zr_{xy}^2 \mathbf{H}}{6mc^2}\right) \quad (7.30)$$

For a diamagnetic material there is no net magnetic moment in the absence of a field. So, the magnetic moment acquired per unit volume, which is the magnetization **M**, is given as equation (7.30).

Thus, the susceptibility χ, given by **M/H**, is

$$\chi = \frac{\mathbf{M}}{\mathbf{H}} = -\left(\frac{N\rho}{A}\right)\left(\frac{e^2 Zr_{xy}^2}{6mc^2}\right) \quad \frac{\text{emu}}{\text{cm}^3 \text{ Oe}} \quad (7.31)$$

Problem 7.1 applies equation (7.31) to carbon, which results in a calculated value for the magnetic susceptibility of

$$\chi_{(\text{carbon})} = -1.5 \times 10^{-6} \quad \frac{\text{emu}}{\text{cm}^3 \text{ Oe}} \quad (7.32)$$

whereas the experimental value is -1.1×10^{-6}.

Agreement between calculated and experimental values for other diamagnetic substances is not quite as good, but are accurate within a factor of 10.

7.4 PARAMAGNETISM

For paramagnetism, the spin moments of unpaired electrons can align themselves parallel or antiparallel to the magnetic field **H**. The antiparallel alignments have higher energy, so a parallel alignment is preferred. This leads to an induced field that aids the applied field, so the magnetic susceptibility is positive:

$$\chi = +\frac{\mathbf{M}}{\mathbf{H}} > 0 \quad (7.33)$$

Many metals are weakly paramagnetic and temperature independent, because the antiparallel alignment of electron spins is unstable in metals. This is because the energy change between the spin alignments is much smaller than kT, the thermal energy.

Atoms must have electrons in partially filled electron shells to be paramagnetic. The electrons in these unfilled orbitals must also have unpaired spin quantum numbers. Thus, the transition metal series with unfilled d orbitals and rare earth elements with unfilled f orbitals are paramagnetic.

The temperature dependence of many paramagnetic materials was found empirically by P. Curie to be:

$$\chi = \frac{C}{T} \qquad (7.34)$$

where $C \equiv$ Curie constant
$T =$ temperature

The more general relation is known as the *Curie–Weiss law* and it has the form

$$\chi = \frac{C}{T - T_C} \qquad (7.35)$$

where $T_C \equiv$ Curie temperature

The *Curie temperature* is the temperature above which the magnetization **M** disappears. Langevin used statistical mechanics to show that the Curie constant can be expressed as $C = \mu_0 N_0 \bar{\mu}_m^2 / 3k$, where N_0 is Avogadro's number, μ_0 is the magnetic permittivity of free space ($4\pi \times 10^{-7}$ H/m), $\bar{\mu}_m$ is the average magnetic moment per molecule, and k is Boltzmann's constant.

7.5 FERROMAGNETISM

The response of a material to an external magnetic field can be studied in a coil as shown in Figure 7.3. If the magnetization **M** of a ferromagnetic material such as iron is measured while the external magnetic field **H** is applied, the result is a hysteresis loop as shown in Figure 7.4. Curve 1 in Figure 7.4 shows the response to the virgin (unpoled) ferromagnetic material with no previous magnetization. As the external magnetic field is increased, the magnetization increases to saturation at

$$\mathbf{M} = \mathbf{M}_s \qquad (7.36)$$

If the external field is reduced, the magnetization remains until

$$\mathbf{M} = \mathbf{M}_r \quad \text{at} \quad \mathbf{H} = 0 \qquad (7.37)$$

This is called the *remanent magnetization*, $\mathbf{M}_r$. If nothing more was done, the material would remain magnetizied and thus would be called a permanent magnet. However, if the external field is reversed and increased,

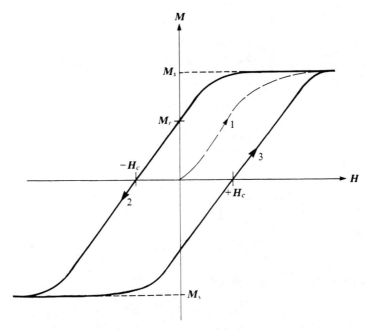

$M_s \equiv$ Saturated magnetization
$M_r \equiv$ Remanent magnetization at $H = 0$
$H_c \equiv$ Coercive magnetic field

FIGURE 7.4 Ferromagnetic hysteresis.

the magnetization finally goes to zero at the coercive magnetic field, H_c. At this point, a further increase in H increases M until saturation occurs in the opposite direction. If the field is again reversed the magnetization follows and a hysteresis loop is created.

It was stated earlier that the magnetic spins in a ferromagnetic material spontaneously align to create a net magnetization. Figure 7.4 clearly shows that the magnetization was zero for virgin magnetic materials. This behavior is explained by introducing the concept of magnetic domains. For virgin material, the magnetization of each domain is still spontaneous and a net magnetic moment exists within each domain. However, the domains are not aligned in a virgin, or unpoled material, and therefore the bulk magnetization for the material is zero.

Thus, for ferromagnetic materials, a spontaneous magnetization within the domains exists in the material without the presence of an external field. This is due to the interaction of the magnetic moment of unpaired electrons which are in a lower energy state when aligned parallel to each other. For example, consider the spin alignments of the ferromagnetic metals Fe, Co, and Ni as shown in Figure 7.5. The net magnetic moment on each atom, due to unpaired spins, is aligned with each neighboring atom within a domain. The difference between the predicted net atomic moment of four

(a) Single atoms

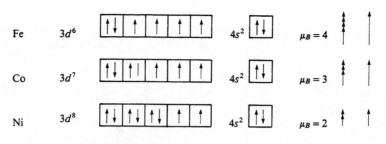

(b) Ferromagnetic crystals

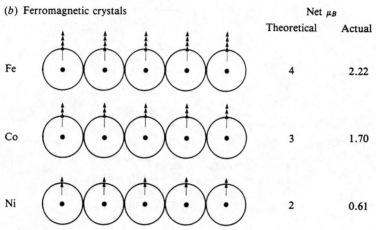

FIGURE 7.5 Spin alignments of elemental transition metals.

Bohr magnetons for Fe, three for Co, and two for Ni and the real values is due to overlap of $4s$ orbitals with $3d$ orbitals in the material. For a more accurate theoretical description of ferromagnetism in metals, see Cullity (1972).

The spontaneous alignment of the magnetic moments illustrated in Figure 7.5 occurs only below the critical Curie temperature T_c. The thermal energy is greater than the spin alignment energy and the spins are randomized above T_c. This behavior is described by the Curie–Weiss law (equation 7.35) and is illustrated in Figure 7.6. Transitions occur between ferromagnetic properties and paramagnetic properties as the temperature is increased. Figure 7.6c shows the boundary between these phases at $T = T_c$.

The saturated magnetization at absolute zero for ferromagnetic materials is defined as $\mathbf{M}_0$. The ratio of the saturated magnetization to that at absolute zero and the ratio of temperature to the Curie temperature clearly define a phase boundary between ferromagnetic behavior and paramagnetic behavior.

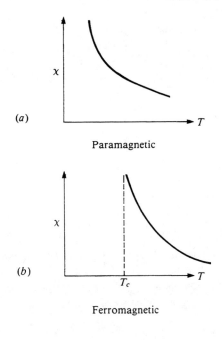

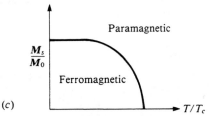

FIGURE 7.6 Ferromagnetic to Paramagnetic phase changes as a function of temperature.

7.6 ANTIFERROMAGNETISM

Antiferromagnetism occurs when there is an ordered antiparallel arrangement of magnetic spins. A comparison between ferromagnetic, antiferromagnetic, and ferrimagnetic spin alignments is shown in Figure 7.7.

Above a critical temperature, T_c, antiferromagnetic spin alignments are also randomized and paramagnetic behavior takes over (Figure 7.8). This phenomenon is equivalent to two ferromagnetic lattices of opposite spin orientations interacting in such a way that their spontaneous magnetizations cancel.

In Figure 7.7, an antiferromagnetic spin alignment is shown. If spin alignment A represents one component of a lattice and B represents the

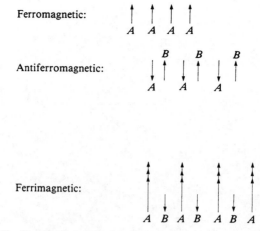

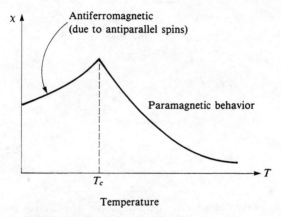

FIGURE 7.7 Comparison of spin alignments between magnetic materials.

FIGURE 7.8 Antiferromagnetic below T_c.

other, then their magnetic moments cancel because the number of spin states match in sublattice A and sublattice B.

If the lattice was of type A only or of type B only, the material would be ferromagnetic.

7.7 FERRIMAGNETISM

As in the case of antiferromagnetism, ferrimagnetism results from the combined contribution of different atoms or sublattices within a lattice. Referring again to Figure 7.7, the ferrimagnetic spin alignment of the sublattice A is opposite to that of sublattice B. There is one important difference, however, between the antiparallel spin alignments of antiferromagnetic and ferrimagnetic lattices. For a ferrimagnetic crystal, there is a mismatch in the number of magnetic moments of the atoms on the A and B sublattice. There are more magnetic moments on the A sublattice than on the B. Therefore, a ferrimagnetic material is actually an antiferromagnetic material with unbalanced magnetic spins.

Once again, we can count the nonpaired spins on each sublattice as in the electron orbital case. The count here is the nonpaired atoms with a net magnetic moment. The number of Bohr magnetons agree closely with experimental values. Small errors are due to the distribution of the A and B atoms within each sublattice and the perturbations to the electron orbitals caused by lattice interactions.

7.8 EXCHANGE AND INDIRECT EXCHANGE INTERACTIONS

Let us now consider why ferromagnetic and ferrimagnetic materials spontaneously align within magnetic domains. As we have discussed in Chapter 2, any particle, such as an electron, or any system tends toward the lowest possible energy. Matter does not like to be in an excited state. Thus, for the model of aligned spins to be correct for ferromagnetism and ferrimagnetism the spin-ordered state must be the lower energy state.

In order to account for this phenomenon, Weiss postulated an *internal magnetic field* that acts as an alignment mechanism for ferromagnetic behavior. He assumed that the total magnetic field, $\mathbf{H}_T$, is the sum of the external field, $\mathbf{H}_e$, and a "molecular" or internal field, $\mathbf{H}_m$:

$$\mathbf{H}_T = \mathbf{H}_e + \mathbf{H}_m \tag{7.38}$$

where $\mathbf{H}_m = \gamma \mathbf{M}$
γ = molecular field constant

Then we can substitute $\mathbf{H}_T$ for $\mathbf{H}$ in the equation for the susceptibility:

$$\chi = \frac{\mathbf{M}}{\mathbf{H}_T} \tag{7.39}$$

$$\chi = \frac{\mathbf{M}}{\mathbf{H}_e + \mathbf{H}_m} \tag{7.40}$$

$$\chi = \frac{\mathbf{M}}{\mathbf{H}_e + \gamma \mathbf{M}} \tag{7.41}$$

Using equation (7.34) for the temperature dependence of the susceptibility:

$$\chi = \frac{C}{T} \tag{7.42}$$

we can set equation (7.42) equal to (7.41) and solve for **M**, as

$$\frac{C}{T} = \frac{M}{H_e + \gamma M} \tag{7.43}$$

or

$$M = \frac{C}{T}(H_e + \gamma M) \tag{7.44}$$

$$M - \frac{C}{T}\gamma M = \frac{C}{T}H_e \tag{7.45}$$

or

$$M = \frac{\frac{C}{T}H_e}{\left(1 - \frac{C}{T}\gamma\right)} \tag{7.46}$$

yielding

$$M = \frac{H_e C}{T - \gamma C} \tag{7.47}$$

This relation (7.47) can be used for the magnetization in equation (7.39) yielding

$$\chi = \frac{H_e C}{T - \gamma C}\left(\frac{1}{H_e}\right) \tag{7.48}$$

and finally

$$\chi = \frac{C}{T - \gamma C} \tag{7.49}$$

which has exactly the form of the Curie–Weiss law. The problem with this result is that the magnitude of the internal field is very large (on the order of 10^7 Oe). On the other hand, this simple theory accounts for the existence of both paramagnetic and ferromagnetic properties. These properties differ in the magnitude of the internal field, H_m.

The quantum-mechanical explanation for Weiss's molecular magnetic field arises from the exchange of energy from one atom to another. This exchange energy is a short-range effect and reduces the overall energy state of an election if its spin is aligned *with* adjacent atomic electrons. The exchange energy is released when the electrons pair with aligning magnetic moments.

7.8 EXCHANGE AND INDIRECT EXCHANGE INTERACTIONS

The classical coupled oscillator has two frequency modes

$$\omega^+ = \frac{\omega_1 + \omega_2}{2} \tag{7.50}$$

and

$$\omega^- = \frac{\omega_1 - \omega_2}{2} \tag{7.51}$$

where ω_1 = angular frequency of the particles
ω_2 = angular frequency of the coupling

The energy of a de Broglie particle was determined in Chapter 2 as

$$E = h\nu \tag{7.52}$$

where h = Planck's constant
ν = frequency

The frequency ν is related to the angular frequency ω by 2π:

$$2\pi\nu = \omega \tag{7.53}$$

Therefore, we can compare the energy of the particles in a coupled oscillator with that of the energy of two "free" particles. The energy of a pair of noncoupled, E_{nc}, particle is

$$E_{nc} = 2E = \frac{2h\omega}{2\pi} = 2\hbar\omega \tag{7.54}$$

Likewise, the energy for the coupled pair, E_c, is

$$E_c = E_1 + E_2 \tag{7.55}$$
$$E_c = \hbar\omega_1 + \hbar\omega_2 \tag{7.56}$$
$$E_c = \hbar(\omega_1 + \omega_2) \tag{7.57}$$

But the frequency of the coupling ω_2 is much less than the frequency of the individual particle ω_1:

$$\omega = \omega_1 > \omega_2 \tag{7.58}$$

Therefore, the coupling reduces the energy of the pair:

$$E_c < E_{nc} \tag{7.59}$$

This can be seen by observing a coupled oscillator. The energy of the particles is slowly transferred back and forth through the coupling spring or force. Neither particle has all of its energy; the energy is shared by the particles. The coupling mechanism has only a small portion of the energy of one particle. The total energy of the system is less than that of two free particles.

The exchange energy between magnetic moments acts in a similar manner. This exchange energy is given by

$$E_{ex} = -2J S_1 S_2 \cos \phi \qquad (7.60)$$

where J = exchange integral
S_1 = spin vector of electron no. 1
S_2 = spin vector of electron No. 2
ϕ = angle between the spin vectors

The exchange interaction (coupling) is an extremely short-range effect. Consequently, only neighboring atoms need be considered. The exchange integral J has the form

$$J = \int \Psi_a(1) \Psi_b(2) \Psi_a(2) \Psi_b(1) \left(\frac{1}{a} - \frac{1}{a_2} - \frac{1}{a_1} + \frac{1}{r_{12}} \right) dt \qquad (7.61)$$

where a = lattice separation
a_2 = distance between electron no. 2 and atom no. 1
a_1 = distance between electron no. 1 and atom no. 2
r_{12} = distance between electrons

If the exchange integral is plotted as a function of atomic separation of the 3d orbital radius for given elements it is known as a Slater–Bethe curve as shown in Figure 7.9.

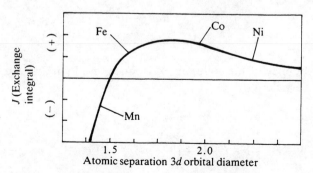

FIGURE 7.9 Slater–Bethe curve.

7.8 EXCHANGE AND INDIRECT EXCHANGE INTERACTIONS

TABLE 7.2 Exchange Integral for Several Metals

Unfilled orbital	Element	E_{ex}	Magnetic property
d	Fe	$E_{ex} < 0$	Ferromagnetic
d	Co	$E_{ex} < 0$	Ferromagnetic
d	Ni	$E_{ex} < 0$	Ferromagnetic
d	Mn	$E_{ex} > 0$	Antiferromagnetic
d and s	Cr	$E_{ex} > 0$	Antiferromagnetic
f	Rare earths	$E_{ex} < 0$	Ferromagnetic

When the exchange integral is *positive,* then the exchange energy is negative and *parallel* spin alignment is favorable:

$$\text{If} \quad J > 0 \quad \text{then} \quad E_{ex} < 0 \qquad (7.62)$$

When the exchange integral is *negative,* then the exchange energy is positive and *antiparallel* spin alignment is favorable:

$$\text{If} \quad J < 0 \quad \text{then} \quad E_{ex} > 0 \qquad (7.63)$$

Table 7.2 summarizes this effect for various elemental materials.

Indirect exchange interactions also provide the origin of ferrimagnetism. As we have seen in ferromagnetism, when J is positive, a parallel orientation of spins occurs. In ionic crystals, such as oxides, with negative values of J an antiparallel alignment is favored. However, in ferrimagnetic oxides, an oxygen ion with an inert shell of paired electron spins lies in between the metal ions with unpaired spins.

Various interaction mechanisms have been proposed to account for the indirect (or super) exchange interaction. In the ground state, with the ions in an inert gas configuration, the ion is inert and can produce no spin coupling. The surrounding ions disturb this state somewhat, so there is partial sharing of electrons with a neighboring ion. The electron goes back to the metal ion from where it originally came. There is a strong interaction with the other metal electrons and the electrons associated with the oxygen orbital can remain only if its spin has a given orientation with respect to the metal electron spins.

The remaining now-unpaired electrons can interact with another metal ion with a general intermingling of excited states between the two metal ions (see Figure 7.10). Therefore, there has been an effective coupling between the spin moments of the two metal ions, because if their moments had been parallel there would have been no intermingling and no reduction

302 CHAPTER 7: MAGNETIC CERAMICS

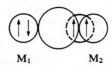

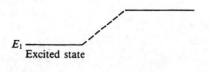

FIGURE 7.10 Indirect exchange.

in energy. Thus:

$$J > 0 \quad \text{for} \quad d^1 \text{ to } d^4 \quad \text{configuration} \ (\uparrow\uparrow) \tag{7.64}$$
$$J < 0 \quad \text{for} \quad d^5 \text{ to } d^9 \quad \text{configuration} \ (\uparrow\downarrow) \tag{7.65}$$

The interaction will depend strongly on the distance between the M and O ions and will rapidly diminish with distance.

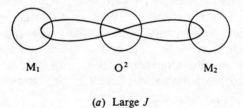

(a) Large J

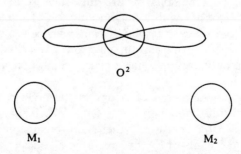

(b) Low J

FIGURE 7.11 (a) Large and (b) low overlap of electron orbitals.

The orientation dependence is strong, because the oxygen electrons are stretched $2p$ orbitals. Since overlap is required, it is most favorable if the oxygen lies online between two metal ions, as shown in Figure 7.11a. If there is little orbital overlap, Figure 7.11b, then the exchange interaction will be small.

7.9 SPIN ORDER

For negative values of J, we expect antiparallel ($\uparrow\downarrow$) spins for neighboring metal ions which have a net magnetic moment. Spin order depends on crystal structure and the ratio of the magnitudes of the interactions. The order occurring will be that of lowest energy (see Figure 7.10) or highest value of J. When an anion is present in the center, only the distances between the cation and anion are important, not the metal–metal distances (Figure 7.11).

Two alternative spin configurations of a M—O lattice are depicted in Figure 7.12. In Figure 7.12a, the A—B coupling is stronger than A—C. Therefore, A—B is antiparallel. In Figure 7.12b, the A—C bond is strongest because of the 180°C angle between connecting ions although the distance between the A—C metal ions is greater than the distance between A—B ions.

This spin ordering is reflected in the MnO antiferromagnetic lattice, as shown in Figure 7.13. As in the cases of Figure 7.12a and b, the lattice can be split into two sublattices, each possessing uniform magnetization. If the sublattices are occupied by identical ions, which is the case for MnO, the resultant moment will be zero (i.e., *antiferromagnetism*). If the magnetic moments of both lattices differ as is the case of ferrite spinels, discussed below, a net moment results (i.e., *ferrimagnetism*).

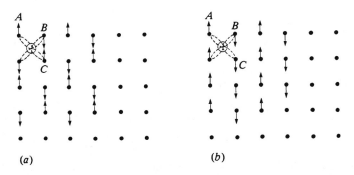

FIGURE 7.12 Spin coupling. (a) A–B coupling stronger than A–C. (b) A–B coupling weaker than A–C.

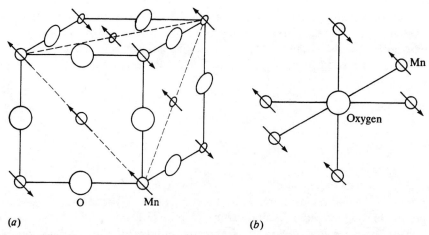

FIGURE 7.13 Antiparallel magnetic spin lattices of MnO (*a*) cubic structure (*b*) octahedral coordination.

7.10 LATTICE INTERACTIONS

There will be interactions between the spin-ordered sublattices. Interactions occur between spins within each sublattice, which are usually negative, and they will try to upset the parallel alighment of the spins within this sublattice. However, if the interactions with other sublattices are stronger, this will not happen.

The ion configuration in a typical ferrimagnetic crystal is shown in Figure 7.14. As long as $\bar{J}_{AB} > 2\bar{J}_{B'B''}$ then B' and B'' will be antiparallel with A. At the same time B' and B'' will be parallel with each other.

Consequently, the magnetization versus temperature of a ferrimagnet can vary considerably if the ratio of the two sublattice interactions are changed. Total magnetization is obtained by subtraction of the two magnetizations, so M_{total} can pass through zero or a maximum as a function of temperature as shown in Figure 7.15.

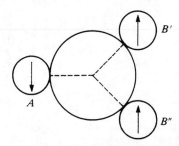

FIGURE 7.14 Lattice interaction. A belongs to one lattice, B' and B'' belong to another.

7.11 FERRIMAGNETIC AND FERROMAGNETIC DOMAINS

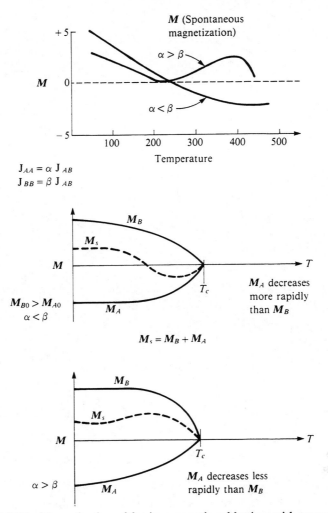

FIGURE 7.15 Magnetization of ferrite magnetic sublattices with temperature.

7.11 FERRIMAGNETIC AND FERROMAGNETIC DOMAINS AND DOMAIN MOTION

The spontaneous parallel alignment of magnetic moments within a crystal results in a large saturation magnetization, M_s, of the crystal. As shown in Figure 7.16a the aligned magnetic moments also produce a large external magnetic field since the lines of magnetic force must close. There is a substantial energy associated with this external field. This energy can be reduced by closure of the magnetic lines of force within the crystal. Formation of magnetic domains within the crystal eliminates the external field and thereby lowers the overall energy.

As illustrated in Figure 7.16b and c, formation of domains of opposite

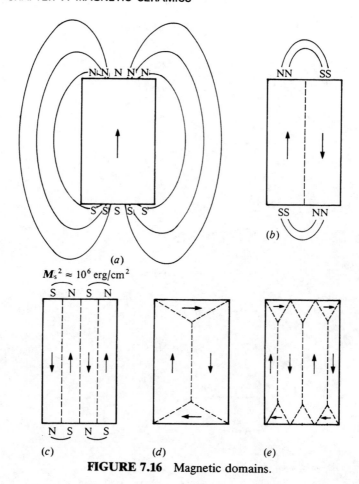

FIGURE 7.16 Magnetic domains.

magnetization substantially lower the external force lines to the ends of the crystal.

Domain arrangements leading to a zero external magnetic field M_s can be formed using *domains of closure*, Figure 7.16d, which make angles of 45° with respect to the magnetization in rectangular domains. Consequently, since there are no poles in the material, there is no external field associated with the magnetization and the flux is complete within the crystals. However, domains of closure are also opposed by *magnetostrictive energy*. The crystal increases in length in the direction of magnetization. Therefore, domains of closure are opposed by these elastic forces.

The energy required for domains of closure comes from crystalline anisotropy energy. This energy tends to make the magnetization of a domain line up along certain crystal directions. These axes are known as preferred axes, or axes of easy magnetization (see Figure 7.17). The Miller indices of these axes are shown in Figure 7.18.

7.11 FERRIMAGNETIC AND FERROMAGNETIC DOMAINS

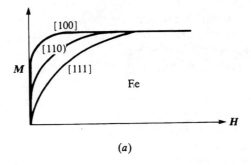

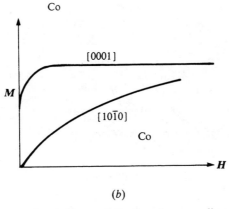

FIGURE 7.17 Iron and cobalt magnetization in easy directions (*a*) iron (cubic axes) (*b*) cobalt (hexagonal axes).

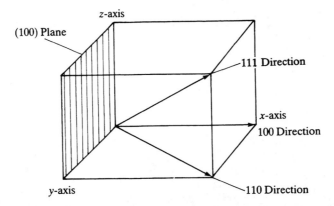

FIGURE 7.18 Miller indices of easy directions of magnetization.

308 CHAPTER 7: MAGNETIC CERAMICS

If, in hexagonal cobalt, the rectangular domains are in the easy direction, the domains of closure will be in the hard direction. In cubic Fe both basic domains and domains of closure will be along different easy axes. The energy comes from magnetostriction, (i.e., the change of length with magnetization direction).

In order to form domains of closure, work must be done against these elastic forces. The transition layer separating adjacent domains is called the *Bloch wall*. The exchange energy is lower for a gradual change of spin. The thickness and energy of the wall is a balance of exchange energy favoring thickness and anisotropy energy. This favors a decreasing thickness because of magnetostriction.

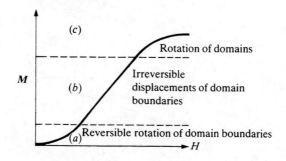

Representative magnetization curve

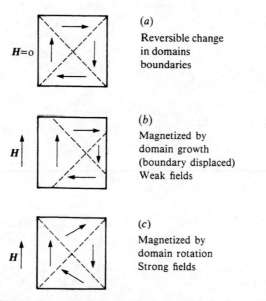

FIGURE 7.19 Magnetization by growth or rotation of domains.

The width of the domains is also a balance of energy. Wall energy tends to increase domain width; anisotropy energy tends to decrease width (due to domains of closure).

The subdivision into the smaller domains, Figure 7.16e continues until the energy required for an additional domain wall is greater than the reduction in magnetic field energy.

The boundary layer of a domain is due to the following energies. Exchange forces favor parallel and oppose antiparallel orientations of the magnetization. Thus, energy will be required to establish a boundary. The value of the boundary energy is, $E \cong 1 \text{ erg/cm}^2$, so if there are $N > 10^3$ domains/cm the total boundary energy will be $\simeq 10^3$ ergs and the magnetic energy will be $\simeq 10^3$ ergs. This represents an equilibrium condition.

As the external magnetic field, **H**, is increased the magnetic domains change in a ferromagnetic or ferrimagnetic material. Figure 7.19 shows the progressive nature of the domain motion. For a small **H**, the domain change is small and reversible. For larger values of **H** the boundaries shift or are displaced to enhance domains with magnetic moments aligned with the external field. Finally, near saturation the domains rotate to align themselves with the external field.

7.12 CLASSES OF MAGNETIC CERAMICS

Magnetic ceramics are subdivided into three classes based upon their crystal structures: spinels or cubic ferrites, garnets or rare earth ferrites, and magnetoplumbites or hexagonal ferrites (Table 7.3). The molecular composition of each class can be optimized or tailored for specific applications. The cubic spinels, also called *ferrospinels*, are used as soft magnetic materials because of their very low coercive force (**H**$_c$) of 4×10^{-5} weber/m^2 and high saturation magnetization (**B**$_s$) 0.3–0.4 weber/m^2. Hexagonal ferrites are hard magnetic materials with **H**$_c$ = 0.2 to 0.4 weber/m^2 and a large resistance to demagnetization, measured by (**BH**)$_{max}$. Values of (**BH**$_{max}$) range from 2 to 3 J/m^3. The **BH** hysteresis loops for hard and soft ferrites are compared in Figure 7.20.

The prime requirement for a soft magnetic material is that a high saturation magnetiziation be produced by a small applied field. Consequently only a small field, $(\mu_v\mathbf{H})_c$, is necessary to cause demagnetization and very small energy losses occur per cycle of the hysteresis loop. This is important for applications such as transformers used in touch tone telephones or inductors or magnetic memory cores where continuing changes in the direction of magnetization occur. During use a soft ferrite has its magnetic domains rapidly and easily realigned by the changing magnetic field.

TABLE 7.3 Classes of Magnetic Ceramics

Structure	Composition	Applications
Spinel (cubic ferrites)	$1\text{MeO}:1\text{Fe}_2\text{O}_3$ (where MeO = transition metal oxide, e.g., Ni, Co, Mn, Zn)	Soft magnets
Garnet (rare earth ferrites)	$3\text{Me}_2\text{O}_3:5\text{Fe}_2\text{O}_3$ (where Me_2O_3 = rare earth metal oxide, e.g., Y_2O_3, Gd_2O_3)	Microwave devices
Magnetoplumbite (hexagonal ferrites)	$1\text{MeO}:6\text{Fe}_2\text{O}_3$ (where MeO = divalent metal oxide from group IIA; e.g., BaO, CaO, SrO)	Hard magnets

In contrast, a hard (or permanent) ceramic magnet achieves its magnetization during manufacture. The magnetic domains of an hexagonal ferrite are "frozen in" by poling in an applied magnetic field as the material is cooled through its Curie temperature. Because of the anisotropy of the crystals and their small grain size it is extremely difficult to nucleate

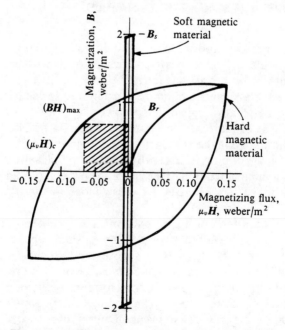

FIGURE 7.20 Representative curves of initial magnetization and hysteresis loops for soft and hard magnetic materials.

magnetic domains in hexagonal ferrites. Consequently, the materials are magnetically very hard and will retain in service the residual flux density, B_r, that remains after the strong magnetizing field has been removed. Hard ferrites are used in loudspeakers, magnetic latches, motors, magnetos, separators for ore benefication or tramp iron removal.

The third type of magnetic ceramics, the rare earth garnets, are especially suited for high frequency microwave applications due to the ability to tailor properties such as magnetization, line width, g-factor, T_c, and temperature stability. The versatility of molecular tailoring of magnetic properties is due to the presence of *three* sites in the garnet lattice that can be occupied by magnetic or nonmagnetic cations. Thus, many combinations of interacting spin lattices are possible.

7.13 SPINEL FERRITES

The compositions of the magnetically soft spinel ferrites, also called ferrospinels, are based upon the mineral spinel ($MgAl_2O_4$) but have the chemical formula $MeFe_2O_4$ where Me stands for a divalent transition metal ion, such as Ni^{2+}, Fe^{2+}, Co^{2+}. The divalent metal ion can also be replaced with an electronically neutral mixture of trivalent and monovalent metal ions. For example, in lithium ferrite the divalent metal is $\frac{1}{2}Li^{+1} + \frac{1}{2}Fe^{3+}$.

The spinel crystal structure is cubic with the O^{2-} ions forming a fcc lattice. There are two types of interstitial sites in this close-packed structure. In a unit cell containing eight formula equivalents of $MeFe_2O_4$ there are 32 O^{2-} ions and 64 sites with tetrahedral coordination of O^{2-} ions around the cations, called A sites. There are 32 sites with octahedral coordination of O^{2-} around metal ions, called B sites. As shown in Table 7.4 only eight of the 64 A sites and 16 of the 32 B sites are occuped in a ferrite unit cell. In the mineral spinel, the Mg^{2+} ions occupy A sites and Al^{3+} ions occupy B sites. This arrangement is called a *"normal"* spinel. Some ferrites are also normal spinels, as shown in Table 7.5, with Me^{2+} in A sites and Fe^{3+} in B sites.

TABLE 7.4 Arrangements of Metal Ions in the Unit Cell of Ferrite $MeFe_2O_4$

Kind of site	Size of site (Å)	Number available	Number occupied	Occupants	
				Normal spinel	Inverse spinel
Tetrahedral (A)	0.3–0.6	64	8	$8Me^{2+}$	$8Fe^{3+}$
Octahedral (B)	0.6–1.0	32	16	$16Fe^{3+}$	$8Fe^{3+}$ $8Me^{2+}$

TABLE 7.5 Ion Distribution of Net Moment/Molecule of Spinel Ferrites[a]

Spinel ferrite	Structure	Tetrahedral A sites	Octahedral B sites	Net moment (μ_B/molecule)
Fe_3O_4	Inverse	Fe^{3+} 5 →	Fe^{2+}, Fe^{3+} 4 5 ← ←	4
$NiO \cdot Fe_2O_3$	Inverse	Fe^{3+} 5 →	Ni^{2+}, Fe^{3+} 2 5 ← ←	2
$ZnO \cdot Fe_2O_3$	Normal	Zn^{2+} 0	Fe^{3+}, Fe^{3+} 5 5 ← →	0
$MgO \cdot Fe_2O_3$	Mostly inverse	Mg^{2+}, Fe^{3+} 0 4.5 →	Mg^{2+}, Fe^{3+} 0 5.5 ←	1
$0.9NiO \cdot Fe_2O_3$ and $0.1ZnO \cdot Fe_2O_3$	Inverse	Fe^{3+} 4.5 →	Ni^{2+}, Fe^{3+} 1.8 4.5 ← ←	1.8
	Normal	Zn^{2+} 0	Fe^{3+}, Fe^{3+} 0.5 0.5 ← ←	1.0
		4.5 →	7.3 ←	2.8

[a] After Cullity (1975).

When the divalent cations are on *B* sites and the trivalent ions are equally divided between *A* and *B* sites the structure is called an *"inverse"* spinel. Examples of inverse spinels include iron, cobalt, and nickel ferrites (Table 7.5). Intermediate spinel structures also exist where the divalent cations are split between *A* and *B* sites. Manganese ferrite, for example has

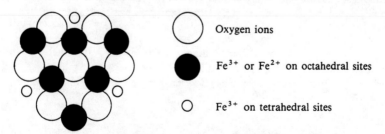

FIGURE 7.21 Fe_3O_4 lattice.

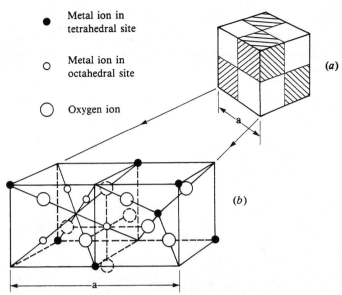

FIGURE 7.22 Schematic of two segments of a spinel structure. (After Smit and Wijn, 1959.)

0.8 of the Mn^{2+} ions on A sites and 0.2 of the Mn^{2+} ions on B sites. Mixtures of two different types of divalent cations, such as $(Ni, Zn)O:Fe_2O_3$ are also possible. Such compositions, called a mixed ferrite, are solid solutions of $NiO:Fe_2O_3$ and $ZnO:Fe_2O_3$. The magnetic properties of mixed ferrites vary greatly depending upon the magnetic moment of the cations and their distribution among the octahedral and tetrahedral sites (Figures 7.21 and 7.22). Consequently, the properties of commercial ferrites are "molecularly tailored" by controlling the solid solution of mixed ferrites.

7.14 Fe_3O_4

The simplest magnetic oxide, magnetite, is Fe_3O_4. This is the original "lodestone" which, if supported to float on water, would turn until it pointed approximately north and south. Fe_3O_4 has an inverse spinel structure: for example in Fe_3O_4 (Figure 7.21)

8 Fe^{3+} are in tetrahedral sites,
8 Fe^{3+} are in octahedral sites,
8 Fe^{2+} are in octahedral sites.

The Fe^{3+} ions spins are antiparallel with the Fe^{2+} ions and they are directly connected through a corner oxygen atom so that the magnetic

TABLE 7.6 Magnetic Moments of Spinel Ferrites

Ferrite	Observed moment	Calculated
$Fe^{3+}(Fe^{2+}Fe^{3+})O_4$	4.1	4
$Fe^{3+}(Ni^{2+}Fe^{3+})O_4$	2.3	2
$Fe^{3+}(Co^{2+}Fe^{3+})O_4$	3.7	3
$Fe^{3+}(Mn^{2+}Fe^{3+})O_4$	4.6–5.0	5
$Fe^{3+}(Cu^{2+}Fe^{3+})O_4$	1.3	1
$Fe^{3+}(Zn^{2+}Fe^{3+})O_4$	0	0

moments are cancelled leaving a resultant moment due to the Fe^{2+} ions; for example

$$1Fe^{3+} = -5 \text{ Bohr magnetons (see Figure 7.2)}$$
$$1Fe^{3+} = +5 \text{ Bohr magnetons}$$
$$1Fe^{2+} = +4 \text{ Bohr magnetons (see Figure 7.2)}$$
$$\text{Net moment} = +4 \text{ Bohr magnetons (see Table 7.5)}$$

The saturation moment, M_s, of some simple ferrites are given in Table 7.6. Many 2+ ions can be substituted for the Fe^{2+} ions which changes the total magnetic moment.

7.15 STRUCTURE OF SPINEL FERRITES

A schematic representation of the spinel unit cell is shown in Figure 7.22. The unit cell can be considered to contain eight octants of ions with the configurations shown in Figure 7.22b. The octant on the left of Figure 7.22b, with cell edge dimension of $a/2$, is represented by one of the shaded blocks of the unit cell (Figure 7.22a). The octant on the right is represented by one of the unshaded blocks.

One tetrahedral A site occurs at the center of the right octant and other A sites are at various octant corners, represented by the black circles. Four octahedral B sites are shown in the left octant. The coordination of one B site with six oxygen ions is indicated by six dashed lines. Two of the oxygen ions are in adjacent octants behind and below the ones shown. The oxygen ions are aranged tetrahedrally with respect to the metal cations in all octants.

7.16 SPINEL SPIN LATTICE INTERACTIONS

Figures 7.22, 7.23, and 7.24, show various interionic bonds in a spinel structure where ion A represents metal ions on tetrahedral sites and ion B

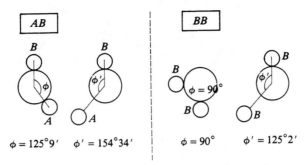

FIGURE 7.23 Spinel lattice exchange interactions.

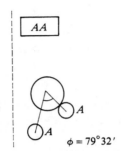

FIGURE 7.24 AA spinel interactions.

represents metal ions on octahedral sites. The AB interactions are the greatest due to short distances and angles approaching 180°. One can expect the spins of the A and B ions in ferrites with the spinel structure to be oppositely oriented so that at $T = 0$ there will be two oppositely magnetized sublattices present.

The resultant magnetization is thus the difference between the magnetization of the octahedral lattice and that of the tetrahedral lattice with the octahedral being the greatest.

Deviations from this result can occur if:

1. The ion distribution is not perfect;
2. The orbital moment is not completely quenched; or
3. Other angles may occur.

7.17 EFFECT OF COMPOSITION IN FERRITES

Ions commonly participating in the spinel lattice include:

Valence	Ion
1	Li
2	Mg, Mn, Fe, Co, Ni, Cu, Zn, Cd
3	Al, Cr, Mn, Fe
4	Ti, Mn

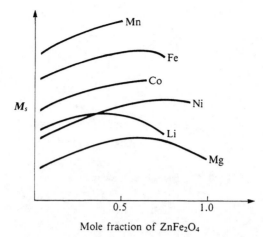

FIGURE 7.25 Saturation magnetization of various ferrospinels with substitution of zinc ferrite.

In general the divalent ions prefer octahedral sites and thus most ferrites form inverse spinels. However, Zn and Cd ions prefer tetrahedral sites forming normal spinels.

For $ZnFe_2O_4$ there is no net magnetization since Zn contributes no magnetic moment (see Figure 7.2) and occupies all the tetrahedral A sites. Thus, there can be no AB interaction and there remains only an antiparallel BB interaction.

The Zn ion preference for tetrahedral sites makes it possible to design ferrites with specific saturation moments. When zinc ferrite is added to an inverse structure, the Zn ions which enter tetrahedral sites force Fe^{3+} ions into octahedral positions. If all the ferrite ions were in a parallel orientation on octahedral positions, the resulting moment would be 10 Bohr magnetons. Small percentages of zinc solid solution produces this arrangement.

At larger percentages of zinc, greater than 50%, the fraction of the magnetic ions in tetrahedral sites decreases. The AB interaction is overcome by the BB antiparallel interaction and the resulting moment tends toward zero for pure zinc ferrite (see Figures 7.2, 7.25, and Table 7.5).

7.18 EFFECT OF THERMAL TREATMENT

Increasing temperature increases the randomization of the ion distribution between the octahedral and tetrahedral sites of a ferrite. Quenching from different temperatures freezes in different percentages of the random structure and thereby changes the saturation magnetization M_s. For example, as the quench temperature is raised for magnesium ferrite the

7.20 MANGANESE AND NICKEL ZINC FERRITES

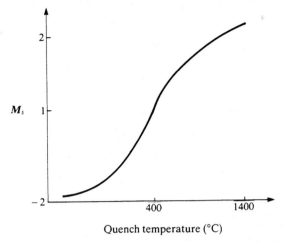

FIGURE 7.26 Effect of quench temperature of magnesium ferrite on saturation magnetization.

increase in fraction of magnesium on tetrahedral sites increases the magnetic moment of the quenched spinel (see Figure 7.26).

7.19 OXIDATION STATES

In addition to affecting cation ratios, oxidation states determine structure and resulting magnetic properties. For Fe^{3+} $(Mn^{2+}Fe^{3+})O_4$ at a larger oxygen content, Fe_2O_3 exists as a solid solution in the ferrite until at sufficient oxygen content, αFe_2O_3 forms. Thus, there exists a mixture of $(MnFe)_3O_4$ and Fe_2O_3 in the ferrite. At higher oxygen levels a $Mn_2O_3 \cdot Fe_2O_3$ solid solution appears and finally MnO_2 is formed.

At lower oxygen partial pressures there is a mixture of $MnFe_2O_4$ plus a solid solution of $MnO \cdot FeO$. At lower oxygen partial pressures the $FeO \cdot MnO$ phase is present and even Fe or Mn metals are present. In each case rapid quenching maintains the high temperature structure and the ferrite must be cooled under controlled atmosphere to achieve a desired oxygen ratio.

7.20 MANGANESE AND NICKEL ZINC FERRITES

Spinel ferrites based upon the $(Mn, Zn, Fe)O_4$ system are important examples of commercial magnetic ceramics. Although spinel ferrites are inferior to magnetic metals for static or power frequency fields, they are superb as cores for high frequency transformers or inductors. (See Lax and Button, 1962 for details.) This is because spinel ferrites combine high electrical resistivity and moderately good magnetic properties. Consequently, spinel ferrites can operate with essentially no eddy current losses at

318 CHAPTER 7: MAGNETIC CERAMICS

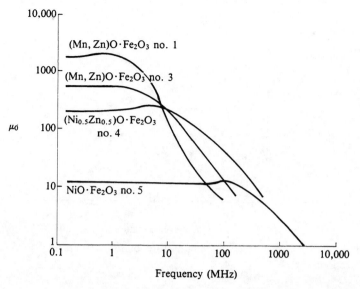

FIGURE 7.27 Variation of initial permeability (μ_0) with frequency for four spinel ferrites.

TABLE 7.7 Properties at 20°C of Some Spinel Ferrites[a]

	Mn–Zn (no. 1)	Mn–Zn (no. 2)	Mn–Zn (no. 3)	Ni–Zn (no. 4)	Ni (no. 5)
Composition (mol%)					
MnO	27	34	30		
NiO				32	50
ZnO	20	14	15	18	
Fe_2O_3	53	52	55	50	50
Initial permeability (μ_i)	2000	1500	650	110	20
Coercivity H_c(Oe)	0.2	0.25	0.7	4	14
Curie temperature	160	220	210	380	440
dc electrical resistivity	20	20	150	10^6	10^6
Applications	Low power-wide band transformers	High power transformers	Medium frequency inductors	High frequency inductors	Misc.

[a] After Snelling (1969).

frequencies where electrically conducting metallic magnets would be useless.

The (Mn, Zn, Fe)O_4 ferrites offer reasonably high permeabilities, on the order of 1000 to 2000 (Figure 7.27 and Table 7.7) and coercivities of less than 1 Oe. As Figure 7.27 shows, the Mn–Zn ferrites are usable without serious losses up to frequencies from 500 KHz (composition no. 1 in Table 7.7) to 2–3 MHz (composition no. 3 in Table 7.7). Resistivities are in the range of 20 to 200 ohm-cm. Applications include use in transformers with the power and frequency of use dependent upon the Mn, Zn, Fe content. A ternary diagram of the phase relations for three oxide components is given in Figure 7.28. The metal ions occupying A sites are given outside the brackets for the various compositions. The metal ions occupying B sites are given within the brackets. Within the portion of the diagram labeled cubic all three components form a continuous series of solid solutions. However, when the Mn content is increased beyond that of the zinc ferrite–manganese ferrite join, Mn^{3+} is introduced into the structure. Mn^{3+} prefers octahedral sites which it distorts as a result of the Jahn–Teller effect. The distortions are cooperative and produce a first order transition to a tetragonal phase.

The room temperature values of magnetocrystalline anisotropy (K_1) (two

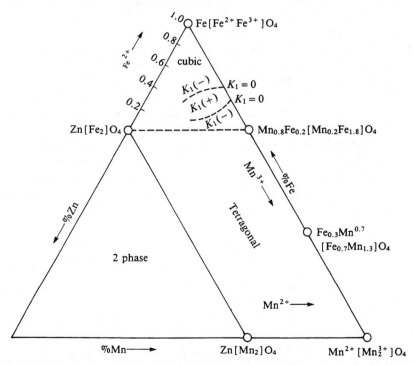

FIGURE 7.28 Cation distributions and crystalline anisotropy constants (K_1) in the (Fe, Mn, Zn)O_4 system. (After Schnettler, 1973.)

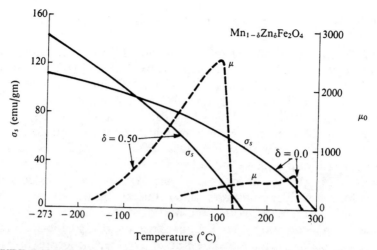

FIGURE 7.29 Variation with temperature of the specific saturation magnetization (σ_s) and permeability (μ) of Mn–Zn ferrites. (After Smit and Wijn, 1959.)

dashed lines) are shown for the cubic Mn–Zn–Fe spinel solid solutions in Figure 7.28. (The scale at the left of the cubic triangle indicates the divalent iron content expressed as atoms per formula unit.) The two dashed lines, drawn through the compositions for which $K_1 = 0$, enclose the region of positive anisotropy. The easy direction of magnetization within this region is [100]. For ferrites with negative anisotropy, outside these boundaries, the direction of easy magnetization is [111].

The effect of varying the Mn–Zn content δ on the temperature dependence of the specific saturation magnetization and permeability is shown in Figure 7.29. It is typical of any magnetic material for the initial permeability μ to increase with temperature to a maximum just below T_c. This is because the crystal anisotropy and magnetostriction normally decrease as temperature increases. Zinc ferrite additions shift T_c and the accompanying peak in μ to lower temperatures. The zinc additions also increase the height of the peak. Consequently, the room temperature value of μ also increases with the zinc content.

Replacement of Mn by Ni in spinel ferrites greatly increases the frequency over which they can be used. As indicated in Figure 7.27 and Table 7.7 the Ni–Zn ferrites have initial permeabilities in the range of 10 to 1000, decreasing with Ni content. Coercivities are several oersteds and resistivities are very high, 10^5 to 10^6 ohm cm.

At very high frequencies losses can be decreased if domain wall motion is inhibited and the magnetization is forced to change by domain rotation. Consequently, some grades of Ni–Zn ferrites are underfired so that the resulting porosity can interfere with wall motion. However, permeability decreases as well as losses.

7.21 HEXAGONAL FERRITES

The commercially most important hexagonal ferromagnetic ceramics are based upon $BaO:6Fe_2O_3$. This class is also called *"magnetoplumbite ferrites"* since barium ferrite is isostructural with the mineral magnetoplumbite of approximate composition $PbFe_{7.5}Mn_{3.5}Al_{0.5}Ti_{0.5}O_{19}$. The Ba ions in barium ferrite occupy the same position as the Pb ions in the magnetoplumbite lattice and the Fe ions in barium ferrite occupy the same positions as the mixture of Fe, Mn, Al, and Ti ions in the mineral.

The hexagonal unit cell of barium ferrite is very large, containing two "molecules" or $2 \times 32 = 64$ atoms. As illustrated in Figure 7.30, the unit cell is elongated along the magnetically active c-direction. The c-axis is 23.2 Å and the a-axis is 5.88 Å. Since Ba^{2+} and O^{2-} ions are large and about the

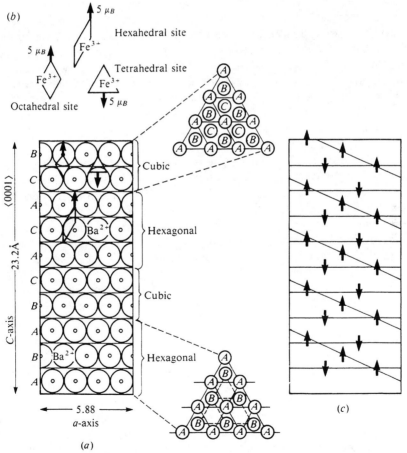

FIGURE 7.30 Schematic of barium ferrite crystal structure: (*a*) stacking of lattice planes; (*b*) Fe^{3+} sites; (*c*) a possible orientation of Fe^{3+} spin moments.

same size (see the Table of Ionic Radii in Chapter 1), they form a close-packed lattice. The smaller Fe^{3+} cations fit into interstitial positions in the Ba–O lattice. There are 10 layers of the close packed Ba^{2+} or O^{2-} ions with four ions per layer (Figure 7.30a). Eight of the layers are entirely O^{2-} ions while two contain one Ba^{2+} each. Thus, the unit of 10 layers can be considered as a stack of four blocks with alternating hexagonal and cubic stacked sequences.

As indicated in Figure 7.30a within each layer the large ions are located at the corners of a network of adjoining equilateral triangles. When the stacking sequence is *ABABAB* . . . , as in the bottom part of the figure, the third layer lies directly over the first and the resulting structure is close packed. When the sequence is *ABCABC* . . . it does not repeat until the fourth layer and the structure is face-centered cubic. This is the packing sequence for the spinel ferrites where the O^{2-} ions are in an *ABCABC* arrangement with the M^{2+} and Fe^{3+} cations in the interstices.

In the barium ferrite lattice, the arrangement of oxygen ions and the occupied tetrahedral sites and occupied octahedral sites in the cubic blocks are the same as described earlier for the spinel ferrite. However, in each hexagonal block a Ba^{2+} ion substitutes for an O^{2-} ion in the center of the three layers, as emphasized in Figure 7.30a. Since the cubic and hexagonal sections overlap, the four layers between those containing the Ba^{2+} ions have cubic packing and the five layers centered on a Ba^{2+} ion have hexagonal packing. The unit cell has hexagonal symmetry. (See Smit and Wijn, 1959, for additional details.)

The Fe^{3+} ions, each with a magnetic moment of 5 μ_B, are located in three different crystallographic sites: tetrahedral, octahedral, and hexahedral. Each is shown schematically in Figure 7.30b. The tetrahedral and octahedral sites are equivalent to those described above for spinels. However, the hexahedral site has the Fe^{3+} ion surrounded by five equidistant O^{2-} ions, arranged at the corners of a bipyramid with a triangular base. One hexahedral site occurs in each Ba^{2+} containing layer.

As illustrated in Figure 7.30c, the magnetic moment of each Fe^{3+} ion is normal to the plane of the O^{2-} layers and parallel or antiparallel to the $+c$-axis, the $\langle 0001 \rangle$ direction. Therefore, the spin moments will have an orientation of either [0001] or [000$\bar{1}$]. Eighteen of the 24 Fe ions per unit cell are in octahedral sites, four are in tetrahedral sites, and two are in hexahedral sites.

The spin directions of the Fe^{3+} ions in the cubic blocks are equivalent to those described earlier for the ferrospinels. Interactions of the indirect exchange force between Fe^{3+} ions in the hexagonal block and the cubic spinel block yield a net of 16 Fe^{3+} ions per unit cell with spins in one direction and eight in the opposite (see problem 7.3). A simplified configuration of the spin alignments for the hexagonal ferrite lattice is shown in Figure 7.30c. (For details of the complex spin alignments see Smit and Wijn, 1959.) The magnetic moment therefore is $(16-8)5\,\mu_B = 40\,\mu_B$

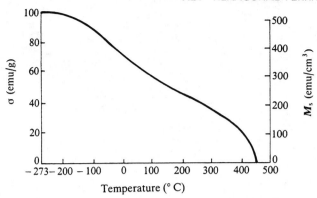

FIGURE 7.31 Temperature dependence of specific saturation magnetiziation (σ_s) and saturation magnetization (M_s) of barium ferrite. (After Smit and Wijn, 1959.)

per unit cell or 20 μ_B per molecule of $BaO:6Fe_2O_3$. This value corresponds to 100 emu/g (530 emu/cm³) exactly matching the measured saturation magnetization M_0 value at 0°K.

Due to thermal randomization of Fe^{3+} spin orientations, the saturation magnetization, M_s, decreases as temperature increases reaching a value of 72 emu/g (380 emu/cm³) at 20°C, as shown in Figure 7.31, with a Curie temperature of 450°C.

Substitutional solid solution of various divalent cations for Ba^{2+} can occur in the barium ferrite lattice. A variation in Ba^{2+} content is also possible in the $BaO:6Fe_2O_3$ lattice before the equimolar $BaO:Fe_2O_3$ phase is formed. These features offer the compositional flexibility, illustrated in Figure 7.32, to tailor the magnetic properties of barium ferrite by substituting a cation, such as Co^{2+} with magnetic moments, for nonmagnetic Ba^{2+} ions. Other hexagonal ferromagnetic oxides include:

$BaO:2MeO:8Fe_2O_3$ (W)
$2(BaO:MeO:3Fe_2O_3)$ (Y)
$3BaO:2MeO:12Fe_2O_3$ (Z)

Me designates a metal ion and the letters (W, Y, Z) are abbreviations for these compounds (see Smit and Wijn, 1959). Consequently, Co_2Z stands for the compound $3BaO:2CoO:12Fe_2O_3$. Figure 7.32 shows the location of these compounds on the $BaO-MeO-Fe_2O_3$ compositional diagram.

Table 7.8 summarizes the physical properties of several hexagonal ferrites. Strontium ferrite has nearly identical properties to barium ferrite, although its crystal anisotropy constant, K_1, is somewhat larger than the 3.3×10^6 ergs/cm³ value for $BaO:6Fe_2O_3$. Both barium and strontium ferrites owe their magnetic hardness to their uniaxial crystal anisotropy fields and their large K values. If the material consisted of aligned,

324 CHAPTER 7: MAGNETIC CERAMICS

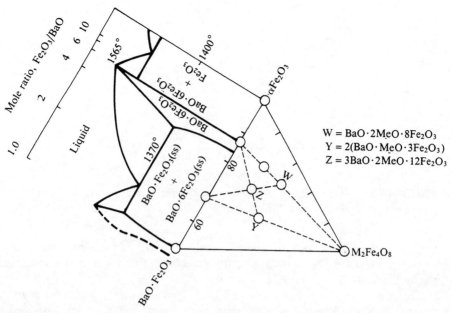

FIGURE 7.32 Relation of phase equilibria of the $BaO:Fe_2O_3-Fe_2O_3$ system with hexagonal ferrites: $BaO:6Fe_2O_3$, W, Y, and Z.

spherical, single-domain particles, its intrinsic coercivity $\mathbf{H}_{ci}$ would be

$$\mathbf{H}_{ci} = \mathbf{H}_K = \frac{2K}{\mathbf{M}_s} = \frac{2(3.3 \times 10^6)}{380} = 17{,}000 \text{ Oe} \qquad (7.66)$$

Since particle interactions do not affect crystal anisotropy this value should apply to both an isolated particle and to a fully dense pressed and sintered ceramic body. However, the hexagonal ferrites have a tabular crystal habit, (i.e., the 1 μm grains crystallize in the form of flat plates with the basal plane of the unit cell, the a-axis of Figure 7.30, parallel to the surface of the plate). The c-axis of the crystal, thus, is at right angles to the tabular grain. The flat plates tend to stack together when the powder is pressed to make the ceramic into its desired shape. Anisotropic grades (nos. 5, 6, 7 in Table 7.8) are made by applying a magnetic field during wet pressing of the compact. This aligns the c-axis of the particles with the field which is along the axis pressing. A cylindrical magnet is produced with the easy axis parallel to the cylinder axis.

The tabular shape, however, is less than ideal with respect to the best permanent magnet properties. This is because the easy axis due to shape is at right angles to the easy axis due to crystal anisotropy, rather than being parallel. Consequently, for an isolated particle, the shape effect reduces $\mathbf{H}_{ci}$

TABLE 7.8 Properties of Hexagonal Ferrites

Material	Saturation magnetization $M_0(0°K)$ emu/cm^3	Saturation magnetization $M_s(20°C)$ emu/cm^3	Curie temperature $T_c(°C)$	Remanence $B_r(G)$	Cercivity $H_c(Oe)$	Max energy product BH_{max}(MGOe)
BaO:6Fe$_2$O$_3$ (isotropic) (no. 1)	530	380	450	2250	1850 4000 (H_{ci})	1.2
BaO:6Fe$_2$O$_3$ (anisotropic) (no. 5)	530	380	450	3950	2400 2470 (H_{ci})	3.5
BaO:6Fe$_2$O$_3$ (anisotropic) (no. 6)	530	380	450	3600	2900 3100 (H_{ci})	3.1
SrO:6Fe$_2$O$_3$ (anisotropic) (no. 7)	530	380	450	3425	3300 4100 (H_{ci})	2.9
2(BaO:CoO:3Fe$_2$O$_3$)	204	—	320	—	—	—
BaFe$_{18}$O$_{27}$	520	—	445	—	—	—

Note: Nos. 4, 5, 6, 7 are designations of the Magnetic Materials Producers Association.

to 12,000. (See Cullity, 1972 for further discussion of particle size and shape effects on coercivity.)

The observed values of H_{ci} (Table 7.8) for the hexagonal ferrites are 1/3 to 1/4 of the theoretical limits of 12,000–17,000 Oe. This is because the 1 μm grains are nearly ten times too large for reversal of magnetization by single domain rotation. Instead the nucleation and movement of domain walls is responsible for the magnetization reversal in hexagonal ferrites. Smaller crystal sizes and fewer crystal imperfections will result in an increase in coercivity. Likewise, the value of the maximum energy product BH_{max} of commercial hexagonal ferrite magnets of 3 to 3.5 MGOe can be increased to about 5.0 MGOe if all particles were single domains and aligned.

A high coercive force is important when making permanent magnets. This is because the function of a permanent magnet is to provide an external field. Therefore, it must have free poles. The free poles create a demagnetizing field, H_d, which makes the induction lower than the remanence value, B_r, found in a closed ring. Certain shapes, such as plates or rings result in large self-demagnetizing fields. Hexagonal ferrites can resist large demagnetization fields due to crystal anisotropy and large coercivity. For example, consider Figure 7.33 which compares the demagnetization curves of an oriented Alnico 5 metallic magnet (8Al–15Ni–24Co–3Cu–50Fe) with four hexagonal ferrites nos. 1, 2, 5, 6 (Table 7.8). The ferrites have a much lower remanence and a much higher coercivity than the metal magnet.

The maximum value of $B \times H$ on the demagnetization curve is designated BH_{max}. The value of BH_{max} is used as a figure of merit for the magnetic material because it indicates the available magnetic energy in the air gap of the magnet. A material with a high value of BH_{max} will yield a smaller magnet for the same available energy. The dashed curves in Figure 7.33 are for constant values of BH_{max}. The value of BH_{max} for an isotropic barium ferrite (no. 1) is about 1.0 MGOe. The anisotropic grades 2, 5, and 6 have BH_{max} values up to 3.5 MGOe.

In contrast, a good metallic permanent magnet (such as Alnico 5) has a typical BH_{max} value of 5 MGOe. Consequently, in order to deliver the same magnetic flux in a gap a ferrite magnet has to have a larger cross-sectional area than the Alnico 5 magnet. This negative feature is offset by several advantages of hexagonal ferrite permanent magnets.

1. Low price and availability; ferrites do not contain expensive Ni or Co.
2. High coercive force (1800–4000 Oe); this makes ferrites attractive for applications where strong demagnetizing influences are present, such as dc motors, magnetic chucks, and magnets for periodic focusing of traveling wave tubes.
3. High resistivity ($\geq 10^8$ Ωm); this makes ferrites suitable for use in high frequency ac fields without giving rise to eddy current losses.

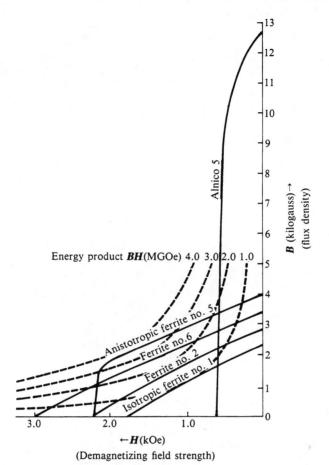

FIGURE 7.33 Comparison of demagnetizing curves of hexagonal ferrites (grades nos. 1, 2, 5, 6) with a metal magnet (Alnico 5).

4. Lightweight ($\rho = 4.9$–5.3 g/cc); compared with Alnico ($\rho = 8.7$ g/cc).
5. Ease of manufacture; isotropic grades of ferrites can be pressed and sintered easily. Anisotropic grades are used where configurations are simple and high performance is important, such as stereo loudspeakers.

7.22 GARNET FERRITES

Magnetic ceramics with the garnet crystal structure are widely used in microwave devices. Garnets are cubic and isostructural with the mineral garnet which has the formula $Ca_3Al_2(SiO_4)_3$. The general formula of garnet ferrites is $3Me_2O_3:5Fe_2O_3$, where Me = yttrium or any other rare earth

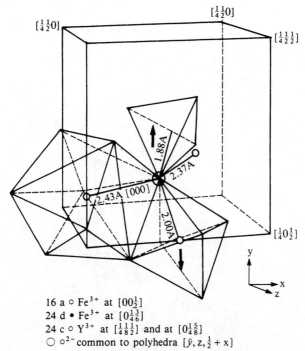

16 a ○ Fe^{3+} at $[00\frac{1}{2}]$
24 d ● Fe^{3+} at $[0\frac{1}{4}\frac{3}{6}]$
24 c ○ Y^{3+} at $[\frac{1}{4}\frac{1}{8}\frac{1}{2}]$ and at $[0\frac{1}{4}\frac{5}{8}]$
◯ o^{2-} common to polyhedra $[\bar{y}, z, \frac{1}{2}+x]$

FIGURE 7.34 Schematic indication of the structure of yttrium-ion garnet. The oxygen coordination of cations and the spin directions of the Fe^{3+} ions are shown. (After Reynolds and Buchanan.)

element. The complicated cubic structure contains three nonequivalent lattice sites, termed 16a, 24c, and 24d sites, which are occupied by the trivalent ions. The 16a sites have an octahedral coordination of oxygen ions, the 24c sites are dodecahedral, and the 24d sites are tetrahedral. Figure 7.34 shows the relationship of these three cation sites and the spin moment of the Fe^{3+} ions. The octahedral a and tetrahedral d sites have slightly canted axes of the octahedra and the tetrahedra.

The most common garnet ferrites are based upon $3Y_2O_3:5Fe_2O_3$ or $Y_3Fe_5O_{12}$ (called YIG as an abbreviation for yttrium–iron–garnet). In the YIG lattice there are eight formula units of $Y_3Fe_5O_{12}$ per unit cell or 160 atoms which gives rise to a large lattice constant of 12.38 Å. There are 24 tetrahedral sites per unit cell (24d) which are the smallest and normally occupied by Fe^{3+}. The 16 octahedral sites (16a) are occuped by Fe^{3+} and the 24 dodecahedral sites (24c) are occupied by Y^{3+} ions. In the YIG lattice the magnetic moments of the d sites are opposite to the moments of the a sites. Since yttrium is a nonmagnetic ion, the magnetization of a garnet ferrite is due to the difference in magnetization between the a and d sites; that is, $24d(5\,\mu_B) - 16a(5\,\mu_B) = 8(5\,\mu_B) = 40\,\mu_B$/unit cell; $40\,\mu_B$/unit cell ÷ 8 formula units/unit cell = $5\,\mu_B$ (Bohr magnetons) per formula unit.

7.22 GARNET FERRITES

TABLE 7.9 Structural Features of Ions Substituted in Garnet Ferrites

Ion	Ionic radii (Å)	↓16a	↓24c	↑24d	Number of unpaired electrons[a]
Y^{3+}	0.95		1.0		0
Gd^{3+}	0.97		1.0		7
Yb^{3+}	0.93		1.0		1
Er^{3+}	0.97		1.0		3
Sm^{3+}	1.00		1.0		5
Dy^{3+}	0.92		1.0		5
Ho^{3+}	0.91		1.0		4
Tb^{3+}	0.93		1.0		6
Nd^{3+}	1.04		1.0		3
Bi^{3+}	0.93		1.0		0
Ca^{2+}	0.99		1.0		0
Fe^{3+}	0.66	0.4		0.6	5
Al^{3+}	0.51			1.0	0
Ga^{3+}	0.62			1.0	0
Cr^{3+}	0.64	1.0			3
V^{5+}	0.59			1.0	

[a] Based upon Harrison and Hodges (1972).

Substitution of trivalent rare earth ions for yttrium or multivalent cations for iron can produce a wide range of magnetic properties for the garnet ferrites. The substitutional sites that are preferred depend upon the radii of the ions. Table 7.9 summarizes the preferred site in the garnet lattice for various cations, their number of unpaired electrons, and their ionic radii.

The preference of lattice sites depends upon ion size; for example, the small V^{5+} cation prefers the small 24d site. A Ca^{2+} cation is substituted along with V^{5+}, in suitable proportions, in order to main electrical neutrality (see Chapter 4). Thus, one V^{5+} and two Ca^{2+} ions replace three Fe^{3+} ions. The Ca^{2+} prefers the larger 24c site, generally referred to as the rare earth or yttrium site in garnet ferrites. Since nonmagnetic V^{5+} locates on the site normally occupied by magnetic Fe^{3+}, the resulting compositions exhibit magnetizations that decrease with increasing V^{5+} content until the 16a site becomes dominant. In other words, a substitution of one V^{5+} per formula unit into the material results in two Fe^{3+} on the 24d site (instead of three). Since the indirect exchange interaction between the a and d sites yields antiparallel magnetic moments of the Fe^{3+} ions, the total magnetization is zero, that is, it is antiferromagnetic. As more V^{5+} is substituted into the lattice, the 16a site becomes the dominant site and the magnetization depends on Fe^{3+} in the 16a site. Thus, substitution of Ca^{2+} and V^{5+} not only can alter the overall magnetization but the lattice site controlling the magnetization.

A similar situation exists when Ga^{3+} and Al^{3+} are substituted in garnet ferrites because nonmagnetic ions replace the highly magnetic Fe^{3+} ions.

As indicated in Table 7.9, trivalent rare earth ions can be substituted for the nonmagnetic yttrium ions on the 24c site. Since all of the rare earth ions listed in Table 7.9 have unpaired electrons in their outermost shells they have magnetic moments. The 24c site is not strongly coupled to the two Fe^{3+} sublattices. Consequently, the magnetic moment of this site relaxes much faster with increasing temperature than does the Fe^{3+} sites. The important consequence of this behavior is the production of *compensation* points in the temperature dependence of magnetization. Ionic substitutions can be used to vary the location of these compensation points and achieve temperature stable magnetizations.

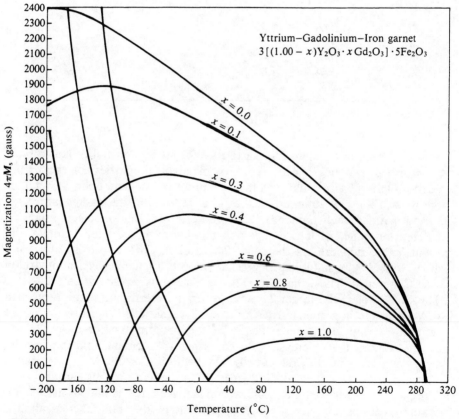

FIGURE 7.35 Variation of saturation magnetization with temperature for mixed yttrium gadolinium iron garnet. The compensation point changes from 15°C to −180°C. (After Harrison and Hodges, 1972.)

7.22 GARNET FERRITES

Substitution of gadolinium Gd^{3+} for yttrium is a good example of compositional tailoring of garnet ferrites. Since Gd^{3+} has seven unpaired electrons one can expect gadolinium iron garnet to have a magnetic moment of 32 Bohr magnetons when Gd^{3+} is located in the 24c sites instead of Y^{3+} (see Neel, 1954). The ionic distribution per formula unit of the $3Gd_2O_3 \cdot 5Fe_2O_3$ material is as follows:

↓24c	↑24d	↓16a
$6Gd^{3+}$	$6Fe^{3+}$	$4Fe^{3+}$

The two iron sublattices are strongly antiferromagnetically coupled together. The resultant magnetic moment of these two sites, ↑24d and ↓16a, will be $(2Fe^{3+})(5\ \mu_B/Fe^{3+}) = 10\ \mu_B$ (Bohr magnetons) in the direction given by the 24d site which contains the excess Fe^{3+} ions.

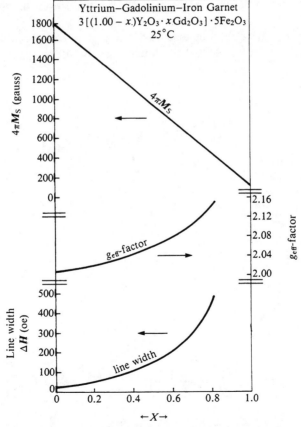

FIGURE 7.36 Variation of linewidth, saturation magnetization, and effective g-factor with Gd content in Y(Gd)IG. (After Harrison and Hodges, 1972.)

However, the magnetic moment of the Gd^{3+} ions relaxes much more with temperature than the Fe^{3+} ions. At some temperature the magnetic moments of the Gd^{3+} and the Fe^{3+} are equal and a compensation point arises. The compensation point, when $\mathbf{M} = 0$, for the gadolinium–iron-garnet is 15°C (Figure 7.35). As the proportion of Gd^{3+} is reduced and Y^{3+} is increased the compensation point changes systematically, as does the magnetization, line width, and g_{eff}-factor (Figure 7.36). The Curie temperature remains at 290°C for all compositions (Figure 7.35) since the number of Fe^{3+}–Fe^{3+} interactions has not been altered and it is the exchange interaction between the Fe^{3+} spin lattices that determines T_c.

This property control is important from an applications standpoint. If one desires a microwave material that has a nearly constant magnetization over a wide temperature range, say from $-30°C$ to $120°C$, then Figure 7.35 shows that 60% gadolinium substituted garnet has this characteristic; its compensation point is $-120°C$. For room temperature applications the magnetization of the $3[(1-x)Y_2O_3 \cdot xGd_2O_3] \cdot 5Fe_2O_3$ system can be varied from as much as 1760 G for the pure yttrium iron garnet to 100 G for the gadolinium iron garnet.

The microwave line widths also vary with temperature and composition, as shown in Figure 7.36. At room temperature the line width of yttrium iron

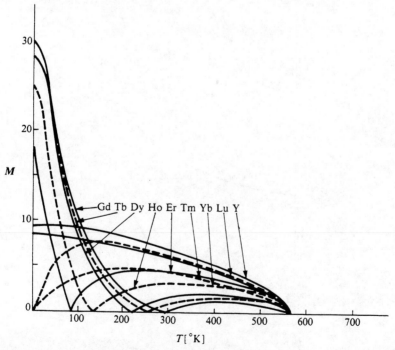

FIGURE 7.37 Magnetization curves of yttrium iron garnet and rare-earth-iron garnets. (After Bertaut and Forrat, 1956.)

garnet is quite narrow (30 Oe) whereas that of the Gd substituted garnet is very broad (500 Oe). The broadening is due to the large anisotropy fields that exist at and near the compensation point for any of these materials. Thus, the mixed yttrium gadolinium iron garnets are useful when application requires low to moderate line widths and in devices where some temperature compensation of the magnetization is required.

As indicated in Table 7.9, in addition to Gd^{3+}, many other trivalent rare earth ions can be substituted for Y^{3+} in the YIG lattice. In each case when the rare earth lattice is occupied by magnetic ions, the magnetization is parallel to the a-sublattice. Also, because of the screening by the $5s$ bonding electrons of the electrons that provide the magnetic moments, the rare earth ions are only weakly coupled with indirect exchange interactions antiparallel to the magnetic d lattice. Consequently, the magnetization of the rare earth substituted garnets all vary rapidly with temperature and have compensation points. Figure 7.37 compares the magnetization-temperature curves for the rare earth iron garnets.

Use of all the substitutional concepts discussed above and the ions listed in Table 7.9 make it possible to "molecularly tailor" this class of ferrites at will. (See Lax and Button, 1962 and Snelling, 1964, and Chapter 15 in *Magnetic Ceramics,* Proceedings of the British Ceramic Society, for discussion of microwave operation of garnet ferrites. See Harrison and Hodges, 1972 for details of microwave properties of a wide range of garnet ferrite compositions.)

PROBLEMS

7.1 Calculate the magnetic susceptibility in emu/cc-Oe for:
 (a) carbon.
 (b) Diamond.
 (c) Iron.
 (d) Cobalt.
 (e) Manganese.
 (f) Magnesium.

7.2 Calculate the magnetic moments in Bohr magnetons for the following list of materials:
 (a) Fe
    Co
    Ni
    $MnOFe_2O_3$
    $FeOFe_2O_3$
    $NiOFe_2O_3$
    $MgOFe_2O_3$

(b) Compare these with the measured values of the magnetic moments.

(c) What is the trend between the metals and the metal-oxides?

(d) What can be said about the electrons "belonging" to the metals in the presence of oxygen?

7.3 Based upon the principles of indirect exchange and the spinel lattice exchange interactions (Figure 7.19) show the magnetic spin orientations in a barium ferrite unit cell that produce a net moment of 20 μ_B.

READING LIST

Azaroff, L. and Brophy, J. J., (1963). *Electronic Processes in Materials*, McGraw-Hill, New York, Chapter: 13.

Bertaut, F. and Forrat, F. (1956). "Compt. Rend.", *Acad. Sci. Paris*, **242**, 382.

Cullity, B. D., (1972). *Introduction to Magnetic Materials*, Addison Wesley, New York.

Goodenough, J. B., (1963). *Magnetism and the Chemical Bond*, Interscience, New York.

Harrison, G. R. and Hodges, L. R. Jr., (1972). "Microwave garnet compounds," *Physics of Electronic Ceramics*, (L. L. Hench and D. B. Dove, eds.), Dekker, New York.

Hench, L. L. and Dove, D. B., (1971). *Physics of Electronic Ceramics*, Part B, Dekker, New York, Chapters: 24, 25.

Kingery, W. D., Bowen, H. K., and Uhlmann, D. R., (1976). *Introduction to Ceramics*, 2nd ed., Wiley, Chapter: 18.

Kittel, C., (1986). *Introduction to Solid State Physics*, 6th ed., Wiley, New York, Chapter: 8.

Lax, B. and Button, K. J., (1962). *Microwave Ferrites and Ferrimagnetics*, McGraw-Hill, New York.

Martin, D. H., (1967). *Magnetism in Solids*, Illife Books, Ltd., London.

Néel, L., (1954). "On the interpretation of the magnetic properties of the rare earth ferrites," *Compt. Rend.*, **239**.

Proceedings of the British Ceramic Society No. 2, "Magnetic ceramics," Stoke-on-Trent, England, Chapters: 1, 6, 7, 10, 11, 12, 15.

Reynolds, T. G. and Buchanan, R. C., (1986). *Ceramic Materials for Electronics*, (R. C. Buchanan, ed.), Dekker, New York, Chapter: 4.

Rosenberg, H. M., (1978). *The Solid State*, 2nd ed., Oxford Physics Series, Oxford University Press, New York, Chapters: 11, 12.

Schnettler, F. J., (1972). "Microstructure and processing of ferrites," *Physics of Electronic Ceramics*, L. L. Hench and D. B. Dove, eds.), Dekker, New York.

Smit, J. and Wijn, H. P. J., (1959). *Ferrites*, Wiley, New York.

Snelling, E. C., (1969). *Soft Ferrites, Properties and Applications*, Iliffe, London.

Standley, K. J., (1962). *Oxide Magnetic Materials*, Clarendon Press, Oxford, England.

Wohlfarth, E. P., (1980). *Ferromagnetic Materials*, North Holland, New York.

8

PHOTONIC CERAMICS

8.1 INTRODUCTION

There are a number of ways in which quanta of light, that is, photons, can interact with crystalline ceramics and amorphous glasses. The type of photon interactions that occur depend considerably upon the composition of the materials and the nature and types of phases and interfaces present within the material and between the material and its ambient media. Figure 8.1 summarizes these various photonic interactions.

The incident radiant flux of photons is split into beams of reflected, transmitted, absorbed, and scattered radiation, i.e.,

$$\phi = \phi_{\rho_r} + \phi_{\rho_d} + \phi_\tau + \phi_\sigma + \phi_\alpha \tag{8.1}$$

where ϕ = total incident flux

ϕ_{ρ_r} = directly reflected flux; ϕ_{ρ_r}/ϕ = coefficient of specular reflection, ρ_r

ϕ_{ρ_d} = diffusely reflected flux; ϕ_{ρ_d}/ϕ = coefficient of diffuse reflection, ρ_d

ϕ_τ = transmitted flux; ϕ_τ/ϕ = coefficient of direct transmission, τ

ϕ_σ = scattered flux; ϕ_σ/ϕ = coefficient of scattered transmission, σ

ϕ_α = absorbed flux; ϕ_α/ϕ = the absorptance or absorption factor, α

Thus

$$(\rho_r + \rho_d) + (\tau + \sigma) + \alpha = 1 \tag{8.2}$$

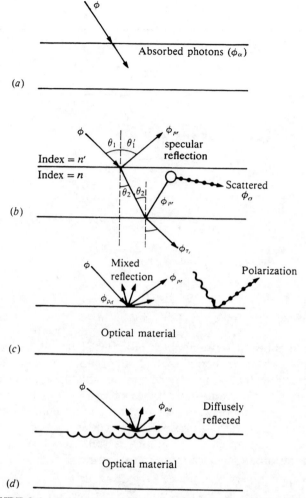

FIGURE 8.1 Loss mechanisms of photons in ceramics and glass.

and

$$\rho + \tau + \sigma + \alpha = 1 \qquad (8.3)$$

where ρ = coefficient of total reflectance
 τ = coefficient of total transmittance
 σ = coefficient of scattering
 α = coefficient of absorption

Other effects such as birefringence and polarization will be dealt with later in this chapter. Phenomena such as spontaneous emission and stimulated emission will be the concern of Chapter 11. Photoelectric and electro-optical effects will be dealt with in Chapter 10. The derivations of various laws describing the interaction of photons with matter will be developed in Chapter 9.

An optical coefficient can be expressed in fractions or in units of decibels. A decibel is defined through the relationship:

$$+/-1 \text{ db} = +/-10 \log \frac{\text{(power output)}}{\text{(power input)}} \qquad (8.4)$$

Engineers tend to express these coefficients in terms of decibels but they must be converted to fractions to calculate the power or intensity transmitted through a photonic material. The plus sign refers to a gain in power while the minus sign refers to a loss of power.

8.2 RADIATION

Photonic interactions with materials depend greatly on the frequency of the incident ratiation. This is because photons are quanta of energy E, with

$$E = h\nu = hc/\lambda \quad \text{(see Chapter 2)} \qquad (8.5)$$

where ν = frequency of the radiation in hertz (cycles per second)
λ = wavelength of the radiation in micrometers (10^{-6} m) or nanometers (10^{-9} m)
c = the velocity of propagation *in vacuo* = $2.99742 \pm 0.00003 \times 10^8$ m/sec

Photons interact with the electrons, ions, and molecules of the material, which also have characteristic energy levels, as discussed in Chapter 2. Consequently, the optical phenomena depicted in Figure 8.1 are by means of specific, quantized photon–electron, photon–ion, photon–molecular or photon–phonon interactions depending upon the relative energies of the particles.

The names ascribed to various portions of the electro-magnetic (EM) spectrum are shown in Table 8.1 with the relevant frequencies and wavelengths. Only the ultraviolet (UV), visible (vis), near infrared (NIR), and infrared (IR) portions of the EM spectrum will be considered in this chapter. Interactions of ceramics and glasses with EM oscillations in the range of 1 to 10^7 Hz are discussed in Chapters 4 and 5.

By substituting the values of h and c into equation (8.5), the energy for one mole of quanta is

$$E = \frac{1232.7}{\lambda} \text{ eV} \qquad (8.6)$$

338 CHAPTER 8: PHOTONIC CERAMICS

TABLE 8.1 Photon Energies and Wavelengths

Energy of quanta (J)	Frequency (Hz)	Description	Wavelength (nanometers)	Energy of quanta (eV)
6.63E–11	1E23	Cosmic rays	1E–6	4.1383E8
6.63E–12	1E22		1E–5	41383185
6.63E–13	1E21		1E–4	4138318.
6.63E–14	1E20	Gamma rays	1E–3	413831.8
6.63E–15	1E19		1E–2	41383.18
6.63E–16	1E18	X-rays	1E–1	4138.318
6.63E–17	1E17		1E 0	413.8318
6.63E–18	1E16		1E 1	41.38318
6.63E–19	1E15	Ultraviolet	1E 2	4.138318
6.63E–20	1E14	Visible	1E 3	.4138318
6.63E–21	1E13		1E 4	.0413832
6.63E–22	1E12	Infrared	1E 5	.0041383
6.63E–23	1E11		1E 6	4.138E–4
6.63E–24	1E10		1E 7	4.138E–5
6.63E–25	1E9	Hertzian	1E 8	4.138E–6
6.63E–26	1E8		1E 9	4.138E–7
6.63E–27	10,000,000		1E 10	4.138E–8
6.63E–28	1,000,000		1E 11	4.138E–9
6.63E–29	100,000	Radio	1E 12	4.14E–10
6.63E–30	10,000		1E 13	4.14E–11
6.63E–31	1,000		1E 14	4.14E–12
6.63E–32	100	Power generators	1E 15	4.14E–13
6.63E–33	10		1E 16	4.14E–14
6.63E–34	1		1E 17	4.14E–15

based upon Einstein's theory of the photoelectric effect where one quantum of radiation acts upon one molecule in a photochemical reaction. One molecule absorbs one light quantum, thereby increasing the energy of the molecule by $h\nu$. The energy of one mole of light quanta in joules is

$$1 \text{ eV} = 1.6021 \times 10^{-19} \text{ J} \qquad (8.7)$$

The energy of one mole of light quanta for various wavelengths is shown in Table 8.1.

8.3 REFLECTION

As indicated in Figure 8.1, the magnitude and character of reflected radiation depends upon the quality of the interface (roughness) and angle of incidence. It also depends upon the difference between the refractive indices of the medium and the glass or ceramic and the wavelength of the radiation.

Upon entering a material of refractive index n, at an angle of incidence θ, from a medium of index n', a beam of monochromatic or single wavelength photons is split into two beams or rays (Figure 8.1b). The reflected ray remains in the n' medium, the second ray is refracted into the material with the angle of refraction, θ_2, dependent upon the index n, as defined by Snell's law of refraction:

$$n \sin \theta_2 = n' \sin \theta_1 \tag{8.8}$$

The angle of reflection is equal to the angle of incidence

$$\theta_1' = \theta_1 \tag{8.9}$$

If the ray enters into a medium of higher optical density, $n' < n_1$ then $\theta_2 < \theta_1$ and the ray is refracted towards the normal. When $n' > n$ then $\theta_2 > \theta_1$ and refraction is away from the normal. At angles larger than a *critical angle* of incidence θ_m, where $\theta_1 = \theta_1' = \theta_m$, the entire radiation is reflected by the interface and is therefore called *total* or *specular reflection*.

The coefficient of reflection ρ on one surface can be calculated from Fresnel's formula for the incident angle perpendicular to the surface (see Section 9.8):

$$\rho = \left(\frac{n - n'}{n + n'}\right)^2 \tag{8.10}$$

where ρ = coefficient of reflection from one surface
  n' = refractive index of the incident medium
  n = refractive index of the material

Figure 8.2 shows the effect on loss of light by increasing the index of refraction of a glass or ceramic and the number of reflecting surfaces m'.

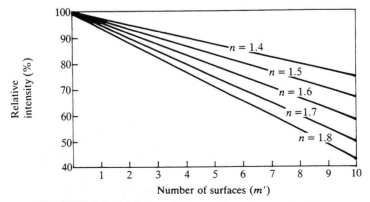

FIGURE 8.2 Light loss because of multiple reflections.

The loss is due to reflection, based upon the following relationship which is valid for normal incidence and zero absorptivity:

$$\rho = 1 - \left[1 - \left(\frac{n-1}{n+1}\right)^2\right]^{m'} \quad (8.11)$$

An important consequence of equation (8.11) and Figure 8.2 is that optical instruments with many boundary surfaces between components can lose as much as 50% light from reflection. Antireflection coatings must be used to prevent such losses. However, such reflections are the basis for decorative glass objects and crystalline jewelry where a large number of facets (surfaces) and high indices of refraction are used to reflect incident light and achieve "sparkle" (see problem 8.1).

When a surface is rough or irregular, the incident radiation is scattered in a multiplicity of directions, resulting in *diffuse reflection* (Figure 8.1). Complete diffuse scattering can occur (Figure 8.1d), or a mixture of specular and diffuse scattering is also possible (Figure 8.1c).

An important consequence of the reflection of light is its *polarization*. The reflected ray oscillates in the plane perpendicular to the plane of incidence (Figure 8.1c). If the angle of incidence corresponds to *Brewster's angle of polarization*, θ_p, the ray is totally polarized. Polarization is discussed in more detail in a later section.

The critical angle, θ_m, where total reflection occurs is a function of the refractive index of the material as shown in Figure 8.3. The critical angle therefore decreases as n increases.

If an optical material absorbs a significant fraction of the incident radiation, Fresnel's formula (8.9) must be modified to include an absorptive factor, k', with

$$k' = \frac{a}{4\pi}\lambda_n \quad (8.12)$$

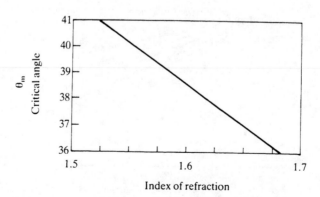

FIGURE 8.3 Critical angle as a function of index of refraction.

where a = linear absorption coefficient
 λ_n = wavelength

yielding for the reflectance of a partially absorbing optical material,

$$\rho = \frac{(n-1)^2 + k'^2}{(n+1)^2 + k'^2} \tag{8.13}$$

For conducting materials like silver, aluminum or other metals the relationship for reflectance is more complicated (see Garbuny, 1967).

8.4 REFRACTION

As shown in Figure 8.1b and Snell's law (equation 8.8), when a ray of light enters a medium of greater optical density, the ray is bent or refracted proportional to the difference in indices of refraction of the two media. Within the denser medium, the phase velocity of light is decreased:

$$n(\lambda) = \frac{\sin \theta_1}{\sin \theta_2} = \frac{c}{v_\lambda} \tag{8.14}$$

where $\sin \theta_1$, θ_2 are defined in Figure 8.1, and
 n_λ = index of refraction of the optical material at a given wavelength λ
 c = velocity of light *in vacuo*
 v_λ = phase velocity of light in the material at wavelength λ

The decrease in v_λ in the material is due to the oscillating electric field of the light causing forced harmonic oscillations of the electron shells of the atoms or ions composing the material. As the polarizability of the electron shells increases, v_λ decreases proportionally causing an equivalent increase in n_λ. When the electrons are in perfect resonance with the frequency of the incident photons absorption occurs, as discussed in the next section.

In order to understand refraction effects we use the model of a forced harmonic oscillator. First, in the absence of an external field, each electron of charge e and mass m can be considered bound to its equilibrium position with an elastic force constant k. Thus, the electron's motion about its equilibrium position will be defined as

$$-kx = m\ddot{x} \tag{8.15}$$

where x = displacement of mass m from its equilibrium position
 $\ddot{x}$ = acceleration
 k = force constant = $m\omega^2$ (ω = angular frequency)

Equation (8.15) has the solution

$$x = x_0 \cos 2\pi v_0 t \qquad (8.16)$$

where v_0 is the natural frequency of the electron vibrations, e.g.,

$$v_0 = \frac{1}{2\pi}\left(\frac{k}{m}\right)^{1/2} \qquad (8.17)$$

The electric field $\mathbf{E}$ of the incident light subjects the electron of charge e to a force $e\mathbf{E}$ which varies with frequency v as

$$\mathbf{E} = \mathbf{E}_0 \cos 2\pi v t \qquad (8.18)$$

thus changing equation (8.15) to

$$m\ddot{x} = -kx + e\mathbf{E}_0 \cos 2\pi v t \qquad (8.19)$$

When t is large, the solution of equation (8.18) is also:

$$x = x_0 \cos 2\pi v_0 t \qquad (8.20)$$

which when substituted into equation (8.18) yields

$$-m4\pi^2 v^2 x_0 \cos 2\pi v t = -kx_0 \cos 2\pi v t + e\mathbf{E}_0 \cos 2\pi v t \qquad (8.21)$$

Reducing equation (8.20) yields for the amplitude of the electron vibrations the following:

$$x_0 = \frac{e}{(k - 4\pi^2)mv^2}\mathbf{E}_0 \qquad (8.22)$$

and

$$x = \frac{e}{(k - 4\pi^2)mv^2}\mathbf{E}_0 \cos 2\pi v t \qquad (8.23)$$

By definition, if the displacement of the electrons x is multiplied by their electric charge e we obtain the induced dipole moment p:

$$p = ex = \frac{e^2}{(k - 4\pi^2)mv^2}\mathbf{E} = \alpha_e \mathbf{E} \qquad (8.24)$$

where α_e is the electronic polarizability.

By substitution of k from equation (8.17), the following expression for α_e

is obtained:

$$\alpha_e = \left(\frac{e^2}{4\pi^2 m}\right)\left(\frac{1}{v_0^2 - v^2}\right) = \frac{e^2}{m(\omega_0^2 - \omega^2)} \quad (8.25)$$

where $\omega_0 = 2\pi v_0$ and $\omega = 2\pi v$ \hfill (8.26)

Equation (8.25) shows that the polarizability of an electron is inversely proportional to the frequency of the incident photons, ω. For the limiting condition when the external field is constant, that is, $\lambda \to \infty$ and $v \to 0$, then the coefficient of static electronic polarizability ($\alpha_{e(0)}$) can be defined, yielding:

$$\alpha_{e(0)} = \frac{e^2}{4\pi^2 m v_0^2} = \frac{e^2}{k} \quad (8.27)$$

For an optical material there will be a very large number of electrons, each of which vibrate at a specific frequency, v_i. Consequently, for a multielectron system we can write equation (8.27) as a sum of all electron frequencies:

$$\alpha_e = \frac{e^2}{4\pi^2 m} \sum_i \frac{f_i}{(v_i^2 - v^2)} \quad (8.28)$$

where f_i represents the oscillator strength, or the extent by which a given electron participates in a particular vibrational frequency. The coefficient of static electronic polarizability will therefore be

$$\alpha_{e(0)} = \frac{e^2}{4\pi^2 m} \sum_i \frac{f_i}{v_i^2} = e^2 \sum_i \frac{f_i}{k_i} \quad (8.29)$$

Values of α_e for a number of ions important to photonic ceramics are given in Table 8.2, with units of 10^{-24} cm^3 (Debye). To turn them into SI units multiply by $(1/6) \times 10^{-15}$ (after Pauling).

Since, according to Maxwell's theory, the relationship

$$n^2_{\lambda \to \infty} = \epsilon \quad (8.30)$$

holds for infinitely long waves, then we can substitute equation (8.29) describing electronic polarizability and equation (8.30) into the Clausius–Mossotti equation (see equation 6.9 in Chapter 6):

$$\frac{N\alpha}{3\epsilon_0} = \frac{\epsilon - 1}{\epsilon + 2} = \frac{n^2 - 1}{n^2 + 2} = N \frac{e^2}{12\epsilon_0 \pi^2 m} \sum_i \frac{f_i}{v_i^2 - v^2} \quad (8.31)$$

where N is the number of electrons in the solid.

TABLE 8.2 Photonic Characteristics of Ionic Species

Species	Polarizability α_e (10^{-24} cm^3)	Ionic refraction R_i(cm^3)	Radius of ion coordination (nm)	Coordination number with respect to O^{2-} ion
Li^+	0.029	0.08	0.068	4, 6
Na^+	0.179	0.47	0.098	6, 8
K^+	0.83	2.24	0.133	6, 10, 12
Rb^+	1.40	3.75	0.149	10, 12
Cs^+	2.42	6.42	0.165	12
Be^{2+}	0.008	0.03	0.034	4
Mg^{2+}	0.094	0.26	0.074	4, 6
Ca^{2+}	0.47	1.39	0.104	6, 8
Sr^{2+}	0.86	2.56	0.120	6, 8, 12
Ba^{2+}	1.55	4.67	0.138	8, 12
B^{3+}	0.003	0.006	0.020	3, 4
Al^{3+}	0.052	0.14	0.57	4, 6
Si^{4+}	—	1.56	0.039	4
Ti^{4+}	0.185	0.6	0.064	4, 6
Zr^{4+}	0.37	1.1	0.082	6, 8
Ce^{4+}	0.73	—	—	—
O^{2-}	3.88	6.95	0.136	—
Cl^-	3.66	—	—	—
F^-	1.04	2.44	1.33	—
S^{2-}	—	22.7	0.182	—
Se^{2-}	—	28.8	0.193	—
OH^-	—	4.85	—	—

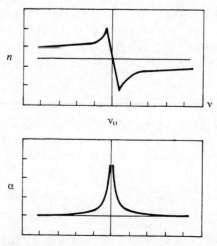

FIGURE 8.4 Incident light and electron resonance.

Equation 8.28, called the *dispersion equation,* describes the frequency dependence of the index of refraction and explains the *anomalous dispersion* that occurs at frequencies where absorption occurs. This resonance effect is shown in Figure 8.4.

Figure 8.4 shows that as the frequency of the incident radiation, v, approaches the natural vibrational frequency of the electrons in the material the denominator of equation (8.31) increases and the refractive index becomes smaller. When $v = v_0$, equation (8.31) predicts that n goes to infinity. This obviously cannot happen due to absorption of energy caused by resonance of the electron frequencies and the incident radiation. The net change in n and α is shown in Figure 8.4 near the region of an absorption band.

The extent of dispersion, that is, variation of index of refraction with wavelength, depends upon the composition of a glass or ceramic. For example, consider the dispersion of SiO_2-based glasses shown in Figure 8.5. At the UV end of the spectrum, the large increase in n is due to resonance with the electrons of the Si—O bonds, whereas in the IR region the large decrease of n is due to the onset of the molecular vibrations of Si—O bonds, as discussed in the next section. Compositions that have a broader distribution of bond energies have lower dispersion (see Figure 8.5).

The extent of dispersion is most commonly defined in terms of the index at two wavelengths, $n_e = 546.07$ nm and $n_d = 587.56$ nm, which can be used to calculate Abbé values, v_e and v_d, as follows:

$$v_e = \frac{n_e - 1}{n_{479.99} - n_{643.85}} = \frac{n_e - 1}{n'_F - n'_C} \tag{8.32}$$

and

$$v_d = \frac{n_d - 1}{n_{486.1} - n_{656.27}} = \frac{n_d - 1}{n_F - n_C} \tag{8.33}$$

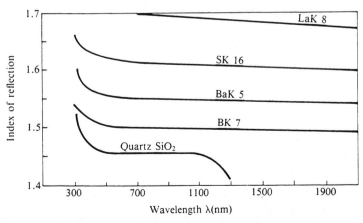

FIGURE 8.5 Dispersion curves of silica based optical glasses. (See Appendix A.)

The values of $n'_F - n'_C$ and $n_F - n_C$ are termed the *mean dispersion of optical glasses*. For other aspects of dispersion of optical materials see Fanderlik (1983), who also reviews the quantum-mechanical treatments of polarization, absorption, and dispersion.

As indicated in equation (8.31), refractive index depends upon composition of an optical material; the more polarizable the outer electrons, the higher is the refraction index. When equation (8.31) is expressed in terms of the specific mass or density ρ, it reduces to

$$R = \frac{n^2 - 1}{n^2 + 2}\left(\frac{1}{\rho}\right) \qquad (8.34)$$

and describes the specific refraction R of the material. The molar refraction, R_M, is

$$R_M = \frac{n^2 - 1}{n^2 + 2}\left(\frac{M}{\rho}\right) \qquad (8.35)$$

where M is the molecular weight of the material and M/ρ is the molar volume. Equations (8.34) and (8.35) are called the *Lorentz–Lorenz equations*.

R_M is proportional to polarizability α that is

$$R_M = \frac{4\pi N_L}{3}\alpha \qquad (8.36)$$

where N_L is Loschmidt's number (the number of molecules of an ideal gas at 0°C and normal atmospheric pressure):

$$N_L = 2.687 \times 10^{19}/cm^3 \qquad (8.37)$$

Consequently, the molar refraction R_M and index of refraction n depend on the polarizability of the material which in turn is determined, approximately, by the sum of ionic refractions R_i. The value of ionic refraction in crystal compounds depends upon at least four factors:

1. The electronic polarizability of the ion (see Table 8.2), which increases as the number of electrons in the ion increases.
2. The coordination number of the ion (see Table 8.2).
3. The polarizability of the first neighbor ions coordinated with it.
4. The field intensity z/a^2 where z is the valence of the ion and a is distance of separation.

8.4 REFRACTION

TABLE 8.3 Ionic Refraction of Silicates

Species	Ionic refraction R_i(cm^3)
Ba$_2$SiO$_4$	4.97
BaSiO$_3$	4.55
Ba$_2$Si$_3$O$_8$	4.30
BaSi$_2$O$_5$	4.10
Be$_2$SiO$_4$	3.35
Ca$_2$SiO$_4$	4.53
LiAlSiO$_4$	4.01
Mg$_2$SiO$_4$	3.83
NaAlSiO$_4$	4.03
SiO$_2$ (quartz)	3.55
Sr$_2$SiO$_4$	4.67

Table 8.2 lists values for the ionic refraction of various cations coordinated with O^{2-} ions. Table 8.3 lists R_i values for several silicate compounds. Note that the anionic refraction of O^{2-} differs considerably depending upon the electronic polarization and field intensity of the cations in the silicate crystals.

The variations in R_i shown in Table 8.3 for the different silicates are in part due to differences in molar volume as well as polarizability. As indicated in equation (8.35), when molar volume M/ρ decreases, and the material gets denser, the refractive index increases. For example, the values of n for SiO$_2$, shown on a Lorentz–Lorenz plot in Figure 8.6, vary from $n_{gel} = 1.28$ to $n_{glass} = 1.46$ to $n_{quartz} = 1.55$, due to the densities varying from 1.2 g/cm^3 to 2.65 g/cm^3. Thus, an ultraporous gel silica has a very low index and vitreous silica has a moderately low refractive index due to low electronic polarizabilities and a high specific molar volumes. An SiO$_2$ crystal with an open network, such as trydimite, has a lower density and a lower index than quartz with a tightly packed crystal structure.

When one substitutes alkali oxides (M$_2$O) for SiO$_2$ in alkali silicate glasses bridging Si—O—Si bonds are broken and nonbridging Si—O—M$^+$ bonds are formed as discussed in Chapter 1. The nonbridging oxygen (NBO) bonds have a much more ionic character and have much lower bond energies, as shown in Figure 8.7. Consequently, the NBO bonds have higher polarizabilities and cation refractions, and the index of refraction of the M$_2$O–SiO$_2$ glass increases proportionally (Figure 8.8). Multicomponent glasses generally behave in a similar manner, for example, n increases proportional to the increase in M$_2$O and MO content in M$_2$O–MO–SiO$_2$ glasses. Also, there is a general trend of n to increase with the value of R_i of the cations added.

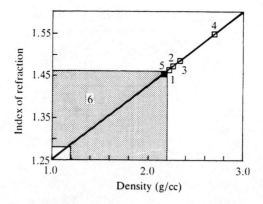

1 Corning 7940
2 Tridymite
3 Cristobalite
4 Quartz
5 Type V gel silica
6 Type VI porous gel silica

FIGURE 8.6 Lorentz–Lorenz plot.

However, calculations of the dependence of n or $dn/d\lambda$ on composition are complicated by the molar volume dependence of n (equation 8.35) discussed above. As shown in Figure 8.8, substitution of Li_2O for SiO_2 in a glass increases n even though Li^+ has a very small value of R_i. This is because Li_2O *decreases* the molar volume of the glass sufficiently to increase the index of refraction.

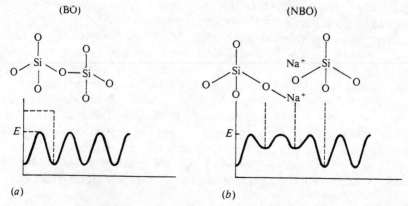

FIGURE 8.7 (*a*) Pure silica. (*b*) Silica plus Na_2O.

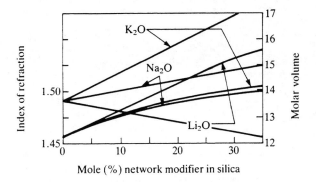

FIGURE 8.8 Index and molar volume as a function of mole% modifier.

The low polarizabilities and ionic refractivities of small nonoxygen anions such as the halides (F^- and Cl^-) and hydroxyls (OH^-), result in lower refractive indices and large Abbé numbers (reciprocal relative dispersion, ν_d) (Figure 8.9). In contrast, elements with high polarizabilities and ionic refractivities such as S^{2-}, Se^{2-}, Te^{2-}, and As^{3+}, result in high indices of refraction and low dispersion.

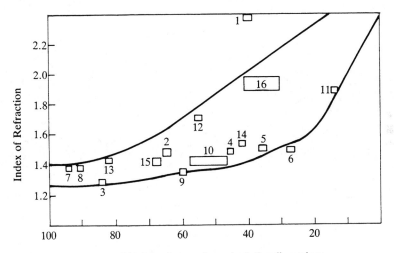

1 Diamond
2 Plexiglass
3 LiF
4 NaCl
5 KBr
6 O_6H_6
7 CaF_2
8 SrF_2
9 B_2O_3
10 Titanium glasses
11 LiI
12 MgO
13 BaF_2
14 KCl
15 SiO_2
16 Borate glasses

FIGURE 8.9 Dispersion of optical glasses and crystals.

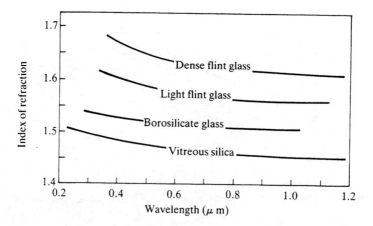

FIGURE 8.10 Dispersion of various silicate glasses.

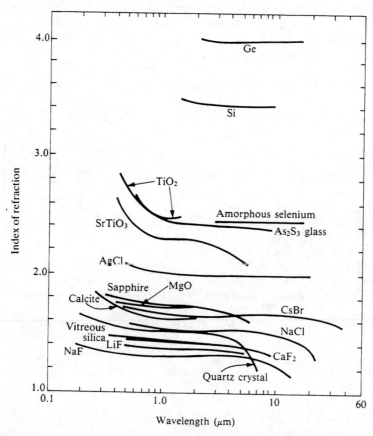

FIGURE 8.11 Change in refractive index with wavelength for several crystals and glasses.

Plots comparing the wavelength dependence of index of refraction of several glasses and crystalline ceramics are given in Figures 8.10 and 8.11. The wavelength dependence of the dispersion of various amorphous and crystal compounds is compared with elemental Si and Ge in Figure 8.12.

It is possible to use the physical–structural factors discussed above to calculate the index, mean dispersion, and Abbé values for glasses by assuming that the property is equal to the sum of the individual components in the material. However, such calculations are only valid for certain ranges of composition, due to the complex effects of specific molar volume, polarizability, and coordination numbers (see Fanderlik, 1983). For compositions and the effects of temperature, annealing, fictive temperature on optical properties of glasses, see Fanderlik, 1983 and Musikant, 1985.

8.5 SCATTERING

Scattering of photons can be considered as an attenuation mechanism. The effect of scattering occurs between the effects of reflection and refraction at interfaces and the effects of electronic absorption. The portion of the photon flux ϕ removed by scattering has been defined as ϕ_σ, the scattered flux. The coefficient of scattering σ is defined as (see Figure 8.1)

$$\sigma = \left(\frac{\phi_\sigma}{\phi}\right)_\lambda \tag{8.38}$$

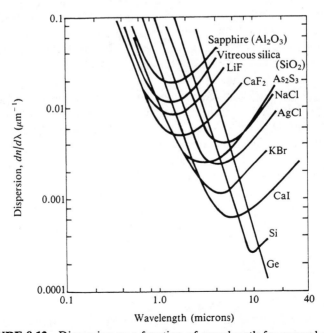

FIGURE 8.12 Dispersion as a function of wavelength for several ceramics.

The dependence of scattering on wavelength is rather strong. The blue sky we see during the day is the direct result of the scattering of sunlight by particles in the atmosphere. If the diameter of particles are of the same order of one-twentieth of the wavelength of light, then the scattering coefficient is a maximum. This is known as *Rayleigh scattering* and has the form

$$\sigma_R = a\left(\frac{d}{\lambda}\right)^4 \tag{8.39}$$

where a = specific scattering coefficient, (i.e., for H_2, N_2, or O_2)
d = particle diameter
λ = wavelength

There is also an angular dependence of scattering. The intensity of the scattered photons follows the relation:

$$I_S \approx (1 + \cos^2 \theta) \tag{8.40}$$

This gives a maximum at right angles to the incident radiation. This basically accounts for the blue sky during the daytime. At sunset, the sky reddens because the blue component has been scattered away and the red (longer wavelength) light is very weakly scattered.

If the momentum of the particle and the photon are approximately the same, then the scattering takes on an unusual form. The photon has sufficient momentum actually to make the particle recoil. Thus, momentum must be transferred and the scattered photon is now a different color. This is known as *Compton scattering*. It usually involves X-ray photons and "free" electrons. This effect is seen in *wide angle X-ray scattering* (WAXS) in crystals, glasses and ceramics. Selective filters are used in pairs such as ruthenium and rhodium to minimize this effect in WAXS analysis.

If the scattering particles are equal to or larger than the wavelength of the photon, then classical refraction and reflection takes place at the interfaces. This type of scattering is not strongly dependent upon wavelength. The general mathematical treatment was developed by Mie; therefore, scattering of spherical particles of any size is commonly referred to as *Mie scattering*. The scattering coefficient for this type of effect of effect is proportional to the cross-sectional area presented by the particles.

8.6 ABSORPTION

In addition to reflection, refraction, and scattering, part of the flux of incident photons (ϕ) may be absorbed (ϕ_α) with the remainder transmitted (ϕ_τ) or scattered (ϕ_σ). The absorption factor α and transmission factor τ at a given wavelength λ is defined as:

$$\alpha = \left(\frac{\phi_\alpha}{\phi}\right)_\lambda \tag{8.41}$$

and

$$\tau = \left(\frac{\phi_\tau}{\phi}\right)_\lambda \tag{8.42}$$

Both α and τ are dependent upon wavelength, angle of incidence, and polarization of the radiation.

When scattering is absent, the ratio of the emergent flux to the incident flux is termed the *internal transmittance*, τ_i and the ratio of the flux absorbed to the incident flux is called the *internal absorptance*, α_i.

The attenuation of the incident flux is proportional to the path traversed l and the concentration of absorbing centers c as described by the *Lambert–Beer law*:

$$\ln\left(\frac{\phi_\tau}{\phi}\right)_\lambda = -acl \tag{8.43}$$

or

$$\log\left(\frac{\phi_\tau}{\phi}\right)_\lambda = -0.434acl = \log \tau \tag{8.44}$$

where a is the linear absorption coefficient.

The optical density D is defined as the common logarithm of the reciprocal of the transmission factor:

$$D = -\log \tau \tag{8.45}$$

Thus, the linear absorption coefficient is

$$a = \frac{D}{0.4343cl} \tag{8.46}$$

The change in intensity, I, can be expressed as

$$dI = -aI\,dl \tag{8.47}$$

Rearranging we obtain

$$\frac{dI}{I} = -a\,dl \tag{8.48}$$

Integrating from the initial intensity, I_0, to the final intensity, I_f:

$$\ln\left(\frac{I_f}{I_0}\right) = -al \tag{8.49}$$

Finally, the intensity is attenuated exponentially

$$I_f = I_0 \exp(-al) \tag{8.50}$$

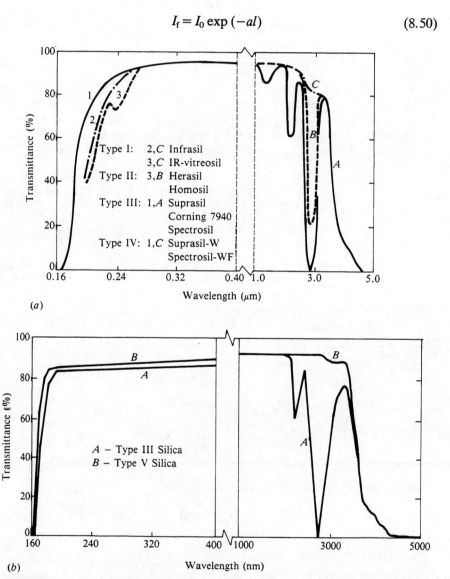

FIGURE 8.13 (a) Typical transmission curves of different types of optical silica. (b) Transmission spectra for types III and V silicas.

8.6 ABSORPTION

When the frequency of the incident radiation approaches resonance of the bonding electrons, as discussed above, absorption occurs. For ultrapure vitreous silica, the absorption edge is in the hard ultraviolet range of the spectrum, near 160 nm (Figure 8.13). Addition of M_2O, MO, or OH^- to SiO_2 increases the concentration of nonbridging oxygens and decreases the resonance frequencies. Consequently the absorption edge increases to longer wavelengths, for example, 300 nm for window glass which is near the visible (Figure 8.14). Cationic and hydroxyl impurities in vitreous quartz increases the absorption edge to 250 nm and also give rise to specific UV absorption losses (Figure 8.13a).

Chalcogenides, that is, those materials containing primarily S, Se, Te or similar elements have very much larger electronic polarizabilities than oxides. Consequently, a chalcogenide glass such as $Ge_{34}As_8Se_{58}$, is opaque to visible light and has an absorption edge shifted all the way to the infrared, ~20 nm. Halide glasses also have excellent IR transmission, as shown in Figure 8.15.

The excellent transmission of optical silica in the UV and visible makes it especially important for optical communications (see Chapter 9). The extent of transmission, however, depends upon the method of manufacture, as

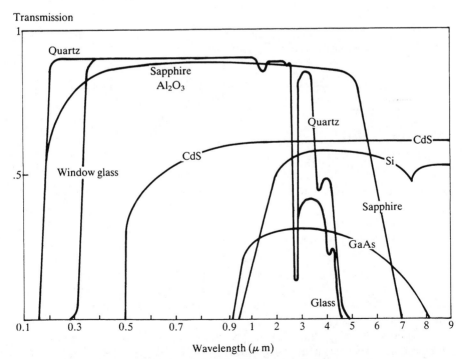

FIGURE 8.14 Spectral transmission.

TABLE 8.4 Types of Vitreous Silicas

Type	Method	Trade Name
	Fused quartz	
I	Electric melting of mineral quartz	Infrasil[c] Vitreosil IR[b] Pursil[c]
II	Flame fusion of mineral quartz	Optosil[a] Homosil[a] Vitreosil OH[b]
	Synthetic fused silica	
III	Vapor-phase hydrolysis of silicon compounds in a flame	Suprasil[a] Tetrasil[c] Spectrosil A, B[b] 7940[d]
IV	Oxidation of silicon compounds subsequently fused electrically or with a plasma, in an OH$^-$-free environment	Suprasil W[a] Spectrosil WF[b]
	Sol–gel silica	
V	Hydrolysis and condensation of silica alkoxide with subsequent dehydration and densification	GELSIL[e]
VI	Hydrolysis and condensation of silica alkoxide with subsequent partial densification and chemical stabilization	Porous GELSIL[e]

[a] Hereaus Amersil.
[b] Thermal Syndicate.
[c] Quartz et Silice.
[d] Corning Glass Works.
[e] GELTECH, Inc.

summarized in Table 8.4. The fused quartz methods of manufacture usually result in some degree of UV absorption (Figure 8.13), whereas the synthetic fused silica processes can lead to residual OH absorption (type III). The highest levels of transmission result from oxidation of silicon compounds such as $SiCl_4$ followed by electrical fusion in an OH$^-$ free environment (type IV) or by low temperature sol–gel processing followed by OH$^-$ free densification at 1150°C (type V). The type III and IV synthetic fused silicas are used extensively in fiber-optical communciations (see Chapter 9 and Midwinter, 1979).

One advantage of the type V sol–gel process, in addition to the high transmission from 160 nm to 4 nm shown in Figure 8.13, is the ease of casting near net shape optical components (Figure 8.16).

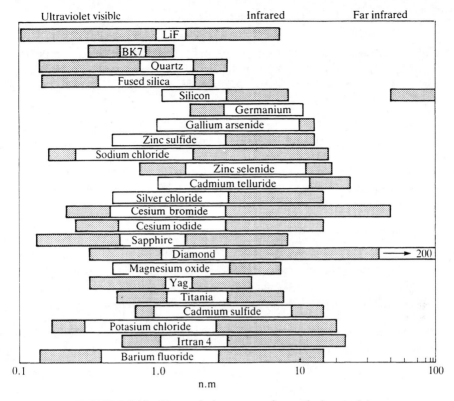

FIGURE 8.15 Transmission ranges for optical materials.

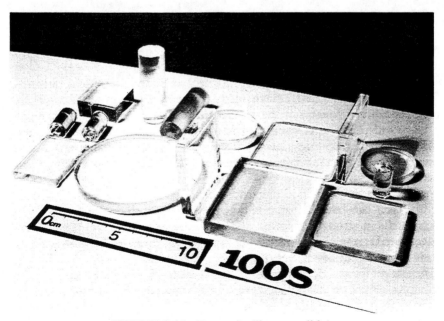

FIGURE 8.16 Pure gel–silica monoliths.

Another feature of the sol–gel silica process is the ability to dope the ultraporous components with elements or organic molecules of differing polarizabilities (for waveguides) or absorption bands (for optical filters or lasers) prior to densification.

8.7 ABSORPTION IN THE VISIBLE

A number of sources can give rise to absorption in the visible range of the EM spectrum, 0.4–0.8 nm, and thereby produce color centers. The four primary sources of color are:

1. Absorption of ions in ligand fields.
2. Absorption by colloidal particles of a second phase.
3. Absorption by microscopic second-phase particles.
4. Absorption by color centers produced by radiation (termed *solarization*).

We will consider each of these sources separately, although more than one may be present in an optical material.

8.8 IONS IN LIGAND FIELDS

Atoms which have filled outer shells, the rare gases, do not produce absorption bands in the visible nor do atoms with a single incomplete electron shell, such as Na, Ca, Ba, and so on. The ionic-covalent bonding of group I and II elements essentially leads to filled outer shells, so that electronic transitions in the range of 1.5–3.0 eV required to produce visible color centers are not possible.

In contrast, atoms with two incomplete shells, the transition elements (Ni, Mn, Cu, Co, Fe, etc.) have many absorption bands or lines in the visible. The electron states of the unfilled $3d$ orbitals are strongly influenced by their local chemical environment and bonding and thus absorption and color centers can vary considerably from material to material. Ligand field theory provides a quantitative explanation for these structure dependent spectral features.

Atoms with three unfilled electron shells, the rare earth elements (Nd, Er, Ce, etc.) can have transitions between levels in the unfilled $4f$ orbital which results in strong absorption and color centers. However, because the $4f$ levels are largely screened from changes during bonding the absorption is largely independent of structural factors.

The concentration dependence of the absorption of crystals and glasses containing either transition elements or rare earths is described by the Lambert–Beer law (equation 8.44). For example, doubling the concentration of the dopant only increases the optical density by 0.3.

8.9 INTRAIONIC ABSORPTION AND LIGAND FIELD THEORY

Table 8.5 summarizes the transitions and hosts for a number of transition elements. The notation used is known as the *Russell–Saunders notation for incomplete d free-ion transitions*:

$$^{(2S+1)}L_J \tag{8.51}$$

where S = net spin
  J = total angular momentum
  L = orbital angular momentum

and the designations for the orbital angular momentum are:

$$L = 0 = S \text{ or } A \tag{8.52}$$

$$L = 1 = P \text{ or } T \tag{8.53}$$

$$L = 2 = D \tag{8.54}$$

$$L = 3 = F \tag{8.55}$$

$$L = 4 = G \tag{8.56}$$

$$L = 5 = H \tag{8.57}$$

$$L = 6 = I \tag{8.58}$$

$$L = 7 = K \tag{8.59}$$

$$L = 8 = L \tag{8.60}$$

TABLE 8.5 Ligand Field Transitions

Ion	Host	Transition	Wavelength (μm)
Cr^{3+}	$BeAl_2O_4$	$^4T_2 \rightarrow {}^4A_2$	0.74–0.788
Co^{2+}	MgF_2	$^4T_2 \rightarrow {}^4T_1$	1.63–2.11
Ni^{2+}	MgF_2	$^3T_2 \rightarrow {}^3A_2$	1.62–1.79
Ni^{2+}	MgO	$^3T_2 \rightarrow {}^3A_2$	1.31–1.41
Ce^{3+}	LiF_4	$5d \rightarrow {}^2F_{5/2}$	0.306–0.315
Nd^{3+}	$Y_3Al_5O_{12}$	$^4F_{3/2} \rightarrow {}^4I_{9/2}$	0.891
Cr^{3+}	Al_2O_3	$^2E(\bar{E}) \rightarrow {}^4A_2$	0.6943(R1)
Cr^{3+}	Al_2O_3	$^2E(2\bar{A}) \rightarrow {}^4A_2$	0.6929(R2)
Er^{3+}	$Y_3Al_5O_{12}$	$^4S_{3/2} \rightarrow {}^4I_{13/2}$	0.863

Ligand field theory, which is a special case of the more general molecular orbital theory, is an alternative to crystal field theory to explain color formation in glasses and crystals. In crystal field theory, bonding is treated as electro-static, derived from the electric field of the ligands viewed as purely ionic species. Thus, in a crystal field model, the chemical compound of a transition metal ion is considered as an aggregate of ions and/or dipolar molecules which symmetrically interact with each other electrostatically, but do not exchange electrons. Consequently, when any covalency is involved, a pure crystal field theory cannot explain the experimental data.

The advantage of the ligand field theory is the mixing between the d electrons of the central ion and the ligands. This feature of mixed ionic covalence successfully explains the coloring phenomena for most situations involving transition elements. In this theory, a ligand presents a negatively charged, nonspherical, partially covalent-bonded, distorted coordination complex towards the positive central transition-metal ion. If the ligand has spherical symmetry (such as an s orbital) then there is no interaction, and the d electron energies are not changed.

For example, in a free ion of a transition metal, the five equivalent d orbitals are depicted spatially as shown in Figure 8.17. The energy level diagram of the five orbitals in a free transition metal ion is also illustrated in Figure 8.18a. The electrons can be found with equal probability in any of these five orbitals (Figures 8.17, 8.18a) and they all have the same energy.

However, place this positive transition metal ion with partly filled d orbitals at the center of a regular (undistorted crystal field) octahedron of six ligands, as shown in Figure 8.19a. An interaction of the five d orbitals of the central ion, Figure 8.17, with the six ligands in the octahedral field will occur. The two lobes of the d_{z^2} orbital point exactly at the two ligands in the $+z$ and $-z$ directions, as indicated in Figure 8.19. Similarly, the four lobes of the $d_{x^2-y^2}$ orbital point exactly at the four ligands in the plus and minus directions of the x and y axes, also shown in Figure 8.19. The electrostatic interaction of the combined d_{z^2} and $d_{x^2-y^2}$ electron orbitals with the negative ligands at the corners of the octahedron is repulsive. Consequently, there is a splitting and raising of the energy of the d_{z^2} and $d_{x^2-y^2}$ orbitals with respect to the three other orbitals of the system, d_{xy}, d_{yz}, and d_{zx}. This leads to the the upper e_g levels of d_{z^2}, $d_{x^2-y^2}$ for the octahedral ligand field configuration as shown in Figure 8.19b.

The remaining three sets of d orbitals of Figure 8.17, the d_{xy}, d_{yz}, and d_{zx} orbitals, have orientations with lobes that protrude halfway between the ligands, as shown in Figure 8.19b. Because there are no repulsive interactions with the ligands, these orbitals will have lower energy levels than the e_g set. Thus, the d_{xy}, d_{yz}, and d_{zx} orbitals are shown as the three equal energy t_{2g} levels in Figure 8.18b.

The energy difference between the e_g and t_{2g} levels, called the *crystal field splitting*, is designated Δ_o. The overall energy cannot be changed. The upward and downward energy levels are inversely proportional to the

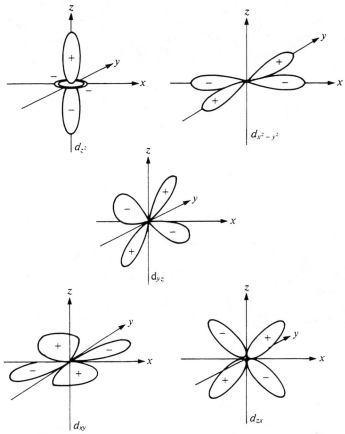

FIGURE 8.17 Electron distribution shapes of the five equivalent d orbitals.

number of equal energy levels (e.g., the degeneracy). Thus in Figure 8.18b, the triply degenerate lower energy t_{2g} level moves down $0.4\Delta_o$, while the upper doubly degenerate e_g level moves up to $0.6\Delta_o$ compared with the unsplit free ion levels (Figure 8.18a). The weighted mean energy must remain unchanged:

$$(2)(0.6\Delta_o) - (3)(0.4\Delta_o) = 0 \tag{8.61}$$

When the transition-metal is in tetrahedral symmetry, as shown in Figure 8.20a and b, the situation is reversed. The lobes of the $d_{x^2-y^2}$ or d_{z^2} oribtals now lie in the direction between the ligands, while the lobes of d_{xy}, d_{yz}, and d_{zx} orbitals, though not pointing directly towards the ligands, lie closer to them. Thus the t_{2g} (d_{xy}, d_{yz}, d_{zx}) orbitals are destabilized with respect to the e_g orbitals. For the same strength ligands, the tetrahedral scheme Δ_t can be

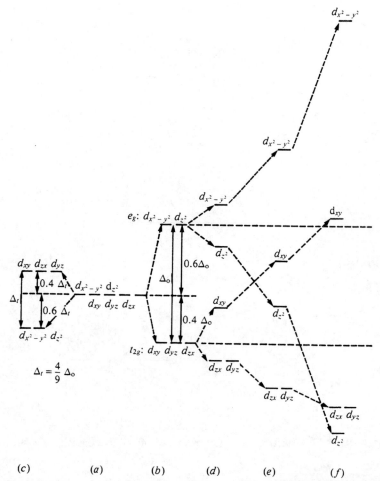

(c)　　　(a)　　(b)　　(d)　　(e)　　　(f)

FIGURE 8.18 Splitting of the five d orbitals in different ligand fields: (a) in a free ion; (b) in an octahedral field; (c) in a tetrahedral field; (d) in a tetragonally distorted octahedral field; (e) as (d), but relatively strongly distorted; (f) in a square planar ligand field.

related to the Δ_o (octahedral) value of the degenerate orbitals:

$$\Delta_t = -\tfrac{4}{9}\Delta_o \qquad (8.62)$$

Relative energy levels for the degenerate tetrahedral orbitals are shown in Figure 8.18c.

The negative sign in equation (8.62) indicates the reversal of the split orbitals in the tetrahedral vs. octahedral configurations.

Practically, octahedral arrangements of the ligands around transition metal ions are often tetragonally distorted. In such a case, the two $+z$ and

8.9 INTRAIONIC ABSORPTION AND LIGAND FIELD THEORY

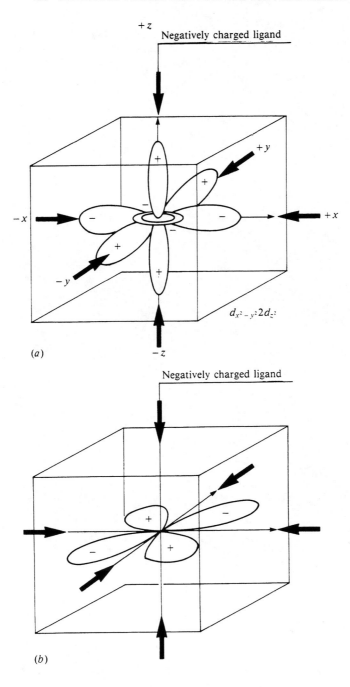

FIGURE 8.19 (a) Head-on interaction of the d_{z^2} and $d_{x^2-y^2}$ orbitals of a central ion with six ligands in an octahedral field. (b) Less interaction of the $d_{xy} d_{yz} d_{zx}$ orbitals of a central ion with six ligands in an octahedral field.

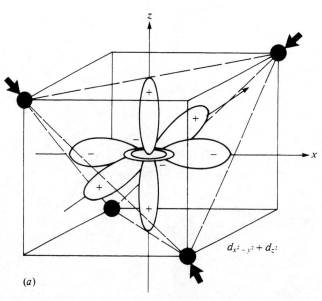

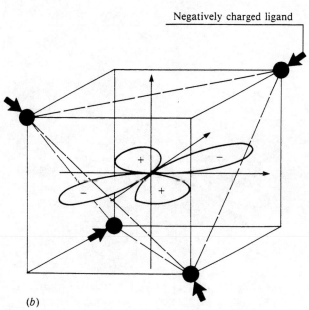

FIGURE 8.20 (a) Interaction of the d_{z^2} and $d_{x^2-y^2}$ orbitals of a central ion with four ligands in a tetrahedral field. (b) Interaction of one of the $d_{xy}d_{yz}d_{zx}$ orbitals of a central ion with four ligands in a tetrahedral field.

$-z$ (d_{z^2}) ligands in Figure 8.19a are gradually moving away from the central transition metal ion and new energy differences arise among the d orbitals. The d_{z^2} level will fall and the $d_{x^2-y^2}$ level will rise equally at the same time. If the two z ligands are completely removed, the d_{z^2} level becomes the lowest energy level in the resulting square planar ligand arrangement, since there are now no energy-raising ligands in that direction and the $d_{x^2-y^2}$ becomes the highest energy level (Figure 8.18f). For a square planar ligand field the location of the d_{yz} and d_{zx} levels will fall and that of the d_{xy} level must rise two times as much. The frequently observed tetragonally distorted octahedral arrangements are shown in Figure 8.18d and e.

Consequently, the kind of energy level arrangement formed by the ligand fields depends on three crucial properties:

1. The orbitals of the central transition metal ion.
2. The surrounding arrangement of ligand fields.
3. The strength of the ligand fields.

For ligand fields involved with individual orbitals of an atom, lower case notations such as a_{1g}, b, e_{1g} are used. Either a or b indicates a nondegenerate orbital with a representing a wavefunction which is symmetric (+) with respect to the rotation axis, whereas b represents a wavefunction which is antisymmetric (−) and changes sign during rotation. The sign in the figures represents symmetric (+) or antisymmetric (−) wavefunctions. The e and t orbitals are symmetrically doubly and triply degenerate. The energy levels within an e level or a t level orbital are equal.

Use of the subscript g designates the presence of a change in sign of the wavefunction on inversion through a center of symmetry. A subscript "1" refers to the presence of mirror planes parallel to the symmetry axis, and a subscript "2" refers to mirror planes normal to this axis. Upper-case designations such as $^2A_{2g}$, $^1B_1$, ^2E_g, $^2T_{2g}$ are generally used to represent the energy levels in the atom, ion, or molecule, with the prefix superscript indicating the $(2S+1)$ multiplicity. The energy states that can accommodate undisturbed or excited electrons in free transition-metal ions having incomplete d orbitals (d^1 to d^9) based on Russell–Saunders coupling were defined earlier.

In a ligand field, the tetrahedral d^9 or d^4 configuration can be viewed as containing one hole; that is, one electron missing from a full d or half-full d shell. This configuration provides a strong analogy with one electron added to an octahedral empty d shell (d^1) or a half-filled d shell (d^6) and, conversely, so do the octahedral d^9 or d^4 and the tetrahedral d^1 or d^6 configurations, as shown in Figure 8.21, except that the highest rather than the lowest split orbital is being occupied. The same applies to all other configurations. For example, the tetrahedral d^2 or d^7 configuration has the same sequence of levels as the octahedral d^8 or d^3 and, conversely, so do the octahedral d^2 or d^7 and the tetrahedral d^8 or d^3 configurations. These

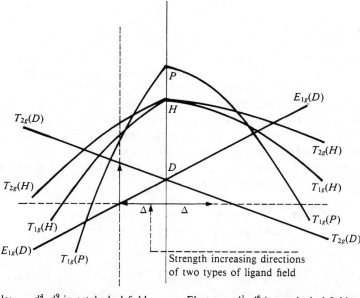

FIGURE 8.21 Three lowest energy levels for d^1, d^6, d^4, d^9 splitting configurations in octahedral and tetrahedral fields.

similarities are shown in Figure 8.22. Consequently, the splitting scheme of d^n (octahedral) configurations is equivalent to that of $d^{(10-n)}$ (tetrahedral) or vice versa. The d^0 and d^{10} configurations corresponding to completely empty or completely full d orbitals cannot show color directly derived from d electronic transitions.

An s^1 orbital is completely symmetrical and hence is unaffected by ligands in an octahedral field such as ^2S or A_{1g}. The p^1 orbitals are not split by octahedral fields such as ^2P or $^2T_{1g}$, since all interact equally as illustrated in Figure 8.23a and b. In an octahedral field the d^1 and $d^9(^2D)$ orbitals, as discussed before, split into $T_{2g}(d_{xy}, d_{yz}, d_{zx})$ and $E_g(d_{x^2-y^2}$ or $d_{z^2})$ levels. The f^1 orbitals are split into three levels in an octahedral field: a $^2T_{1g}$ level at $\frac{1}{3}\Delta$ below, a $^2T_{2g}$ level at $\frac{1}{9}\Delta$ above, and $^2A_{2g}$ level $\frac{2}{3}\Delta$ above the presplit F orbital, as shown in Figure 8.24a.

The two low-energy states ^3F and ^3P, split from either the d^2 or the d^8 configuration, behave in an octahedral field exactly as the F and P states arising from the ^1f and ^1p levels as discussed above. Consequently, the ^3F state is split into $^3T_{1g}(F)$, $^3T_{2g}(F)$, and $^3A_{2g}(F)$ states and the unsplit ^3P becomes the $^3T_{1g}(P)$ state. The d^3 and d^7 states, have ^4P and ^4F orbitals. Under a ligand field, the ^4F splits into $^4T_{1g}(F)$, $^4T_{2g}(F)$, and $^4A_{2g}(F)$ states and the unsplit ^4P becomes the $^4T_1(P)$ state, as shown in Figures 8.22 and

8.9 INTRAIONIC ABSORPTION AND LIGAND FIELD THEORY

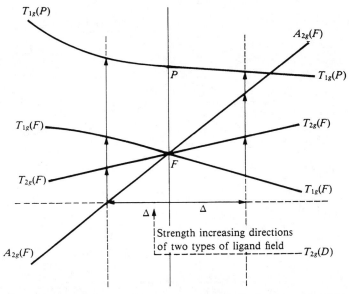

FIGURE 8.22 Two lowest energy levels for d^2, d^3, d^7, d^8 splitting configurations in octahedral and tetrahedral fields.

8.24b. The d^4 and d^6 configurations have a low-energy state ^5D which splits into $^5T_{2g}$ (d_{xy}, d_{yz}, d_{zx}) and ^5E_g ($d_{x^2-y^2}$ or d_{z^2}) in an octahedral field (Figures 8.21 and 8.24c). The d_5 state has an unsplit ^6S or $^6A_1$ level in the octahedral field.

Ligand field theory describes the bonding occurring between center transition-metal ions and the ligands. Molecular orbital theory, which develops the combination of the atomic orbitals of the atoms to form the molecular structure, is used to explain this bonding phenomenon.

The condition for two atoms to form: (1) a bonding molecular orbital ψ_b, (2) a nonbonding molecular orbital, or (3) an antibonding molecular orbital ψ_a, depends on S, the wavefunction overlap integral $\int \psi_A \psi_B \, dt$ of the probability equation

$$\psi_{AB}^2 \, dt = \int \psi_A^2 \, dt + \int \psi_B^2 \, dt + \int \psi_A \psi_B \, dt \tag{8.63}$$

where ψ_A and ψ_B, the wavefunctions of atoms A and B, can be used to describe an electron within a given space (see Chapter 2). Bonding takes place only when the value of S is positive ($S > 0$). The bonding strength (energy) is proportional to the extent of the overlap of the atomic orbitals,

368 CHAPTER 8: PHOTONIC CERAMICS

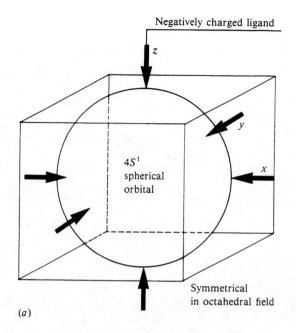

(a)

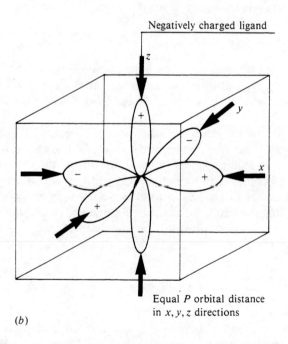

(b)

FIGURE 8.23 (a) All interactions between ligands and $4S^1$ are equal, therefore, no splitting results. (b) All interactions between ligands and $4P^1$ are equal, therefore, no splitting results.

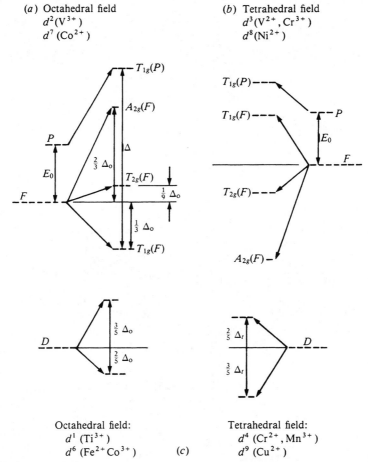

FIGURE 8.24 The splitting of d oribtals (a), (b) for P, F states, (c) for D state in octahedral and tetrahedral ligand field.

and the magnitude of S. The bonding energy level is reduced relative to the level of the free atoms by the same amount as the energy is increased for the antibonding level.

In the case of $3d$ transition-metal ions in an octahedral field, the $d_{x^2-y^2}$ and d_{z^2} configurations are in the direction of the ligands. This results in a positive overlap and a reduction in energy for bonding levels (e_g) and an increase in antibonding (e_g^*) energy levels. The T_{2g} (d_{xy}, d_{yz}, d_{zx}) levels are located between the ligands, and no overlap is observed. As a result, the energy levels of t_{2g} remain unchanged in the presence of the ligand field, as shown in Figure 8.25c.

The $4s$ orbital of transition-metal ions has an A_{1g} spherical symmetry and a corresponding ligand group orbital that is composed of sigma bonds which are cylindrically symmetrical about the internuclear axis (Figure 8.26). The

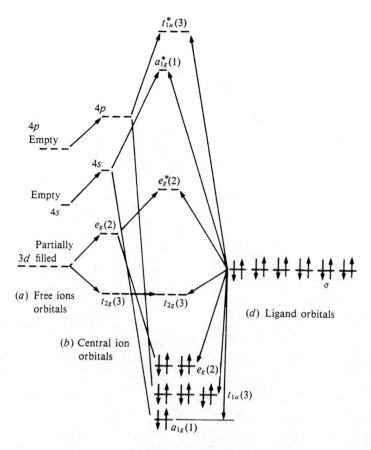

FIGURE 8.25 Molecular orbital splitting levels for a d orbital ion in an octahedral environment with ligands having only sigma bonds.

$4p$ (t_{1u} bonding and t_{1u}^* antibonding) orbitals with the related ligand group orbitals are shown in Figure 8.27.

The 12 electrons from the ligand group orbitals are perfectly paired into three lowest energy molecular orbitals which are $a_{1g}(1)$, $t_{1u}(3)$, and $e_g(2)$, as shown by the vertical arrows in Figure 8.25c. Consequently, the d electrons from the transition-metal ions have to fill the $t_{2g}(3)$ levels first. If there are any remaining electrons, then the $e_g^*(2)$ levels are available. If the energy gap, Δ, is greater than kT, low-spin configurations will be formed.

The low-spin molecular orbitals indicate that the electron pairing energy is less than the split energy levels caused by the ligand. In the case of cobalt with an orthogonally coordinated ligand, the electrons fill the lower three orbitals in pairs (Figure 8.18b). This leaves the electrons in the upper two levels unpaired, and thereby they are called *low-spin levels*.

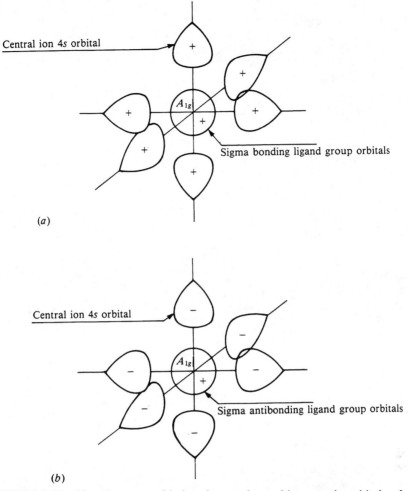

FIGURE 8.26 Ligand group orbital and central matching atomic orbitals of the bonding symmetry: (a) bonding and (b) antibonding.

If a ligand of cobalt was tetrahedrally coordinated then a high-spin state exists. The pairing energy is greater than that of the orbitals because Δ is less than kT. As a result, electrons fill all five states (two lower and three upper) before pairing and filling the two lower states. This leaves three electrons unpaired giving the term *high-spin* coordination or state (see Figure 8.19d).

The gap energy Δ is generally in the visible range of 1 eV (NIR) to 3 eV (UV) energy. If electromagnetic radiation has the same amount of energy as the gap energy, (i.e., $\Delta = h\nu$, photon energy), then the electronic transitions from $t_{2g}(3)$ to $e_g^*(2)$ levels may take place.

These theories of ligand field and molecular orbital transitions are the

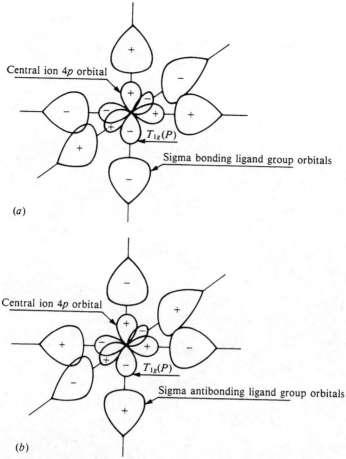

FIGURE 8.27 Ligand group orbital of only sigma bonds and matching atomic orbitals to form molecular orbitals.

basis for interpreting the optical spectra of glasses and crystals containing transition-metal ions, electron population inversion in lasers (Chapter 11), photoluminescence (Chapter 10), and so on. As examples, we will discuss silica gels doped with Co^{2+}, Ni^{2+}, and Cu^{2+} ions and compare them with melt-derived silicate and phosphate glasses. The possible energy level transitions for the Co^{2+} ion in both d^7 octahedral and tetrahedral symmetries are analyzed by means of Figures 8.22a and b. The octahedral Co^{2+} configuration is predicted to have three spin-allowed transitions:

$$^4T_1(F) \rightarrow {}^4T_2(F) \tag{8.64}$$

$$^4T_1(F) \rightarrow {}^4T_1(P) \tag{8.65}$$

$$^4T_1(F) \rightarrow {}^4A_2(F) \tag{8.66}$$

The tetrahedral Co^{2+} configuration is expected to have three major transitions:

$$^4A_2(F) \rightarrow {}^4T_2(F) \tag{8.67}$$

$$^4A_1(F) \rightarrow {}^4T_1(F) \tag{8.68}$$

$$^4A_2(F) \rightarrow {}^4T_1(P) \tag{8.69}$$

The energy level diagram for the Ni^{2+} ion in d^8 tetrahedral and octahedral symmetries is also depicted in Figure 8.22a and b. The tetrahedral Ni^{2+} configuration has three spin-allowed transitions:

$$^3T_1(F) \rightarrow {}^3T_2(F) \tag{8.70}$$

$$^3T_1(F) \rightarrow {}^3A_2(F) \tag{8.71}$$

$$^3T_1(F) \rightarrow {}^3T_1(P) \tag{8.72}$$

The octahedral Ni^{2+} symmetry results in three major transitions:

$$^3A_2(F) \rightarrow {}^3T_2(F) \tag{8.73}$$

$$^3A_1(F) \rightarrow {}^3T_1(F) \tag{8.74}$$

$$^3A_2(F) \rightarrow {}^3T_1(P) \tag{8.75}$$

The Cu^{2+} ion has a $3d^9$ configuration, an inverted d^1 configuration, as shown in Figure 8.21. The major absorption transition is attributed to

$$^2E \rightarrow {}^2T_2 \tag{8.76}$$

as shown in Figure 8.21a.

Silica-gel samples containing 0.25% Co are reddish pink when heated to 160°C, whereas the color of 850°C Co-doped silica gels are deep blue, and Co doped 900°C gels have a greenish black color. The UV–visible spectra characteristic of these Co^{2+}-doped silica-gel samples are shown in Figure 8.28 along with Co-doped phosphate, soda-borate and silicate glasses. The color and absorption spectra of the Co-doped materials depends on the oxidation state and coordination number of the ion. The low-temperature gel shows evidence of a sixfold CN similar to that for Co^{2+} in the metaphosphate and 10 mole% soda-borate glasses, whereas the Co^{2+}-doped silicate glasses and the high-temperature gels have a spectrum and color characteristic of tetrahedrally symmetrical ligands (see Wang, 1988).

The major absorption band of the octahedrally coordinated Co^{2+} ion is due to the $^4T_1(F)$ to $^4T_1(P)$ transition (8.53). The high-energy shoulder at 470 nm is a consequence of spin–orbit coupling in the $^4T_1(P)$ state. The $^4T_1(F)$ to $^4T_2(F)$ transition (8.52) occurs in the infrared region around

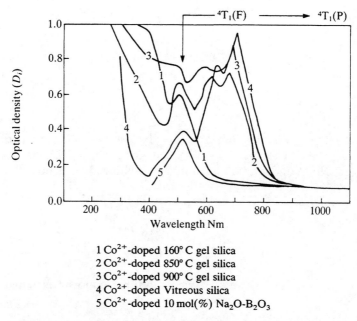

1 Co^{2+}-doped 160° C gel silica
2 Co^{2+}-doped 850° C gel silica
3 Co^{2+}-doped 900° C gel silica
4 Co^{2+}-doped Vitreous silica
5 Co^{2+}-doped 10 mol(%) Na_2O-B_2O_3

FIGURE 8.28 Spectra of Co^{2+}-doped silica (after Wang, 1988).

1250 nm and does not contribute to color formation. The $^4T_1(F)$ to $^4A_2(F)$ transition (8.54) is expected to be at 555 nm. However, this transition is very weak because it involves the forbidden two-electron jump. This weakness combined with the closeness of the major $^4T_1(F)$ to $^4T_1(P)$ transition makes the $^4T_1(F)$ to $^4A_2(F)$ transition unresolved.

The main absorption band in the 550 nm to 700 nm range of the tetrahedrally coordinated Co^{2+} ion is due to the $^4A_2(F)$ to $^4T_1(P)$ transition. The splitting of the $^4A_2(F)$ to $^4T_1(P)$ band (see Figure 8.28), is caused by spin-orbit coupling which splits the $^4T_1(P)$ states and allows the transitions to the neighboring doublet states to gain in intensity. The two other transitions, $^4A_2(F)$ to $^4T_2(F)$, and $^4A_2(F)$ to $^4T_1(F)$ which take place in the infrared region contribute no color chromophores. None of the spectra for the Co^{2+}-doped silica gels is identical to the silicate, borate, or phosphate melt glass spectra in detail. This is because the ligand field strength Δ is varied by the thermal history and density and therefore the specific molar volume of the gels. The OH^- ion is a strong ligand that will contribute to these differences.

The spectrum of a 160°C Ni^{2+}-doped silica gel is similar to that of a 16.2 weight% melt K_2O-borate glass containing Ni ions. It is also similar to that of a $[Ni(H_2O)_6]^{2+}$ octahedral complex in water, as shown in Figure 8.29. The absorption band at 700 nm of Ni^{2+} in an octahedral complex is

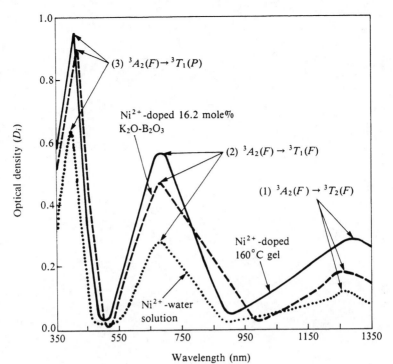

FIGURE 8.29 Absorption spectra of a Ni^{2+}-doped silica–gel sample, a Ni^{2+} water solution and a Ni^{2+}-doped 16.2 mole% K_2O-B_2O_3 glass.

assigned to the

$$^3A_1(F) \rightarrow {}^3T_1(F) \qquad (8.77)$$

transition, and the one at about 400 nm is assigned to the

$$^3A_2(F) \rightarrow {}^3T_1(P) \qquad (8.78)$$

transition. Another band corresponding to a

$$^3A_2(F) \rightarrow {}^3T_2(F) \qquad (8.79)$$

transition is observed in the infrared region at about 1180 nm. In this study, the spectra of these three samples are almost the same, except for the difference in absorption intensity. The similarity in absorption bands of the three curves indicates that the same ligand field strength acts on the Ni^{2+} ion in these three samples.

The absorption spectra of Cu^{2+} in a 160°C gel and three binary sodium boarate Cu^{2+} melt glasses are shown in Figure 8.30. All the absorption spectra consist of a broad band with a maximum at about 780 nm. This

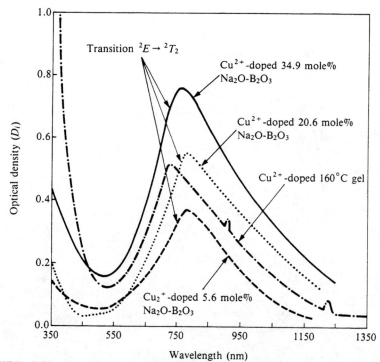

FIGURE 8.30 Absorption spectra of Cu^{2+}-doped silica–gal sample and three Cu^{2+}-doped sodium borate glasses.

absorption is attributed to the transition from ^2E levels to $^2T_2$ levels. The band is asymmetric and departs from Gaussian symmetry, since the $^2T_2$ levels are split by a distorted low-symmetry ligand field.

8.10 OPTICAL FILTERS

Absorption of specific wavelengths is used to filter portions of the optical spectrum. There are many different types of optical filters currently available. The three most common classifications are *neutral filters*, *polarizers*, and *color filters*. Neutral filters are filters that transmit equally across a broad bandwidth, and appear brown or gray in color. They are used to reduce incident intensity without altering the relative intensities of the incident frequencies. Polarizers are used to filter out photons of a given polarization or orientation. Color filters are used to transmit selectively light of a certain frequency or bandwidth with a minimum of attenuation.

Neutral filters can attenuate the light by reflection, absorption, scattering, polarization, or a combination of these methods. Generally any filter using scattering cannot be used in image-forming applications. Reflective filters, and some polarizing filters, offer the advantage of reducing the amount of heating of the filter.

Polarizing filters typically make use of birefringent materials (e.g., $CaCO_3$); other polarizers include diffraction gratings and imbedded wire mesh. These filters cause light of one orientation to be reflected, while the rest of the light is transmitted. The directional alignment of polymers can also be used for polarizing filters. Polarizers can be used singly or in stacked sets. The transmission can be calculated as follows:

$$\tau_0 = 0.5(\tau_{max} + \tau_{min}) \quad \text{for a single filter} \qquad (8.80)$$

$$\tau_0 = 0.5(\tau_{max}^2 + \tau_{min}^2) \quad \text{for two parallel filters} \qquad (8.81)$$

$$\tau_{90} = 0.5(\tau_{max} \times \tau_{min}) \quad \text{for two crossed filters} \qquad (8.82)$$

where τ_{max} is the transmittance through the unpolarized orientation and τ_{min} is the transmittance of the photons orthogonal to the direction of desired transmittance. Typical values for τ_{max} range from 0.60–0.85, while τ_{min} is usually less than 0.001. Polarizers can perform as neutral density filters or as color filters, where τ_{min} is a function of the wavelength λ.

Color filters are the most widely used filters, and are found in a variety of classifications. Short-pass filters transmit at lower wavelength, long-pass filters at higher wavelengths and band-pass filters absorb light at wavelengths above and below a given band. In many cases, band-pass filters are actually manufactured by combining the appropriate short-pass and long-pass filters which yield the desired transmission spectra. Color filters can be made from several materials including glass, gelatin, and plastics or acetates.

Glass filters are the most widely used filters and offer several advantages over plastic and gelatin. The homogeneity and transparency offered by the glass matrix makes them suitable for image-forming beams. However, due to the thickness required and relatively high index of refraction care must be taken in order to prevent a shift in focus in many optical components. Glass filters are relatively durable and impervious to many environments such as moderate heat and moisture.

On the other hand, glass filters are relatively expensive and the high processing temperatures required prevent the use of many compounds as the absorbing media. Polymers and organic compounds can be used to produce absorption profiles which are not possible using classical ionic absorbants. The high processing temperatures used in manufacturing glass filters can destroy many of these new colorants. Sol–gel optics chemically doped with either inorganic elements like Co, Ni, or Cu, such as discussed in Section 8.8 on ligand fields, or organic polymers such as shown in Table 8.6, provide an exciting new alternative.

TABLE 8.6 Organic Dopants in Porous Gel-Silica Optics

Organic polymer	Function
PBT (phenylenebenzobisthiazole)	NLO
MNA (2-methyl-4-Nitroaniline)	NLO
B-PBD (2-(4'-t-Butylphenyl)-5-(4"-Biphenylyl)-1,3,4-Oxadiazole)	Fluorescence
P-TP-p-Terphenyl)	Fluorescence
P-TP(P-Quaterphenyl)	Fluorescence
3-HF(3-Hydroxyflavone)	Wavelength shifter

(After Hench, Wang, and Nogues, 1988.)

8.11 POLARIZATION

Light is composed of electro-magnetic waves which oscillate in directions perpendicular to the direction of propagation of the light. Normally, the orientation of these waves about the propagation direction is random. However, in some circumstances, these oscillations become ordered in time. This is called *polarization*. Normal light is consequently called *unpolarized*.

There are several different types of polarization which can occur: *linear, circular, elliptical,* and *partial*. Linear polarization occurs when the electro-magnetic waves always have the same orientation with the direction of propagation. A common way of representing ordinary, unpolarized light is to draw several arrows of equal length radiating out from a common point (see Figure 8.31a). Each arrow represents one plane of oscillation in which the electro-magnetic waves can move, with the length of the arrow indicating the relative amount of energy in the light when the oscillations are along that direction. This arrow represents the electronic field vector **E**.

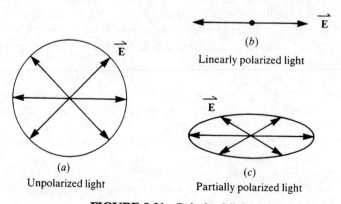

FIGURE 8.31 Polarized light.

Linear polarized light can be drawn in the same representation by a single, double-headed arrow as shown in Figure 8.31b. The diagram shows that all of the energy in the light is contained in oscillations along that single plane. Partially polarized light occurs when there is a mixture of unpolarized light and polarized light. Partial linear polarization can be drawn as in Figure 8.31c. As shown in the diagram, most of the energy in the light is along the polarized direction, but there is still some energy being carried by waves oscillating in different directions.

Circular polarization is a condition wherein the plane in which the electro-magnetic waves oscillate rotates about the direction of propagation of the light beam. A circularly polarized beam can be either right-polarized or left-polarized, depending on the direction of rotation of the electro-magnetic oscillations.

Elliptical polarization occurs when one particular angle is preferred over the others for the transmission of energy. Like circular polarization, elliptical polarization can be either right- or left-polarized. Figure 8.32 shows how elliptical polarization can be represented and how the degree of ellipticity can vary.

Both linear polarization and circular polarization can be considered extremes of elliptical polarization, that is, linear polarization is an ellipse where there is no minor axis, and circular polarization occurs when the minor axis is the same length as the major axis. As with linear polarization, elliptical polarization has an orientation with the direction of propagation.

Linear polarization can be caused in several different ways. One way is by reflection from a surface. When a beam of light which is unpolarized strikes a surface, the electro-magnetic waves will interact with the surface in different ways, depending on their orientation with respect to the surface.

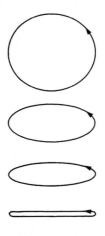

FIGURE 8.32 Elliptically polarized light.

380   CHAPTER 8: PHOTONIC CERAMICS

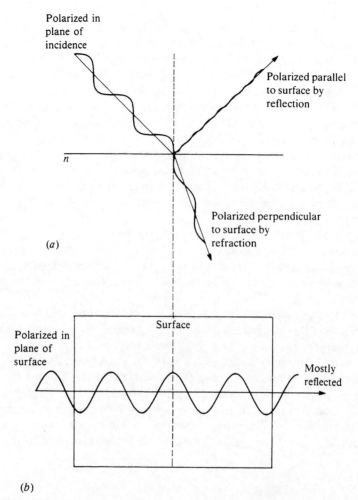

FIGURE 8.33   (*a*) Polarization by reflection, side view; (*b*) Polarized by reflection, top view.

Waves which are *normal* to the surface are not reflected, but pass into the surface. This occurs because the waves cannot oscillate in a continuous manner when reflected (see Figure 8.33). The only component of the light which will be reflected from the surface is that where the waves oscillate parallel to the plane of the surface. Because of this, light reflected from a surface will be linearly polarized parallel to the surface, or normal to the place of incidence.

When light (unpolarized) strikes a transparent surface, part of the light is reflected as polarized light. The remainder of the unpolarized light is then refracted through the material. Since some of the original energy of the light has been reflected along a particular orientation, the energy in the refracted light will be reduced along that same orientation. This causes the refracted

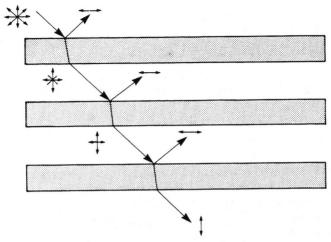

FIGURE 8.34 Polarization by repeated refraction.

light to be partially polarized in the orientation perpendicular to the angle of the polarized reflected light. If the same beam of unpolarized light is refracted through several layers of transparent material, the resulting beam can be strongly polarized, after repeated reduction of the light energy along one orientation due to multiple reflections (see Figure 8.34).

Another way in which light can become linearly polarized is by scattering from suspended particles. As the unpolarized light passes through a volume containing suspended particles, the particles will vibrate in the same directions as the electro-magnetic waves. These vibrations will cause light to be emitted which has its electro-magnetic waves oscillating in the same direction as the oscillating particle, which means the particle will only emit light in the plane perpendicular to the direction of its oscillations. Since the directions of oscillations for the suspended particles are induced by the vibrations of the light, they can only occur along the plane perpendicular to the direction of the propagation. When the two restrictions described above are combined, the result is that an observer will only see scattered light which oscillates in a direction normal to the plane described by the direction of propagation of the original light and that of the scattered light. The degree to which the light is polarized depends on the size of the suspended particles. Smaller particles produce a higher degree of polarization.

In some dichroic crystals, the absorption of light within the crystal depends on the polarization and orientation of the light (see Chapter 10). When unpolarized light passes through the crystal, oscillations along one direction experience very little absorption while oscillations in the perpendicular direction experience a large amount of absorption. If the crystal is thick enough, the light which finally leaves the crystal will have had part of

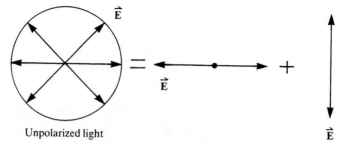

FIGURE 8.35 Resolution of unpolarized light to two polarized rays.

it completely absorbed, leaving the remainder of the light linearly polarized. This is the physical principle of the Polaroid®† material.

In some crystals, the index of refraction depends on the orientation of polarization of the incoming light. If unpolarized light passes through such a crystal, part of the light will be refracted at a different angle from the rest. All of the energy in the unpolarized light can be considered to be contained in two directions of oscillation perpendicular to each other, as shown in Figure 8.35.

After passing through a crystal which has this property, the original unpolarized light will have been split into two rays, with mutually perpendicular orientations of linear polarization. This phenomenon is called *birefringence,* and only occurs in materials which have anisotropic indices of refraction, as discussed in Chapter 10. The amount of birefringence in a material is defined as the difference between the two indices of refraction.

When unpolarized light strikes a birefringent material normal to its surface, the two rays created by the different indices of refraction will not be refracted apart. However, since the velocity of light decreases proportional to the index of refraction, the two rays will travel through the birefringent material with different velocities. When the unpolarized light first strikes the material, both rays are in phase with each other. As the rays pass through the birefringent material, one ray will begin to lag behind the other, causing the two rays to become out of phase. Once out of the material, the two rays will again be moving at the same velocity. However, they will not be out of phase from each other by an amount determined by the wavelength of the light, the indices of refraction for the two rays, and the thickness of the material they passed through. Since the path length of a ray is defined as

$$S = Ln \qquad (8.83)$$

where L = linear length
n = index of refraction

the path difference between the two rays will be

$$S' = Ln_1 - Ln_2 \qquad (8.84)$$

† Registered Trademark of Polaroid Corporation.

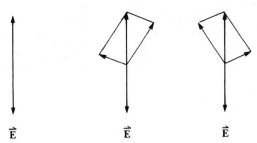

FIGURE 8.36 Perpendicular components of linear polarization.

or

$$S' = L(n_1 - n_2) \tag{8.85}$$

From this we can calculate the phase difference, δ, between the two rays:

$$\delta = S' 2\pi / \lambda \tag{8.86}$$

or

$$\delta = 2\pi L(n_1 - n_2)/\lambda \tag{8.87}$$

This path difference is what causes circular and elliptical polarization.

Linearly polarized light can be represented as consisting of oscillations along two perpendicular directions, with the amplitude of the oscillations depending on their orientation with the final polarized beam. These two oscillations are completely in phase (see Figure 8.36).

When a beam of linearly polarized light passes through a sheet of birefringent material, it can be split into two components lying along the directions with different indices of refraction. After passing through the material, one component will be out of phase with the other. As shown below, when these two components are out of phase, it causes the direction of polarization to change and rotate about the direction of propagation with time.

When the thickness of the material is such that the phase difference induced is exactly one-quarter of a wavelength, the two components will be completely out of phase, giving circularly polarized light. As the path difference increases past this, the circle of polarization becomes compressed into thinner and thinner ellipses, until the light is once again linearly polarized. However, the direction of polarization will have been rotated 90° compared with the orientation of the originally polarized light. A plate of birefringent material that induces a phase difference of one-quarter of a wavelength is called, rather obviously, a *quarter-wave plate*. A *half-wave plate* rotates the orientation of linear polarization 90°.

Any type of polarization can be quantified in terms of its azimuth and ellipticity. The azimuth of elliptically or linearly polarized light is the angle

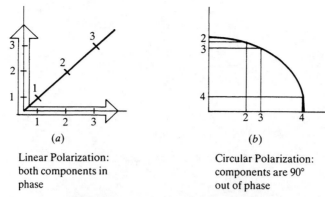

(a) Linear Polarization: both components in phase

(b) Circular Polarization: components are 90° out of phase

FIGURE 8.37 Phase of polarization.

that the major axis makes with the z-axis, when the x-axis is the direction of propagation of the light. The ellipticity of the polarization is defined as the inverse tangent of the ratio between the major and minor axes,

$$\mathbf{E} = \tan^{-1}(b/a) \tag{8.88}$$

Lasers and other devices which depend on efficiently transmitting light to another are constructed with Brewster angle interfaces, so the amount of energy lost due to partial reflection is minimized (see Figures 8.37 and 8.38).

When linearly polarized light is passed through a linearly polarizing substance, the light will have its direction of polarization changed to that of the substance. However, some of the energy contained in the light will be lost with the change of polarization. The greater the angle between the angle of the first polarization and the angle of the material, the less energy will be transmitted. This occurs because linearly polarized light can be considered to be composed of two perpendicular components, one in the

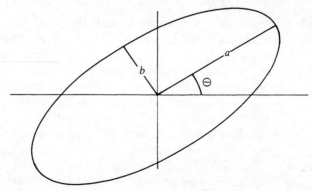

FIGURE 8.38 Major and minor axis with azimuth of elliptically polarized light.

direction of the new polarization, and the other at 90°. The component perpendicular to the new orientation will not be transmitted, so only the parallel component will be transmitted. The amplitude of oscillation of the transmitted component will be $A \cos \theta$, where A is the original amplitude, and θ the difference in directions. Since the intensity of the light, and its energy is the square of the electric field amplitude of the light, the intensity will vary as $\cos^2 \theta$.

In summary, polarization can occur through many different sources with various effects. Linear polarization can be caused by reflection, refraction, scattering, and absorption. Some crystals and stressed glasses can cause light which is linearly polarized to become elliptically polarized by inducing a phase shift between the components of the polarized light. Optically active materials like quartz and sugar solutions can cause the direction of linear polarization to rotate with the length of the material. Application of polarization has been used to reduce glare from surfaces, increase laser efficiency, and determine stress concentrations in materials.

8.12 IONIC POLARIZABILITY

Infrared absorption arises from displacement of ions of opposite sign from their regular lattice sites and also from the deformation of the electronic shells resulting from the relative displacement of the ions.

Consider the response of a diatomic linear crystal to infrared radiation of wavelength $100 \, \mu m$ (10^{-2} cm, frequency $= 3 \times 10^{12}$ Hz, wavevector $k = 2\pi/\lambda \approx 600 \text{ cm}^{-1} \ll k_{max} = \pi/2a \approx 10^8 \text{ cm}^{-1}$). The equations of motion for the lattice are as follows: (see Figure 8.39)

$$m\ddot{u}_{2n} = \beta(u_{2n+1} + u_{2n-1} - 2u_{2n}) \tag{8.89}$$

$$M\ddot{u}_{2n+1} = \beta(u_{2n+2} + u_{2n} - 2u_{2n+1}) \tag{8.90}$$

where β is the force constant. The force required to stretch a single bond is

$$F = \beta(u_n - u_{n-1}) = \beta ae \tag{8.91}$$

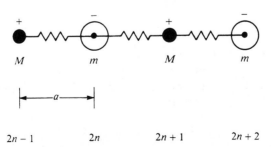

FIGURE 8.39 Diatomic linear lattice.

The solutions have the form of a traveling wave equation:

$$u_{2n} = \xi e^{i(\omega t + 2nka)} \tag{8.92}$$

$$u_{2n+1} = \eta e^{i(\omega t + [2n+1]ka)} \tag{8.93}$$

This leads to a substitution in the equations of motion, which become

$$-\omega^2 m\xi = \beta\eta(e^{i2ka} + e^{-ika}) - 2\beta\xi \tag{8.94}$$

$$-\omega^2 M\eta = \beta\xi(e^{ika} + e^{-ika}) - 2\beta\eta \tag{8.95}$$

Since $k \ll k_{max}$ we can simplify for the infrared region to

$$-\omega^2 m\xi = 2\beta(\eta - \xi) - e\mathbf{E} \tag{8.96}$$

$$-\omega^2 M\eta = -2\beta(y - \xi) + e\mathbf{E} \tag{8.97}$$

where $\mathbf{E}$ = electric field = $\mathbf{E}_0 e^{i\omega t}$ (8.98)

$\mathbf{E}_0$ = amplitude of the electric intensity of the radiation field (8.99)

These equations can be solved for ξ and η:

$$\eta = \frac{(e/M)\mathbf{E}_0}{\omega_0^2 - \omega^2} \tag{8.100}$$

$$\xi = \frac{-(e/m)\mathbf{E}_0}{\omega_0^2 - \omega^2} \tag{8.101}$$

where $\pm e$ represents the ionic charge.

$$\omega_0^2 = 2\beta\left(\frac{1}{m} + \frac{1}{M}\right) \quad \text{corresponding to } k = 0 \tag{8.102}$$

where $\beta = ca$

c = elastic stiffness constants

a = lattice separation

So we may expect infrared absorption to go through a maximum near the frequency ω_0. It is characteristic of ionic crystals that they have an absorption maximum in the infrared associated with the motion of charges of opposite sign toward each other:

$$c_{11} = 5 \times 10^{11} \text{ dyne/cm}^2 \quad \text{for NaCl} \tag{8.103}$$

$$\beta = (3 \times 10^{-8})(5 \times 10^{11}) = 1.5 \times 10^4 \text{ dyne/cm} \tag{8.104}$$

8.12 IONIC POLARIZABILITY

Then

$$\omega_0^2 \approx 2(1.5 \times 10^4)\left(\frac{1}{35.5} + \frac{1}{23.0}\right)\left(\frac{1}{1.67 \times 10^{-24}}\right) \quad (8.105)$$

where $\omega_0 \approx 3.6 \times 10^{13}$ rad/sec
$\lambda_0 \approx 53$ μm (theoretical)
or $\lambda_0 = 25$ μm (experimental)

For ionic polarization, the total relative displacement x of the positive and negative ions is given by setting ω to zero:

$$x = y - \xi = \frac{e\mathbf{E}_0}{\omega_0^2}\left(\frac{1}{m} + \frac{1}{M}\right) \quad (8.106)$$

The ionic polarization is

$$\mathbf{P} = \frac{e(\eta - \xi)}{\Omega} \quad (8.107)$$

where Ω is the volume/molecule
$\Omega = 2a^3$ for NaCl structure $\quad (8.108)$

Then

$$\mathbf{P} = \frac{e^2\mathbf{E}_0}{2\omega_0^2 a^3}\left(\frac{1}{m} + \frac{1}{M}\right) \quad (8.109)$$

so

$$\Delta\epsilon = \frac{2\pi e^2}{\omega_0^2 a^3}\left(\frac{1}{m} + \frac{1}{M}\right) \quad \text{(Born equation)} \quad (8.110)$$

Therefore, the difference between static and optical dielectric constants for NaCl can be calculated as

$$\Delta\epsilon = \frac{(6.28)(4.80 \times 10^{-10})^2 \text{ esu}}{(3.6 \times 10^{13} \text{ rad})^2(2.81 \times 10^{-8} \text{ cm})^3(1.66 \times 10^{-24} \text{ mol/g})}\left(\frac{1}{23} + \frac{1}{35.5}\right) \quad (8.111)$$

$$\Delta\epsilon = 2.17 \quad (8.112)$$

$$\Delta\epsilon \text{ measured} = 3.4 \quad (8.113)$$

This agreement indicates that a large contribution to the high-frequency dielectric constant k'_∞ is satisfactorily predicted on the basis of a hard-sphere model for sodium chloride (see Chapter 5). This is true for most silicates

TABLE 8.7 Index of Refraction for Photonic Materials

Photonic material	Name	Index n
AlGaAs		3.344
Al_2O_3 (parallel)	Sapphire	1.760
alpha-SiC		2.850
BaF_2		1.476
$BaTiO_3$	Barium Titanate	2.400
C	Diamond	2.424
$CaCO_3$	Calcite	1.650
CaF_2	Fluorite	1.434
CdF_2		1.576
CdS		2.500
CdTe		2.740
GaAs		3.630
Ge		4.000
KBr		1.564
KCl		1.509
LiF		1.394
MgF_2		1.391
MgO	Periclase	1.740
NaCl	Salt	1.552
NaF		1.326
Na_2O–CaO–SiO_2 glass	Soda-Lime Glass	1.510
PbF_2	Litharge	1.782
PbO	Galena	2.610
PbS		3.910
Na_2O–B_2O_3–SiO_2	Pyrex®†	1.470
Si		3.490
SiO_2	Fused Quartz	1.550
Na_2O–CaO–SiO_2	Dense Flint Glass	1.458
SiO_2	Quartz Crystal	1.650
SrF_2		1.442
$SrTiO_3$	Strontium Titanate	2.490
TiO_2	Rutile	2.710
TlBr		2.371
96% Silica	Vycor®†	1.458
Y_2O_3	Yttria	1.920
$ZiSiO_4$	Zircon	1.950
ZnS		2.200
AsSe		2.620

† Pyrex® and Vycor® are Registered Trademark of Corning Glass Works.

and aluminates of interest in ceramics for which the dielectric constant is in the range of 5–15 and results from this kind of ion displacement.

For highly polarizable ions, the deformation of electronic shells following displacements of the ions is an additional contribution to k'_∞. It is qualitatively seen to be of greater importance for ions that are highly polarizable and for structures where considerable ion deformations are possible. The ratio between low-frequency k and n^2 increases as the magnitude of these values increase:

$$\frac{k'}{k'_e} = 2.5 \quad \text{for NaCl} \tag{8.114}$$

and

$$\frac{k'}{k'_e} = 8.5 \quad \text{for BaO} \tag{8.115}$$

Table 8.7 lists a selection of useful photonic materials with their index of refraction. If the index has a large dispersion then the value shown is the index near 500 nanometers wavelength.

PROBLEMS

8.1 A Fabry–Perot interferometer can be made from an optically flat lens. A light ray of a particular wavelength is transmitted through a 3 mm piece of silica glass. The glass is tilted by a small angle very nearly perpendicular to the ray to produce the interference pattern. Estimate the reflection loses for the primary ray and each of the next five transmitted rays. (See Chapter 11 if necessary.)

(a) Plot the reflection loss coefficient as a function of the number of reflections.

(b) If the transmission losses for the above interferometer is 0.01 db/cm, calculate and plot the total losses as a function of the number of the transmitted ray.

(c) Which is the greatest contributor to the total losses?

8.2 The measured absorption for an optical fiber is 0.01 db per kilometer. What is the Lambert–Beer transmission factor for a:

(a) 10 kilometer length?

(b) 20 kilometer length?

(c) 50 kilometer length?

(d) Compare this with commercial optical cables. (Note that $a*c*1 = 1 - Po/Pi$)

8.3 Rayleigh scattering is strongly dependent upon the wavelength of the light. Plot the scattering coefficient as function of wavelength for $a = 1$:

(a) Particles of 1 micron diameter to 100 microns at 630 nm wavelength.

(b) Particles of 10 microns between 100 nm and 3000 nm wavelengths.

8.4 Draw an electron spin/energy level diagram for an octahedrally coordinated ligand around cobalt.

8.5 Draw an electron spin/energy level diagram for a tetrahedrally coordinated ligand around cobalt.

8.6 Assume that delta t for the d-orbital splitting of tetrahedrally coordinated cobalt is 344/cm. This absorption is called a high spin mode.

(a) What would be the wavenumber for octahedral coordination if the ligand and bond lengths remain the same?

(b) Which coordination should have the higher magnetic susceptibility?

8.7 If the concept of lattice vibrations is assumed to be the only mechanism for infrared absorption in photonic materials, then calculate the infrared absorption edges for the following materials. For this exercise, assume the lattice separations and the elastic stiffness constants are the same as for NaCl. Express the absorption edge in wavelength (microns):

(a) LiF.
(b) MgO.
(c) NaCl.
(d) Al_2O_3.
(e) TlBr.

8.8 (a) Compare the results in problem 8.7 with experimental values for the IR absorption edge.

(b) What can be said about the assumptions made to calculate the IR absorption edge using constants for NaCl?

8.9 What is the "IR" lattice force constants for each of the above materials. Use the hard sphere ionic model from Chapter 1 and the experimental absorption edges from problem 8.8.

8.10 (a) Use the difference in lattice force function for these materials with sodium chloride as the basis to compare their relative rigidity.

(b) Tabulate the coefficients of thermal expansion (CTE) for these materials.

(c) Compare the ratios of the CTEs with the ratios of the lattice force constants for this group of materials.

READING LIST

Azaroff, L. and Brophy, J. J., (1963). *Electronic Processes in Materials,* McGraw-Hill, New York, Chapter: 14.

Fanderlik, I., (1983). "Optical properties of glass," *Glass Sci. Tech.,* **5,** Elsevier, Amsterdam.

Garbuny, M., (1967). *Optical Physics,* Academic Press, New York.

Hench, L. L. and Dove, D. B., (1971). *Physics of Electronic Ceramics,* Part B, Dekker, New York, Chapters: 27, 28, 29.

Hench, L. L., Wang, S. H., and Nogues, J. L., (1988). In *Multifunctional Materials,* **878,** (R. L. Gunshor, ed.), SPIE, Bellingham, Washington, pp. 76–85.

Kingery, W. D., Bowen, H. K., and Uhlmann, D. R. (1976). *Introduction to Ceramics,* 2nd ed., Wiley, New York, Chapter: 13.

Kittel, C., (1986). *Introduction to Solid State Physics,* 6th ed., Wiley, New York, Chapter: 18.

Midwinter, J. E., (1979). *Optical Fibers for Transmission,* Wiley, New York.

Musikant, S., (1985). *Optical Materials: An Introduction to Selection and Application,* Dekker, New York.

Wang, S. H. (1988). "Sol-Gel derived silica optics," Ph.D. Dissertation, University of Florida, Gainesville, Florida.

9

OPTICAL WAVEGUIDES

9.1 INTRODUCTION

Just over 100 years ago, Alexander Graham Bell transmitted a telephone signal over 200 m using a beam of sunlight as a carrier. In other words, Bell invented the first optical communication system. At the receiver, the modulated sunlight fell on a photoconducting selenium cell, which converted the message to electrical current. A telephone receiver completed the system. Although the photophone worked rather well, it never achieved commercial success due to the lack of a reliable, intense light source and a dependable, low-loss transmission medium.

A major breakthrough that led to high-capacity optical communications was the invention of the laser, the first one of which was constructed in 1960. The laser was immediately recognized as the long-awaited carrier source. Extensive research began on optical devices, components, signal-processing techniques and subsystems, as well as on transmission media such as line-of-sight atmospheric paths and beam waveguides that employ periodic focusing elements. During the early 1960s, work on dielectric waveguides (*optical fibers*) was mainly theoretical, as the available glass fibers exhibited transmission losses in the vicinity of 100 dB/Km, about two orders of magnitude too large for telecommunications applications. In 1970, the first truly low-loss fiber was developed and fiber-optics communications became practical.

Although the initial research in integrated optics was primarily directed toward the optical communications area, other potential applications for combining the unique properties of light into extremely small packages were apparent. The development of micron and submicron fabrication technology in the semiconductor areas made a significant impact on the pace of progress in integrated optics. These advances make possible the combining of optical, electro-optical, and electrical components on the same chip. Progress in all areas of optical fiber communications has been both rapid and abundant. Examples of recent laboratory achievements include fibers with losses as low as 0.16 dB/Km, lasers with threshold current of a few milliamperes, and experimental systems that operate at 2 Gbit/sec over repeater systems as long as 130 Km. These and other laboratory demonstrations are presently transformed quickly into commercial realities. Indeed, optical fiber transmission systems are planned and deployed at a phenomenally fast pace in almost all areas of telecommunications.

The fiber-optics industry is now gearing up for greater activity in nontelecommunications applications. These growing markets include military programs, fiber-optics sensors, laser beam delivery systems, automotive emission and control, process control links, and other short-haul communications, and even optical computing.

9.2 OPTICAL FIBERS

Optical fibers are made from amorphous SiO_2 dielectric. They are the transmission media. Therefore the link between source and detector is nonmetallic, which avoids ground-loop pick-up problems resulting from electro-magnetic interference. Such immunity is important for military applications, since it permits communications cables to be located near wirings for electrical power systems in vehicles for land, sea, air, and space.

An optical fiber is a cylindrical structure, consisting of a central core of radius a and permittivity, ϵ_2, surrounded by a concentric cladding region of slightly lower dielectric constant, ϵ_1. Lowest loss fibers are generally fabricated of a silica (SiO_2) core doped with either GeO_2 or P_2O_5 to increase the refractive index and are surrounded by silica (or P_2O_5-doped silica) claddings. For applications in which attenuation is not a primary concern, such as for short-distance communications and short links, higher loss plastic-clad or totally plastic fibers can be used.

In theory, guidance along such fibers is possible without claddings. However, a cladding serves several useful purposes. It protects the core from external surface contaminants, improves mechanical strength, and reduces scattering losses resulting from dielectric discontinuities at the core surface. Further, by varying the value of the refractive index of the cladding relative to that of the core, an additional degree of freedom in controlling waveguide propagation characteristics is obtained. Frequently, one or more additional plastic coating layers surround the claddings to provide further cable strength, as well as mechanical protection from external shock and perturbations of the fiber geometry.

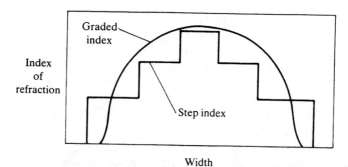

FIGURE 9.1 Alternative index profiles of optical waveguides.

A number of different fiber geometries are presently used for various communications applications. These configurations may be classified according to the variation in the dielectric constant or refractive index along the radial direction of the fiber. Waveguides having a uniform refractive index throughout, but exhibiting an abrupt step at the core-cladding interface are referred to as *step-index fibers*. Also, fibers whose refractive index varies in some continuous fashion as a function of radial distance are known as *graded-index fibers* (see Figure 9.1). All of the above fiber geometries can be divided into two classes, *monomode* and *multimode*. Monomode fibers will support only one propagating mode, while multimode fibers will generally support hundreds of modes.

Although optical fibers can be made from different materials, all-glass fibers have the lowest losses and the smallest intermodal pulse spread. Because of these properties, they are useful at moderately high information rates or fairly long lengths. The low numerical aperture, NA:

$$\text{NA} = (n_2 - n_1)^{1/2} \tag{9.1}$$

of glass fiber results in large losses when coupled to a light source. However, the low transmission loss partially compensates for this problem. Typical core diameters are 50 μm, and 200 μm.

Plastic clad silica (PCS) fibers have higher losses and larger pulse spreads than all-glass fibers. Consequently, they are suitable for shorter links. The PCS fibers have higher numerical apertures which increase the efficiency of coupling with the light, but this advantage is lost in a long fiber due to increased absorption. Core diameters of 200 μm are typical for PCS fibers. The large core diameter improves the source coupling efficiency. *Polymeric clad* fibers are normally suitable choices when the path lengths are less than 1 Km.

All-plastic fibers are limited to path lengths less than a few tens of meters by their high propagation losses. Large cores and large numerical apertures make plastic fibers usable because of the resulting high coupling efficiencies. Core diameters as large as 1 mm are typical.

TABLE 9.1 Optical Waveguide Characteristics

Type	n_2	n_1	NA	α_0 (degrees)
All-glass	1.48	1.46	0.1414	8.13
PCS	1.46	1.4	0.2449	14.17
All-plastic	1.49	1.39	0.3162	18.43

Typical numerical apertures (NA) of the form

$$\text{NA} = (n_2 - n_1)^{1/2} = \sin \alpha_0 \tag{9.2}$$

having acceptance angles α_0 are shown in Table 9.1.

Only rays emitted with a core having an acceptance angle α_0 will be trapped by the fiber. Typical LEDs and laser diodes (see Chapter 11) emit light over a wide angular range, often larger than the acceptance angles.

9.3 GENERATIONS OF OPTICAL COMMUNICATIONS

As with any new technology, the changes in fiber-optics communications systems and equipment are often discussed in terms of *generations*. Multimode technology at 820 nm or 850 nm represents the first generation. The second generation is the transmission of 1300 nm photons into a multimode fiber. However, this technology has been dramatically overshadowed by 1300 nm single-mode transmissions. Whether 1300 nm single-mode technology is considered the second or third generation, it was the next to receive widespread commercial acceptance. This generation should be more specifically defined as using conventional 1300 nm single-mode fiber and Fabry–Perot (multifrequency) diode lasers. The next technology to find a large market will be *dispersion-shifted* (D–S) fiber, optimized for 1550 nm transmission. This has been available since 1984. In 1986 *Distributed feedback* (DFB) lasers came on the market. They are believed to offer high-speed operation at 1550 nm over conventional single-mode fiber. (A DFB laser has part of its output fed back to its input, using more than one propagation mode in the feedback for controlling mode generations and mode conversion. It usually uses periodically inhomogeneous thin films, periodically inhomogeneous substrate guides, or thin-film waveguides with periodic surfaces. The DFB operates more efficiently when feedback and direct waves are in the same mode.)

There is considerable interest in the use of photons for computing since they potentially would eliminate the von Neumann bottleneck in computer architecture. The development of an optical computer requires arrays of optical waveguides. In current practice there are three basic functions of a computer: arithmetic operations, logical operations and the memory. These are all done by devices that have two stable states: on or off. Electro-optical devices, described in Chapter 10, can be used to achieve optical switching, since the index of refraction of electro-optical materials is changed by the application of electric fields. Optical waveguides are required to connect the optical switches in an array.

9.4 HOLLOW CONDUCTING WAVEGUIDES

The description of optical waveguide behavior can best be visualized by studying an ideal case of an electro-magnetic wave in a hollow conducting waveguide.

If we assume that the medium of propagation is linear and homogeneous, then

$$\mathbf{D} = K_e \epsilon_0 \mathbf{E} = \epsilon \mathbf{E} \tag{9.3}$$

$$\mathbf{H} = \frac{\mathbf{B}}{K_m \mu_0} = \frac{\mathbf{B}}{\mu} \tag{9.4}$$

Where ϵ and μ are constant and

$\mathbf{E}$ = electric field
$\mathbf{D}$ = dielectric displacement
$\mathbf{B}$ = magnetic induction
$\mathbf{H}$ = magnetic field
ϵ = electric permittivity
μ = magnetic permeability
K_m = relative permeability
n = index of refraction
K_e = dielectric constant $\approx n^2$

We must also define the conductivity σ of the medium of propagation equal to zero and the conductivity of the walls of the waveguide equal to infinity. Electro-magnetic waves are very highly attenuated within conductors, Consequently, they cannot be used as media for waveguides. Also, the charge density of the media, ρ, is assumed to be zero. As a final assumption, we define the propagation of the wave to be along a single axis (i.e., the z-axis).

The sinusoidal solution for an electro-magnetic plane wave with both transverse and longitudinal components is given by

$$\mathbf{E} = \mathbf{E}_0 \exp\left[j(\omega t - k_g z)\right] \tag{9.5}$$

$$\mathbf{H} = \mathbf{H}_0 \exp\left[j(\omega t - k_g z)\right] \tag{9.6}$$

9.4 HOLLOW CONDUCTING WAVEGUIDES

where the wavenumber

$$k_g = \frac{1}{\lambda_g} \tag{9.7}$$

is that for the guided wave; it is not necessarily equal to

$$k = \frac{1}{\lambda} = \frac{\omega}{u} = \omega(\epsilon\mu)^{1/2} = \omega(K_e\epsilon_0 K_m\mu_0)^{1/2} = \omega n \tag{9.8}$$

which is the wavenumber for a plane wave in the medium under consideration and which depends on the circular frequency ω and velocity u. The wavenumber k_g is in the general case complex:

$$k_g = k_{gr} - jk_{gi} \tag{9.9}$$

As an exercise, the student is requested to develop the wave equation (9.10) from Maxwell's equations (see Problem 9.1)

$$\nabla^2 \mathbf{E} = \epsilon\mu \frac{\partial^2 \mathbf{E}}{\partial t^2} = -k^2 \mathbf{E} \tag{9.10}$$

and show that equations (9.5) and (9.6) are indeed solutions to the wave equation. Maxwell's equations for the field become

$$\nabla \cdot \mathbf{D} = 0 \tag{9.11}$$

with ϵ = constant.

$$\nabla \cdot \mathbf{E} = 0 \tag{9.12}$$

Similarly

$$\nabla \cdot \mathbf{B} = 0 \tag{9.13}$$

with μ = constant,

$$\nabla \cdot \mathbf{H} = 0 \tag{9.14}$$

Substitution of equations (9.5) and (9.6) into equations (9.12) and (9.14) yields:

$$\nabla \cdot (\mathbf{E}_0 \exp[j(\omega t - k_g z)]) = 0 \tag{9.15}$$

and

$$\nabla \cdot (\mathbf{H}_0 \exp[j(\omega t - k_g z)]) = 0 \tag{9.16}$$

expanding:

$$\left(\frac{\partial}{\partial x}\hat{\mathbf{i}} + \frac{\partial}{\partial y}\hat{\mathbf{j}} + \frac{\partial}{\partial z}\hat{\mathbf{k}}\right) \cdot (\mathbf{E}_0 \exp[j\omega t - jk_g z]) = 0 \tag{9.17}$$

CHAPTER 9: OPTICAL WAVEGUIDES

and

$$\left(\frac{\partial}{\partial x}\hat{\mathbf{i}} + \frac{\partial}{\partial y}\hat{\mathbf{j}} + \frac{\partial}{\partial z}\hat{\mathbf{k}}\right) \cdot [\mathbf{H}_0 \exp[j\omega t - jk_g z]] = 0 \quad (9.18)$$

Note that

$$\frac{\partial}{\partial x}(E_{0x} \exp[j\omega t - jk_g z]) = E_{0x} \frac{\partial}{\partial x} \exp[j\omega t - jk_g z]$$

$$+ \left(\frac{\partial}{\partial x} E_{0x}\right)(\exp[j\omega t - jk_g z])$$

$$= \exp[j\omega t - jk_g z] \frac{\partial}{\partial x} E_{0x} \quad (9.19)$$

Therefore

$$\exp[j\omega t - jk_g z]\frac{\partial}{\partial x} E_{0x} + \exp[j\omega t - jk_g z]\frac{\partial}{\partial y} E_{0y}$$

$$+ \exp[j\omega t - jk_g z]\frac{\partial}{\partial z} E_{0z} + E_{0z}(-jk_g) \exp[j\omega t - jk_g z] = 0 \quad (9.20)$$

because we have specified the z-dependence of the solution (see equations (9.5) and (9.6)), $\frac{\partial}{\partial z} E_{0z} = 0$. Divide through by $\exp(j\omega t - jk_g z)$

$$\frac{\partial}{\partial x} E_{0x} + \frac{\partial}{\partial y} E_{0y} - jk_g E_{0z} = 0 \quad (9.21)$$

Likewise,

$$\frac{\partial}{\partial x} H_{0x} + \frac{\partial}{\partial y} H_{0y} - jk_g H_{0z} = 0 \quad (9.22)$$

Using Maxwell's third equation

$$\nabla \times \mathbf{E} = -\frac{\partial}{\partial t}\mathbf{B} \quad (9.23)$$

and Maxwell's fourth equation

$$\nabla \times \mathbf{H} = \frac{\partial}{\partial t}\mathbf{D} \quad (9.24)$$

Substitute equations (9.5) and (9.6) into equations (9.23) and (9.24), respectively:

$$\left(\frac{\partial}{\partial x}\hat{\mathbf{i}} + \frac{\partial}{\partial y}\hat{\mathbf{j}} + \frac{\partial}{\partial z}\hat{\mathbf{k}}\right) \times \mathbf{E}_0 \exp[j\omega t - jk_g z]$$

$$= -\frac{\partial}{\partial t}(\mu \mathbf{H}_0 \exp[j\omega t - jk_g z]) \quad (9.25)$$

$$= -j\omega\mu\mathbf{H}_0 \exp[j\omega t - jk_g z] \quad (9.26)$$

and

$$\left(\frac{\partial}{\partial x}\hat{\mathbf{i}} + \frac{\partial}{\partial y}\hat{\mathbf{j}} + \frac{\partial}{\partial z}j\hat{\mathbf{k}}\right) \times \mathbf{H}_0 \exp[j\omega t - jk_g z]$$

$$= \frac{\partial}{\partial t}(\epsilon \mathbf{E}_0 \exp[j\omega t - jk_g z]) \quad (9.27)$$

$$= ij\omega\epsilon\mathbf{E}_0 \exp[j\omega t - jk_g z] \quad (9.28)$$

Expanding equation (9.26) and letting the exponential be represented by exp [—]:

$$\frac{\partial}{\partial x}E_{0y}\hat{\mathbf{k}}\exp[—] + \frac{\partial}{\partial x}E_{0z}(-\hat{\mathbf{j}})\exp[—] + \frac{\partial}{\partial y}E_{0x}(-\hat{\mathbf{k}})\exp[—]$$

$$+ \frac{\partial}{\partial y}E_{0z}\hat{\mathbf{i}}\exp[—] + \frac{\partial}{\partial z}E_{0x}\hat{\mathbf{j}}\exp[—] + \frac{\partial}{\partial z}E_{0y}(-\hat{\mathbf{i}})\exp[—]$$

$$+ jk_g E_{0x}\exp[—](-\hat{\mathbf{j}}) + jk_g E_{0y}(+\hat{\mathbf{i}})\exp[—]$$

$$= -j\omega\mu\mathbf{H}_0\exp[j\omega t - jk_g z] \quad (9.29)$$

The extra two terms arise from the z dependence of the exponential. Divide through by exp [—] and collect by unit vectors $\hat{\mathbf{i}}$, $\hat{\mathbf{j}}$, and $\hat{\mathbf{k}}$:

$$jk_g E_{0y} + \frac{\partial}{\partial y}E_{0z} - \frac{\partial}{\partial z}E_{0y} = -j\omega\mu H_{0x} \quad (9.30)$$

$$-jk_g E_{0x} - \frac{\partial}{\partial x}E_{0z} + \frac{\partial}{\partial z}E_{0x} = -j\omega\mu H_{0y} \quad (9.31)$$

$$\frac{\partial}{\partial x}E_{0y} - \frac{\partial}{\partial y}E_{0x} = -j\omega\mu H_{0z} \quad (9.32)$$

but the vectors $\mathbf{E}_0$ and $\mathbf{H}_0$ are functions of x and y only because they represent a plane wave. Therefore,

$$\frac{\partial}{\partial z}E_{0x} = \frac{\partial}{\partial z}E_{0y} = 0 \quad (9.33)$$

Likewise, from $\nabla \times \mathbf{H} = \dfrac{\partial}{\partial t} D$:

$$\frac{\partial}{\partial y} H_{0z} + jk_g H_{0y} = j\omega\epsilon E_{0x} \tag{9.34}$$

$$jk_g H_{0x} + \frac{\partial}{\partial x} H_{0z} = j\omega\epsilon E_{0y} \tag{9.35}$$

$$\frac{\partial}{\partial x} H_{0y} - \frac{\partial}{\partial y} H_{0x} = -j\omega\epsilon E_{0z} \tag{9.36}$$

Combine equations (9.31) and (9.34) and eliminate H_{0y}:

$$\frac{\partial}{\partial y} H_{0z} + jk_g\left(\frac{k_g}{\omega\mu} E_{0x} + \frac{1}{j\omega\mu}\frac{\partial}{\partial x} E_{0z}\right) = j\omega\epsilon E_{0x} \tag{9.37}$$

Solve for E_{0x}:

$$\frac{\partial}{\partial y} H_{0z} + \frac{jk_g^2}{\omega\mu} E_{0x} + \frac{jk_g}{j\omega\mu}\frac{\partial}{\partial x} E_{0z} = j\omega\epsilon E_{0x} \tag{9.38}$$

$$-j\omega\epsilon E_{0x} + \frac{jk_g^2}{\omega\mu} E_{0x} = -\frac{\partial}{\partial y} H_{0z} - \frac{k_g}{\omega\mu}\frac{\partial}{\partial x} E_{0z} \tag{9.39}$$

$$E_{0x}\left(j\omega\epsilon - \frac{jk_g^2}{\omega\mu}\right) = \left(\frac{k_g}{\omega\mu}\frac{\partial}{\partial x} E_{0z} + \frac{\partial}{\partial y} H_{0z}\right) \tag{9.40}$$

From equation (9.8): $\dfrac{1}{\lambda} = \omega(\epsilon\mu)^{1/2}$, so

$$\frac{1}{\lambda^2} = \omega^2\epsilon\mu \tag{9.41}$$

Note that $\lambda^2 \neq \lambda_g^2$ for the following analysis

$$E_{0x} j\left(\frac{1}{\omega\mu}(\omega^2\epsilon\mu - k_g^2)\right) = \left(\frac{k_g}{\omega\mu}\frac{\partial}{\partial x} E_0 + \frac{\partial}{\partial y} H_{0z}\right) \tag{9.42}$$

$$E_{0x}\frac{j}{\omega\mu}\left(\frac{1}{\lambda^2} - \frac{1}{\lambda_g^2}\right) = \left(\frac{k_g}{\omega\mu}\frac{\partial}{\partial x} E_{0z} + \frac{\partial}{\partial y} H_{0z}\right) \tag{9.43}$$

$$E_{0x} = \frac{-j\omega\mu}{\left(\dfrac{1}{\lambda^2} - \dfrac{1}{\lambda_g^2}\right)}\left(\frac{k_g}{\omega\mu}\frac{\partial}{\partial x} E_{0z} + \frac{\partial}{\partial y} H_{0z}\right) \tag{9.44}$$

9.4 HOLLOW CONDUCTING WAVEGUIDES

Likewise:

$$E_{0y} = \frac{-j\omega\mu}{\left(\frac{1}{\lambda^2} - \frac{1}{\lambda_g^2}\right)} \left(-\frac{k_g}{\omega\mu}\frac{\partial}{\partial x}E_{0z} + \frac{\partial}{\partial y}H_{0z}\right) \quad (9.45)$$

$$H_{0x} = \frac{-j\omega\epsilon}{\left(\frac{1}{\lambda^2} - \frac{1}{\lambda_g^2}\right)} \left(\frac{\partial}{\partial y}E_{0z} - \frac{k_g}{\omega\epsilon}\frac{\partial}{\partial x}H_{0z}\right) \quad (9.46)$$

$$H_{0y} = \frac{-j\omega\epsilon}{\left(\frac{1}{\lambda^2} - \frac{1}{\lambda_g^2}\right)} \left(\frac{\partial}{\partial y}E_{0z} + \frac{k_g}{\omega\epsilon}\frac{\partial}{\partial x}H_{0z}\right) \quad (9.47)$$

The above equations (9.44) to (9.47) show that the guided wave is completely determined once E_{0z} and H_{0z} are known.

The wave equation (9.10) can be used to supply a second-order differential equation for E_{0z}:

$$\nabla^2 \mathbf{E} = -k^2 \mathbf{E} \quad (9.48)$$

Substituting $E_{0z} \exp[j\omega t - jk_g z]$ into equation (9.10) for the z component only:

$$\left(\frac{\partial^2}{\partial x^2} + \frac{\partial^2}{\partial y^2} + \frac{\partial^2}{\partial z^2}\right) \cdot (E_{0z} \exp[j\omega t - jk_g z]) = -k^2 E_{0z} \exp[j\omega t - jk_g z] \quad (9.49)$$

$$\left(\frac{\partial^2}{\partial x^2} E_{0z}\right) \exp[j\omega t - jk_g z] + \left(\frac{\partial^2}{\partial y^2} E_{0z}\right) \exp[j\omega t - jk_g z]$$
$$+ \left(\frac{\partial^2}{\partial z^2} E_{0z}\right) \exp[j\omega t - jk_g z] + E_{0z}(jk_g)^2 \exp[j\omega t - jk_g z]$$
$$= -k^2 E_{0z} \exp[j\omega t - jk_g z] \quad (9.50)$$

divide through by $\exp[j\omega t - jk_g z]$:

$$\frac{\partial^2}{\partial x^2} E_{0z} + \frac{\partial^2}{\partial y^2} E_{0z} + \frac{\partial^2}{\partial z^2} E_{0z} - k_g^2 E_{0z} = -k^2 E_{0z} \quad (9.51)$$

remembering for our solution $\frac{\partial}{\partial z} E_{0z} = 0$ we obtain

$$\frac{\partial}{\partial x^2} E_{0z} + \frac{\partial^2}{\partial y^2} E_{0z} = (k_g^2 - k^2) E_{0z} \quad (9.52)$$

$$\frac{\partial}{\partial x^2} E_{0z} + \frac{\partial^2}{\partial y^2} E_{0z} = \left(\frac{1}{\lambda_g^2} - \frac{1}{\lambda^2}\right) E_{0z} \quad (9.53)$$

Likewise:

$$\frac{\partial^2}{\partial x^2} H_{0z} + \frac{\partial^2}{\partial y^2} H_{0z} = \left(\frac{1}{\lambda_g^2} - \frac{1}{\lambda^2}\right) H_{0z} \tag{9.54}$$

Equations (9.53) and (9.54) represent two second-order partial differential equations for E_{0z} and H_{0z} that must be solved by application of specific boundary conditions of the waveguide.

The radian length of the guided wave λ_g is still unspecified. The radian length must be selected as discrete eigenvalues in order that E_{0z} and H_{0z} satisfy the above equations (9.53) and (9.54). These eigenvalues are functions of the wave frequency, of the electrical characteristic of the materials of construction for the waveguide, and of the geometry of the waveguide:

$$\lambda_g = f(\nu, K_e, K_m, \sigma, K'_e, K'_m) \tag{9.55}$$

The procedure for calculating the field vectors **E** and **H** is as follows:

1. Define the boundary conditions of the waveguide.
2. Solve the wave equations (9.53) and (9.54) for E_{0z} and H_{0z}.
3. Calculate the transverse components for **E** and **H** from equations (9.44)–(9.47).

9.5 TEM WAVES IN A METALLIC WAVEGUIDE

A TEM or *transverse electric and magnetic* wave, where both **E** and **H** are transverse, is a plane wave traveling exactly perpendicular to the direction of propagation. There are different modes of TEM waves indicated by the indices n, m, and q. These integers represent the number of times the intensity goes to zero between the intensity peaks:

$$\text{TEM}_{nmq} \tag{9.56}$$

The different modes represent various modes of standing waves as shown in Figure 9.2. There are an infinite number of possible oscillatory longitudinal cavity modes. They each have a distinctive frequency ν_m:

$$\nu_m = \frac{mv}{2d} \tag{9.57}$$

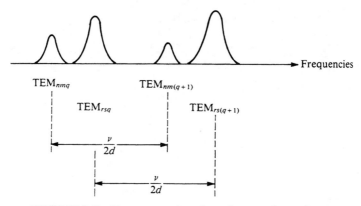

FIGURE 9.2 Transverse electric and magnetic modes.

where n = integer modes along x-axis
  v = wave velocity
  d = length of optical cavity which contains the standing wave
  m = integer modes along y-axis
  q = longitudinal mode number

Figure 9.3 shows near field projections of various TEM_{nm} modes.

Only the case of TEM_{00} will be considered for the solution in a hollow metallic waveguide. This is really the most interesting case, considering it can be focused to the smallest spot in an optical system. It has the least divergence and is the mode most widely used in general applications.

For a TEM wave, the electric and magnetic field vectors **E** and **H** have no longitudinal components:

$$E_{0z} = 0 \tag{9.58}$$

and

$$H_{0z} = 0 \tag{9.59}$$

Because a TEM wave is a plane wave perpendicular to the axis of propagation:

$$\lambda_g = \lambda \tag{9.60}$$

The phase velocity u

$$u = \omega\lambda = \frac{1}{(\epsilon\mu)^{1/2}} \tag{9.61}$$

is the same as that of a plane wave in the medium of propagation.

404 CHAPTER 9: OPTICAL WAVEGUIDES

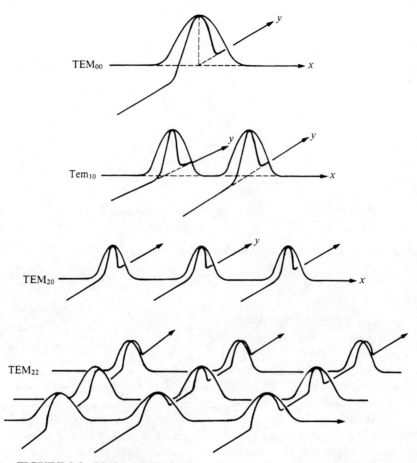

FIGURE 9.3 Various TEM Modes representing near field projections.

For a moment we need to discuss some basic relationships of electrostatics to continue the evaluation of TEM waves. The electro-static field intensity **E** can be derived from the potential V through the relation

$$\mathbf{E} = -\nabla V \tag{9.62}$$

Likewise, the magnetic induction **B** can be related to a vector potential **A** by the relation:

$$\mathbf{B} = \nabla \times \mathbf{A} \tag{9.63}$$

9.5 TEM WAVES IN A METALLIC WAVEGUIDE

From the third of Maxwell's relations:

$$\nabla \times \mathbf{E} = -\frac{\partial}{\partial t}\mathbf{B} \tag{9.64}$$

$$\nabla \times \mathbf{E} = -\frac{\partial}{\partial t}(\nabla \times \mathbf{A}) \tag{9.65}$$

$$\nabla \times \mathbf{E} = -\nabla \times \frac{\partial}{\partial t}\mathbf{A} \tag{9.66}$$

$$\nabla \times \left(\mathbf{E} + \frac{\partial}{\partial t}\mathbf{A}\right) = 0 \tag{9.67}$$

Therefore,

$$\mathbf{E} = -\frac{\partial}{\partial t}\mathbf{A} \tag{9.68}$$

for induced electric fields.

Remembering that the curl of a gradient of a scalar function is zero:

$$\nabla \times (-\nabla V) = 0 \tag{9.69}$$

then the static potential $\mathbf{E} = -\nabla V$ can be added to equation (9.68):

$$\mathbf{E} = -\frac{\partial}{\partial t}\mathbf{A} - \nabla V \tag{9.70}$$

where the first term arises from changes in the magnetic field, and the second from the accumulation of charge. Since the currents for a TEM wave are all longitudinal as well as $\partial \mathbf{A}/\partial t$, then

$$\mathbf{A} = A\hat{\mathbf{k}} \tag{9.71}$$

Noting that

$$\mathbf{E} = E_{0x}\hat{\mathbf{i}} + E_{0y}\hat{\mathbf{j}} + E_{0z}\hat{\mathbf{k}} \tag{9.72}$$

$$\mathbf{E} = -\frac{\partial V}{\partial x}\hat{\mathbf{i}} - \frac{\partial V}{\partial y}\hat{\mathbf{j}} - \left(\frac{\partial V}{\partial z} + \frac{\partial A}{\partial t}\right)\hat{\mathbf{k}} \tag{9.73}$$

But $\mathbf{E}$ is transverse, therefore the longitudinal component of ∇V must cancel that of $\partial \mathbf{A}/\partial t$ exactly at all points:

$$E_{0z} = 0 = \frac{\partial}{\partial z} V + \frac{\partial}{\partial t} \qquad (9.74)$$

$$\frac{\partial}{\partial z} V = -\frac{\partial A}{\partial t} \qquad (9.75)$$

and

$$\mathbf{E} = -\frac{\partial}{\partial x} \hat{\mathbf{i}} \frac{\partial}{\partial y} V \hat{\mathbf{j}} \qquad (9.76)$$

$$E_{0x} = -\frac{\partial V}{\partial x} \qquad (9.77)$$

and

$$E_{0y} = -\frac{\partial V}{\partial y} \qquad (9.78)$$

because we are dealing with a plane wave

$$V = V_0 \exp[j(\omega - kz)] \qquad (9.79)$$

then

$$E_{0x} = -\frac{\partial}{\partial x} V_0 \qquad (9.80)$$

and

$$E_{0y} = -\frac{\partial}{\partial y} V_0 \qquad (9.81)$$

$$E_{0z} = 0 \qquad (9.82)$$

If the waveguide is a hollow perfectly conducting tube, the tangential component of $\mathbf{E}$ at its surface is zero. Therefore V_0 is a constant all around the tube and the only possible solution is $V_0 =$ constant inside as well. Therefore the partial differentials become

$$E_{0x} = 0 \qquad (9.83)$$
$$E_{0y} = 0 \qquad (9.84)$$
$$E_{0z} = 0 \qquad (9.85)$$

with

$$\nabla \times \mathbf{E} = 0 = -\frac{\partial}{\partial t} \mathbf{B} \qquad (9.86)$$

9.6 HOLLOW RECTANGULAR WAVEGUIDE FOR A TE WAVE ($E_{0z} = 0$)

There are no **H** components either. Therefore, a TEM wave cannot be transmitted inside a hollow conducting tube. This result indicates that either **E** or **H** can be transverse for propagation to occur but not both. In these cases $\lambda \neq \lambda_g$, (i.e., the radian length λ_g is different from wavelength λ).

9.6 HOLLOW RECTANGULAR WAVEGUIDE FOR A TE WAVE ($E_{0z} = 0$)

E_0 is always transverse and is always perpendicular to the direction of propagation (see Figure 9.4).

From equation (9.30)

$$\frac{\partial^2}{\partial x^2} H_{0z} + \frac{\partial^2}{\partial y^2} H_{0z} = \left(\frac{1}{\lambda_g^2} - \frac{1}{\lambda_0^2}\right) H_{0z} \tag{9.87}$$

In a TE wave, the field H is independent of the x-coordinate:

$$\frac{\partial^2}{\partial y^2} H_{0z} = \left(\frac{1}{\lambda_g^2} - \frac{1}{\lambda_0^2}\right) H_{0z} = -\alpha^2 H_{0z} \tag{9.88}$$

Therefore the following represents a solution:

$$H_{0z} = X \sin \alpha y + Y \cos \alpha y \tag{9.89}$$

The student is requested to verify that equation (9.89) represents the solution to equation (9.88) (see Problem 9.2).

Now, at the inside surface of the waveguide, E_{0z} is zero everywhere for TE waves. **E** is normal to the inside surface.

$$\mathbf{E} = E_{0x}\hat{\mathbf{i}} + E_{0y}\hat{\mathbf{j}} \tag{9.90}$$

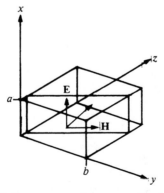

FIGURE 9.4 Hollow rectangular waveguide for a TE wave ($E_{0z} = 0$).

408 CHAPTER 9: OPTICAL WAVEGUIDES

using equation (9.44) and (9.45) with $\frac{\partial}{\partial x} E_{0z} = \frac{\partial}{\partial y} E_{0z} = 0$:

$$\mathbf{E} = \frac{-j\omega\mu}{\left(\frac{1}{\lambda^2} - \frac{1}{\lambda_g^2}\right)} \frac{\partial}{\partial y} H_{0z} \hat{\mathbf{i}} + \frac{j\omega\mu}{\left(\frac{1}{\lambda^2} - \frac{1}{\lambda_g^2}\right)} \frac{\partial}{\partial x} H_{0z} \hat{\mathbf{j}} \qquad (9.91)$$

Combining terms:

$$\mathbf{E}_0 = \frac{-j\omega\mu}{\left(\frac{1}{\lambda^2} - \frac{1}{\lambda_g^2}\right)} \left(-\frac{\partial}{\partial y} H_{0z}\hat{\mathbf{i}} + \frac{\partial}{\partial x} H_{0z}\hat{\mathbf{j}}\right) \qquad (9.92)$$

$$= \frac{j\omega\mu}{\left(\frac{1}{\lambda^2} - \frac{1}{\lambda_g^2}\right)} (\hat{\mathbf{k}} \times \nabla H_{0z}) \qquad (9.93)$$

The three vectors $\mathbf{E}$, $\hat{\mathbf{k}}$, and ∇H_{0z} are mutually perpendicular. This leaves ∇H_{0z} tangent to the conducting wall. Therefore

$$\frac{\partial}{\partial x} H_{0z} = 0 \quad \text{at} \quad x = 0 \quad \text{and} \quad x = a \qquad (9.94)$$

and

$$\frac{\partial}{\partial y} H_{0z} = 0 \quad \text{at} \quad y = 0 \quad \text{and} \quad y = b \qquad (9.95)$$

By differentiating the solution (equation 9.89) and applying the above boundary conditions:

$$\frac{\partial}{\partial y} H_{0z} = \alpha(X \cos \alpha y - Y \sin \alpha y) = 0 \qquad (9.96)$$

At $y = 0$, $X = 0$. At $y = b$ we get

$$0 = -\alpha \sin \alpha b \qquad (9.97)$$

$$\sin \alpha b = 0 \qquad (9.98)$$

and $\alpha = n\pi/b$, where n is an integer that cannot be zero; therefore

$$H_{0z} = Y \cos\left(\frac{n\pi y}{b}\right) \qquad (9.99)$$

9.6 HOLLOW RECTANGULAR WAVEGUIDE FOR A TE WAVE ($E_{0z}=0$)

differentiate H_{0z} twice:

$$\frac{\partial}{\partial y} H_{0z} = Y\left(\frac{n\pi}{b}\right) \sin\left(\frac{n\pi y}{b}\right) \tag{9.100}$$

$$\frac{\partial^2}{\partial y^2} H_{0z} = Y\left(-\frac{n^2\pi^2}{b^2}\right) \cos\left(\frac{n\pi y}{b}\right) \tag{9.101}$$

substituting into equation (9.54):

$$Y\left(-\frac{n^2\pi^2}{b^2}\right) \cos\left(\frac{n\pi y}{b}\right) = \left(\frac{1}{\lambda_g^2} - \frac{1}{\lambda_0^2}\right) Y \cos\left(\frac{n\pi y}{b}\right) \tag{9.102}$$

$$-\frac{n^2\pi^2}{b^2} = \left(\frac{1}{\lambda_g^2} - \frac{1}{\lambda_0^2}\right) \tag{9.103}$$

for $n = 1, 2, 3, \ldots$. This allows only eigenvalues of λ_0 corresponding to integer values of n.

Also notice that the left-hand side of equation (9.103) is less than zero. This requires the radian length to be greater than the plane wave length of the same frequency propagating in the same medium, i.e., a zig-zag path:

$$\lambda_g > \lambda_0 \tag{9.104}$$

There are two remaining quantities to determine:

$$E_{0x} \text{ and } H_{0y} \tag{9.105}$$

The student is requested to combine the above analysis with equations (9.5), (9.6), (9.44), (9.47), and equations (9.99) and (9.103) to determine the complete solution to the TE waves in a hollow metallic waveguide (see Problem 9.9).

$$\mathbf{E} = \frac{j\omega\mu bY}{n\pi} \sin\left(\frac{n\pi}{b} y\right) \exp[j(wt - k_g z)] \tag{9.106}$$

and

$$\mathbf{H} = \frac{jbY}{\lambda_g} \sin\left(\frac{n\pi}{b} y\right) \exp[j(\omega t - k_g z)]$$

$$+ Yn\pi \cos\left(\frac{n\pi}{b} y\right) \exp[j(\omega t - k_g z)]\hat{\mathbf{k}} \tag{9.107}$$

These waves have longitudinal modes. The waves zig-zag down the waveguide. Each higher mode represents a smaller angle of incidence, more reflections, a longer time of travel from beginning to end, and a lower

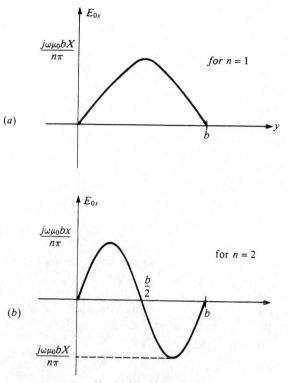

FIGURE 9.5 Electric field intensity for (a) $n = 1$ and (b) $n = 2$.

intensity. This difference in propagation time for the different modes is called *modal dispersion*. It occurs both in metallic and dielectric waveguides.

The amplitude of **E** for a TE wave in a hollow rectangular waveguide can be represented as in Figure 9.5. The different values of n correspond to different angles of incidence θ_i inside the guide. If $\lambda_0 > 2b/n$ there will be no propagation. Waveguides pass high frequencies and reject low frequencies. This is called the cutoff wavelength for the waveguide.

At this time, we need to discuss the character of the incident and reflected waves. We will use a more general case of a plane wave solution incident on a dielectric interface.

9.7 REFRACTION AND REFLECTION

For a plane wave incident on the interface at angle θ_i (using equation 9.5)

$$\mathbf{E}_i = \mathbf{E}_{0i} \exp[j(\omega t - k_1 \hat{\mathbf{n}}_i \cdot \mathbf{r})] \tag{9.108}$$

9.7 REFRACTION AND REFLECTION

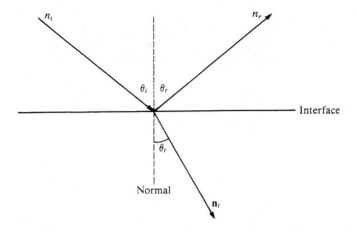

FIGURE 9.6 Refraction and reflection at an interface.

and the reflected and transmitted waves respectively:

$$\mathbf{E}_r = \mathbf{E}_{0r} \exp\left[j(\omega t - k_2 \hat{\mathbf{n}}_r \cdot \mathbf{r} + A)\right] \tag{9.109}$$

$$\mathbf{E}_t = \mathbf{E}_{0t} \exp\left[j(\omega t - k_2 \hat{\mathbf{n}}_t \cdot \mathbf{r} + B)\right] \tag{9.110}$$

where A and B are possible changes in phase (see Figure 9.6).

The tangential components of $\mathbf{E}$ and $\mathbf{H}$ must be continuous across the interface I. Therefore, for $\mathbf{r} = \mathbf{r}_I$ at the interface:

$$\omega_i t - k_1 \hat{\mathbf{n}}_i \cdot \hat{\mathbf{r}}_I = \omega_r t - k_1 \hat{\mathbf{n}}_r \cdot \hat{\mathbf{r}}_I + A \tag{9.111}$$

$$= \omega_t t - k_2 \hat{\mathbf{n}}_t \cdot \hat{\mathbf{r}}_I + B \tag{9.112}$$

for all t and for all $\hat{\mathbf{r}}_I$. Therefore

$$\omega_i = \omega_r = \omega_t \tag{9.113}$$

and for all $\mathbf{r}_I$

$$k_1 \hat{\mathbf{n}}_i \cdot \hat{\mathbf{r}}_I = k_1 \hat{\mathbf{n}}_r \cdot \hat{\mathbf{r}}_I - A = k_2 \hat{\mathbf{n}}_t \cdot \hat{\mathbf{r}}_I - B \tag{9.114}$$

$$(\hat{\mathbf{n}}_i - \hat{\mathbf{n}}_r) \cdot \hat{\mathbf{r}}_I = -\frac{A}{k_1} = \text{constant} \tag{9.115}$$

for all $\hat{\mathbf{r}}_I$. Therefore, $(\hat{\mathbf{n}}_i - \hat{\mathbf{n}}_r)$ must be normal to the interface and tangential components must be equal and opposite in sign. Therefore

$$\theta_i = \theta_r \tag{9.116}$$

The angle of reflection is equal to the angle of incidence and the plane of incidence is normal to the interface. Also

$$(k_1\hat{\mathbf{n}}_i - k_2\hat{\mathbf{n}}_t) \cdot \hat{\mathbf{r}}_I = -B = \text{constant} \tag{9.117}$$

Therefore, $(k_1\hat{\mathbf{n}}_i - k_2\hat{\mathbf{n}}_t)$ is also normal to the interface and the tangential components must be equal:

$$k_1 \sin \theta_i = k_2 \sin \theta_t \tag{9.118}$$

$$\frac{n_1}{\lambda_0} \sin \theta_i = \frac{n_2}{\lambda_0} \sin \theta_t \tag{9.119}$$

$$n_1 \sin \theta_i = n_2 \sin \theta_t \tag{9.120}$$

The values n_1 and n_2 are the index of refraction for materials 1 and 2. Equation (9.120) is known as *Snell's law of refraction*.

We also need to know the intensities of the reflected and transmitted waves. These are determined by equations known as *Fresnel's equations*. They define the amplitude relationships between $\mathbf{E}_i$, $\mathbf{E}_r$, and $\mathbf{E}_t$. For continuity of magnetic field intensity normal to the interface (Figure 9.7):

$$H_{0i} \cos \theta_i - H_{0r} \cos \theta_i = H_{0t} \cos \theta_t \tag{9.121}$$

The electric field normal (N) to the interface must be conserved (Figure 9.8)

$$(E_{0i})_N - (-E_{0r})_N = (E_{0t})_N \tag{9.122}$$

For a plane wave:

$$-\frac{E_y}{H_x} = \frac{E_x}{H_y} = \frac{\omega \mu}{k} \tag{9.123}$$

Therefore, using equation (9.121):

$$\frac{k_1}{\omega \mu_1} (E_{0i} - E_{0r})_N \cos \theta_i = \frac{k_2}{\omega \mu_2} (E_{0t})_N \cos \theta_t \tag{9.124}$$

Divide by E_{0i}:

$$\frac{k_1}{\omega \mu_1} \left(1 - \frac{E_{0r}}{E_{0i}}\right)_N \cos \theta_i = \frac{k_2}{\omega \mu_2} \left(\frac{E_{0t}}{E_{0i}}\right)_N \cos \theta_t \tag{9.125}$$

and rearranging equation (9.122) we obtain:

$$\left(1 + \frac{E_{0r}}{E_{0i}}\right)_N = \left(\frac{E_{0t}}{E_{0i}}\right)_N \tag{9.126}$$

$$\frac{k_1}{\omega \mu_1} \left(2 - \frac{E_{0t}}{E_{0i}}\right)_N \cos \theta_i = \frac{k_2}{\omega \mu_2} \left(\frac{E_{0t}}{E_{0i}}\right)_N \cos \theta_t \tag{9.127}$$

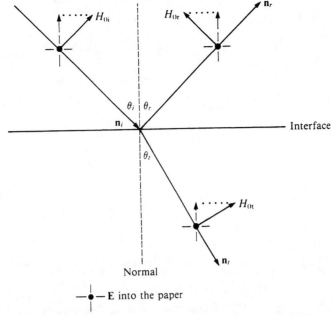

FIGURE 9.7 Continuity of **E** parallel (P) to the interface.

Therefore

$$2\frac{k_1}{\omega\mu_1}\cos\theta_i = \left(\frac{E_{0t}}{E_{0i}}\right)_N \left(\frac{k_1}{\omega\mu_1}\cos\theta_i + \frac{k_2}{\omega\mu_2}\cos\theta_t\right) \qquad (9.128)$$

$$\left(\frac{E_{0t}}{E_{0i}}\right)_N = \frac{\dfrac{2k_1}{\omega\mu_1}\cos\theta_i}{\dfrac{k_1}{\omega\mu_1}\cos\theta_i + \dfrac{k_2}{\omega\mu_2}\cos\theta_t} \qquad (9.129)$$

and

$$\left(\frac{E_{0r}}{E_{0i}}\right)_N = \frac{\dfrac{k_1}{\omega\mu_1}\cos\theta_i - \dfrac{k_2}{\omega\mu_2}\cos\theta_t}{\dfrac{k_i}{\omega\mu_1}\cos\theta_i + \dfrac{k_2}{\omega\mu_2}\cos\theta_t} \qquad (9.130)$$

Repeat the procedure for **H** normal to the plane of incidence (see Figure 9.7).

Therefore the magnetic field must be conserved through the interface for $(E)_P$

$$(H_{0i} - H_{0r})_P = (H_{0t})_P \qquad (9.131)$$

with

$$-\frac{E_y}{H_x} = \frac{E_x}{H_y} = \frac{\omega\mu}{k} \qquad (9.132)$$

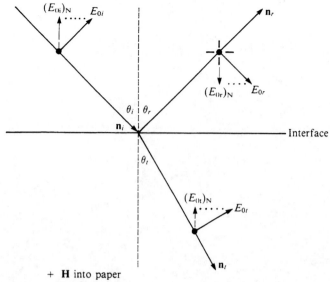

+ **H** into paper
• **H** out of paper

FIGURE 9.8 Continuity of **H** with component of **E** normal to interface.

Therefore

$$\frac{k_1}{\omega\mu_1}(E_{0i} - E_{0r})_P = \frac{k_2}{\omega\mu_2}(E_{0t})_P \tag{9.133}$$

Divide by E_{0i}:

$$\frac{k_1}{\omega\mu_1}\left(1 - \frac{E_{0r}}{E_{0i}}\right)_P = \frac{k_2}{\omega\mu_2}\left(\frac{E_{0t}}{E_{0i}}\right)_P \tag{9.134}$$

$$\left(\frac{E_{0r}}{E_{0i}}\right)_P = -\frac{\omega\mu_1}{k_1}\frac{k_2}{\omega\mu_2}\left(\frac{E_{0t}}{E_{0i}}\right)_P + 1 \tag{9.135}$$

Also

$$E_{0i}\cos\theta_i + E_{0r}\cos\theta_i = E_{0t}\cos\theta_t \tag{9.136}$$

$$\left(1 - \frac{E_{0r}}{E_{0i}}\right)_P \cos\theta_i = \left(\frac{E_{0t}}{E_{0i}}\right)_P \cos\theta_t \tag{9.137}$$

$$\left[2 - \frac{\omega\mu_1}{k_1}\frac{k_2}{\omega\mu_2}\left(\frac{E_{0t}}{E_{0i}}\right)_P\right]\cos\theta_i = \left(\frac{E_{0t}}{E_{0i}}\right)_P \cos\theta_t \tag{9.138}$$

$$\frac{2k_1}{\omega\mu_1}\cos\theta_i - \frac{k_2}{\omega\mu_2}\left(\frac{E_{0t}}{E_{0i}}\right)_P \cos\theta_i = \frac{k_1}{\omega\mu_1}\left(\frac{E_{0t}}{E_{0i}}\right)_P \cos\theta_t \tag{9.139}$$

$$\left(\frac{E_{0t}}{E_{0i}}\right)_P = \frac{\dfrac{2k_1}{\omega\mu_1}\cos\theta_i}{\dfrac{k_1}{\omega\mu_1}\cos\theta_t + \dfrac{k_2}{\omega\mu_2}\cos\theta_i} \tag{9.140}$$

likewise:

$$\left(\frac{E_{0r}}{E_{0i}}\right)_P = \left(\frac{E_{0r}}{E_{0i}}\right)_P \frac{\cos\theta_t}{\cos\theta_i} - 1 \qquad (9.141)$$

$$\left(\frac{E_{0r}}{E_{0i}}\right)_P = \frac{\dfrac{k_1}{\omega\mu_1}\cos\theta_t - \dfrac{k_2}{\omega\mu_2}\cos\theta_i}{\dfrac{k_1}{\omega\mu_1}\cos\theta_t + \dfrac{k_2}{\omega\mu_2}\cos\theta_i} \qquad (9.142)$$

In summary, the following represent the four Fresnel equations that relate the intensities of the incident, reflected, and refracted waves:

$$\left(\frac{E_{0t}}{E_{0i}}\right)_N = \frac{\left(\dfrac{2k_1}{\omega\mu_1}\cos\theta_i\right)}{\left(\dfrac{k_1}{\omega\mu_1}\cos\theta_i + \dfrac{k_2}{\omega\mu_2}\cos\theta_t\right)} \qquad (9.143)$$

E *normal*:

$$\left(\frac{E_{0r}}{E_{0i}}\right)_N = \frac{\left(\dfrac{k_1}{\omega\mu_1}\cos\theta_i - \dfrac{k_2}{\omega\mu_2}\cos\theta_t\right)}{\left(\dfrac{k_1}{\omega\mu_1}\cos\theta_i + \dfrac{k_2}{\omega\mu_2}\cos\theta_t\right)} \qquad (9.144)$$

$$\left(\frac{E_{0t}}{E_{0i}}\right)_P = \frac{\left(\dfrac{2k_1}{\omega\mu_1}\cos\theta_i\right)}{\left(\dfrac{k_1}{\omega\mu_1}\cos\theta_t + \dfrac{k_2}{\omega\mu_2}\cos\theta_i\right)} \qquad (9.145)$$

E *parallel*:

$$\left(\frac{E_{0r}}{E_{0i}}\right)_P = \frac{\left(\dfrac{k_1}{\omega\mu_1}\cos\theta_t - \dfrac{k_2}{\omega\mu_2}\cos\theta_i\right)}{\left(\dfrac{k_1}{\omega\mu_1}\cos\theta_t + \dfrac{k_2}{\omega\mu_2}\cos\theta_i\right)} \qquad (9.146)$$

There is a critical characteristic of the reflection of a wave from a dielectric interface known as the *Brewster angle*. The Brewster angle can be determined from Fresnel's equations with $\mu_1 \approx \mu_2 \approx 1$ and $k = n/\lambda_0$, with **E** parallel to the plane of incidence:

$$\left(\frac{E_{0r}}{E_{0i}}\right)_P = \frac{n_1 \cos\theta_t - n_2 \cos\theta_i}{n_1 \cos\theta_t + n_2 \cos\theta_i} \qquad (9.147)$$

or

$$\left(\frac{E_{0r}}{E_{0i}}\right)_P = \theta \quad \text{if} \quad \cos\theta_t - \frac{n_2}{n_1}\cos\theta_i = 0 \qquad (9.148)$$

from Snell's law and from $\left(\dfrac{E_{0r}}{E_{0i}}\right)_P$:

$$n_1 \sin \theta_i = n_2 \sin \theta_t \tag{9.149}$$

Therefore

$$\cos \theta_t - \dfrac{\sin \theta_i}{\sin \theta_t} \cos \theta_i = 0 \tag{9.150}$$

$$\cos \theta_t \sin \theta_t - \sin \theta_i \cos \theta_i = 0 \tag{9.151}$$

$$\sin 2\theta_t - \sin 2\theta_i = 0 \tag{9.152}$$

because

$$\sin 2x = 2 \cos x \sin x \tag{9.153}$$

$$\tfrac{1}{2}[\sin(\theta_t + \theta_t) - \sin(\theta_i + \theta_i)] = \cos(\theta_t + \theta_i) \sin(\theta_t - \theta_i) = 0 \tag{9.154}$$

because

$$\cos x \sin y = \tfrac{1}{2}[\sin(x+y) - \sin(x-y)] \tag{9.155}$$

while

$$x = \theta_t + \theta_i \quad \text{and} \quad y = \theta_t - \theta_i \tag{9.156}$$

then no reflectance occurs when

$$(\theta_t + \theta_i) = \dfrac{\pi}{2} \quad \text{for} \quad \left(\dfrac{E_{0r}}{E_{0i}}\right)_P = 0 \tag{9.157}$$

then at the incident Brewster angle

$$\dfrac{n_1}{n_2} = \dfrac{\sin \theta_t}{\sin \theta_{iB}} = \dfrac{\sin\left(\dfrac{\pi}{2} - \theta_{iB}\right)}{\sin \theta_{iB}} \tag{9.158}$$

$$\dfrac{n_1}{n_2} = \dfrac{\cos \theta_{iB}}{\sin \theta_{iB}} = \cot \theta_{iB} \tag{9.159}$$

But $\left(\dfrac{E_{0r}}{E_{0i}}\right)_N$ is nonzero for $\theta_i + \theta_t = \dfrac{\pi}{2}$.

Therefore, for unpolarized waves only the normal component of the electric field intensity will be reflected at the Brewster angle θ_{iB}. Thereby reflection polarizes reflected light at this angle, with the **E** vector normal to the plane of incidence. (See Chapter 8 for additional discussion of polarization of light.) This condition gives us additional understanding of why only TE or TM waves will propagate in a waveguide. The zig-zag path can only allow TE or TM waves.

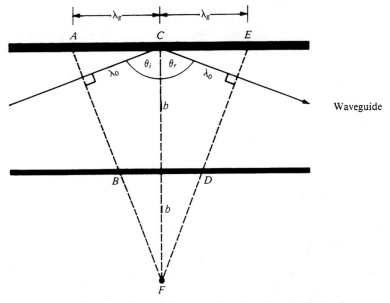

FIGURE 9.9 Waveguide geometry showing zig-zag path.

In the zig-zag path of the wave in the guide, the angle of incidence θ_i and reflection θ_r are shown in Figure 9.9. AB and ED represent incident wave and reflected wavefronts of the TE wave and $\theta_i = \theta_r$ (see Figure 9.10).

$$\cos \theta_i = \frac{\lambda_0}{2b} \tag{9.160}$$

or for a high-mode m:

$$\cos \theta_i = \frac{m\lambda_0}{2b} \tag{9.161}$$

Therefore the angle θ_i can have only discrete values. They simply represent a geometric interpretation of the modes found from the wave equations.

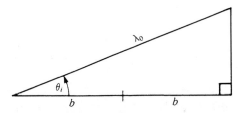

FIGURE 9.10 Constructing the triangle.

In the limiting conditions, θ_i approaches 90°, e.g.,

$$\theta_i \to \frac{\pi}{2} \tag{9.162}$$

Then

$$\cos \theta_i \to 0 \tag{9.163}$$

which prevents propagation because

$$\text{TE waves} \to \text{TEM waves} \tag{9.164}$$

That is, the wavefronts are parallel to the x and y axes of the guide. The internal reflections of the ray then tend to polarize the light. This is true because the normal component is reflected and the parallel component is transmitted through the outside edge of the waveguide. This explains why only TE or TM modes propagate. The propagation creates polarized light, selecting only transverse (polarized) modes that are internally reflected.

9.8 REFLECTANCE

The reflectance, as discussed in Chapter 8, can be calculated from the Fresnel equations. Using equation (9.146) once more:

$$\left(\frac{E_{0r}}{E_{0i}}\right)_P = \frac{n_1 \cos \theta_t - N_2 \cos \theta_i}{n_1 \cos \theta + N \cos \theta} \tag{9.165}$$

$$\theta_i = 0 = \theta_t \tag{9.166}$$

$$\cos \theta_i = \cos \theta_t = 1 \tag{9.167}$$

$$\left(\frac{E_{0r}}{E_{0i}}\right)_P = \frac{n_1 - n_2}{n_1 + n_2} \tag{9.168}$$

The reflectance R is defined as

$$R \equiv \left(\frac{E_{0r}}{E_{0i}}\right)^2 \tag{9.169}$$

Therefore for $\theta_i = \theta_t = \text{zero}$:

$$R = \left(\frac{n_1 - n_2}{n_1 + n_2}\right)^2 \tag{9.170}$$

By inspecting the Fresnel equation (9.144) for the normal component of the polarization, we can see that an identical result for the reflectance can be calculated.

9.9 DIELECTRIC WAVEGUIDES

Optical waveguides and fibers are designed such that the center of the guide has a higher index of refraction than the outside. The analysis for the hollow metal waveguide gives a satisfactory picture of the waveguide mechanisms. The same aproach can be applied to *dielectric waveguides*.

We found that TE or TM waves are the only waves that can propagate in a metallic waveguide; TEM waves attenuate rapidly. This is also true for dielectric waveguides. However, there are characteristic differences that arise because of the very different boundary conditions at the inner surface of the dielectric waveguide. In the case of a dielectric waveguide, there are sets of additional wave equations. There are the wave equations inside the guide (just like the hollow metallic analysis) and a set of wave equations outside the dielectric guide, i.e., in the cladding.

Fortunately, there is also a characteristic equivalence of dielectric waveguides and metallic waveguides. The TM mode in a dielectric guide is equivalent to the TE mode in a metal waveguide. This symmetry can be traced to Maxwell's equations. Thus, **E** and **B** can be interchanged with the appropriate change in sign and change in the factor $[\epsilon\mu]^{1/2}$ in the equations. Secondly, the normal component of **B** almost vanishes for a metallic surface, whereas the normal **E** almost vanishes for the dielectric interface.

The experimental decay of the **E** and **B** vectors inside and outside of the dielectric waveguide has the following form:

$$\mathbf{E}_i = E_{0i} \exp(-jk_1 x) \exp(j(\omega t - k_g z)] \tag{9.171}$$

$$\mathbf{E}_r = E_{0e} \exp(jk_1 x) \exp(j(\omega t - k_g z)] \tag{9.172}$$

$$\mathbf{E}_t = E_{0i} \exp(-jk_2 x) \exp(j(\omega t - k_g z)] \tag{9.173}$$

where

$$k_1^2 = n_1^2 k_0^2 - k_g^2 = -\delta_1^2 \tag{9.174}$$

$$k_2^2 = n_2^2 k_0^2 - k_g^2 = -\delta_2^2 \tag{9.175}$$

Then the exponential decay of the $\mathbf{E}_t$ and $\mathbf{B}_t$ vectors in the cladding becomes

$$\mathbf{E}_t = E_{ot} \exp(-\delta_2 x) \exp(j(\omega t - k_g z)] \tag{9.176}$$

The sign of δ_2 could have been chosen positive, but this would yield an exponentially growing field outside the waveguide. This decaying field (see Figure 9.11) accounts for the losses in dielectric waveguides. Both **E** and **B** exist for short distances. The cross product or pointing vector represents the power removed from the guided wave as the guided wave propagates.

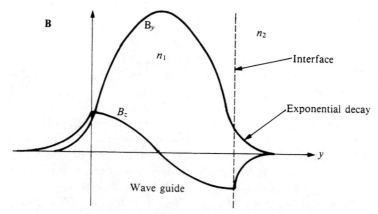

FIGURE 9.11 Dielectric waveguide solution.

9.10 GOOS–HAENCHEN SHIFT

In real systems, the incident wave and the reflected wave make the same angle, $\theta_i = \theta_r$. The amplitude for the reflected wave is slightly lower, due to the losses from the propagating wave. Finally, equation (9.109) and (9.110) indicates a possible phase shift between the incident and reflected wave. In actuality, this phase shift does occur and is known as the *Goos–Haenchen shift*. This shift can be thought of as a physical shift of the point of reflection in the direction of propagation, Δz (Figure 9.12).

9.11 GRADED-INDEX WAVEGUIDES

Optical waveguides are designed such that the core has a slightly higher index of refraction than the cladding. This keeps most of the light rays inside the waveguide by reflection from the interface. Unless the guide or fiber takes a sharp bend, the rays will zig-zag from one side to the other.

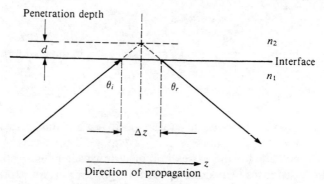

FIGURE 9.12 Goos–Haenchen Shift

9.11 GRADED-INDEX WAVEGUIDES

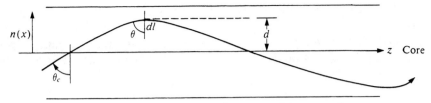

FIGURE 9.13 Graded-index waveguide ray path.

Only rays that enter the guide at small angles to the normal can escape. If the core has a uniform index of refraction, we have seen that multiple modes of reflection exist. Rays that have the higher modes make many more reflections and follow a longer path than those of lower modes. Rays with higher modes will arrive behind rays that make fewer reflections. This defect, called *modal dispersion*, is partially remedied by producing a waveguide whose index of refraction increases near the central axis of the waveguide. In a *graded-index waveguide*, rays that deviate from the central axis travel faster (because of the lower index) than rays that deviate less. This tends to reduce the effects of the longer path length.

We can analyze the graded-index waveguide (Figure 9.13), from Snell's law.

$$n(x) \sin \theta = n_c \sin \theta_c \qquad (9.177)$$

Differentiating with respect to dl (ray path),

$$\left(\frac{d}{dl} n(x)\right) \sin \theta + n(x) \frac{d}{dl} \sin \theta = \left(\frac{d}{dl} n_c\right) \sin \theta_c + n_c \frac{d}{dl} \sin \theta_c \qquad (9.178)$$

$$\sin \theta \frac{dn(x)}{dl} + n(x) \cos \theta \frac{d\theta}{dl} = n_c \cos \theta_c \frac{d\theta_c}{dl} = 0 \qquad (9.179)$$

$$n(x) \frac{d\theta}{dl} = -\frac{\sin \theta}{\cos \theta} \frac{d\theta}{dl} \qquad (9.180)$$

$$\frac{d\theta}{dl} = \frac{-1}{n(x)} \frac{dn(x)}{dl} \tan \theta \qquad (9.181)$$

Also $d\theta/dl$ is the reciprocal of the radius of curvature R:

$$\frac{1}{R} = -\frac{1}{n(x)} \frac{dn(x)}{dl} \tan \theta \qquad (9.182)$$

This shows that the ray bends more sharply as the change in $n(x)$ increases. If $n(x)$ decreases then the ray is bent back toward the core. If $n(x)$ increases, then refraction occurs and the ray escapes from the waveguide.

9.12 RAY DEVIATIONS FROM CENTRAL CORE

The maximum deviation from the center of the waveguide is indicated by the distance d. At this point, the slope of the ray with respect to the z-axis is zero. At $x = d$:

$$\sin \theta = 1 \tag{9.183}$$

substituting into equation (9.69):

$$n(x = d) = n_c \sin \theta_c \tag{9.184}$$

Therefore

$$\frac{n(d)}{n_c} = \sin \theta_c \tag{9.185}$$

Define a parabolic gradient-index profile to be

$$n(x) \equiv n_c(1 - Ax^2) \tag{9.186}$$

Substituting into equation (9.73):

$$(1 - Ax^2)_{x=d} = \sin \theta_c \tag{9.187}$$

$$1 - Ad^2 = \sin \theta_c \tag{9.188}$$

$$-Ad^2 = \sin \theta_c - 1 \tag{9.189}$$

$$d = \left[\frac{1 - \sin \theta_c}{A}\right]^{1/2} \tag{9.190}$$

Remembering that the possible modes, m can be represented by equation (9.67):

$$\cos \theta_i = \frac{m\lambda_0}{2b} \tag{9.191}$$

where b becomes the radius of the waveguide. Remembering also the trigonometric relationship between $\sin \theta$ and $\cos \theta$:

$$\sin \theta = (1 - \cos^2 \theta)^{1/2} \tag{9.192}$$

Letting $\theta_i = \theta_c$ and combining terms:

$$\sin \theta_c = \left(1 - \frac{m^2 \lambda_0^2}{4b^2}\right)^{1/2} \tag{9.193}$$

Then substituting this into equation (9.75):

$$d = \left[\frac{1}{A}\left(1 - \left(1 - \frac{m^2\lambda_0^2}{4b^2}\right)^{1/2}\right)\right]^{1/2} \quad (9.194)$$

Then by using a binomial expansion the expression can be simplified:

$$\left(1 - \frac{m^2\lambda_0^2}{4b^2}\right)^{1/2} \approx 1 - \frac{1}{2}\left(\frac{m^2\lambda_0^2}{4b^2}\right) + \ldots \quad (9.195)$$

Then the distance above the central axis of the waveguide can be expressed as

$$d = \left[\frac{1}{A}\left(\frac{1}{2}\frac{m^2\lambda_0^2}{4b^2}\right)\right]^{1/2} \quad (9.196)$$

Then

$$d = \frac{1}{2\sqrt{2A}}\left(\frac{m\lambda_0}{b}\right) \quad (9.197)$$

Therefore, we can see from equation (9.197) that the distance d from the central axis increases proportionally with the mode m. Likewise, the longer wavelengths λ_0 will also give greater distances from the central core.

Figure 9.14 represents typical ray paths within a graded-index waveguide. These different paths clearly have different path lengths. The analysis of the time of travel for these various modes is complex and is treated by Midwinter (1979).

The dispersion measured through a waveguide can be expressed by the widening of a guided pulse of light after traveling the length of the guide. For a graded-index waveguide, the index can be expressed by

$$n(x) = n_0\left[1 - 2A\left(\frac{x}{b}\right)^\alpha\right]^{1/2} \quad \text{for} \quad x < b \quad (9.198)$$

inside the waveguide and

$$n(x) = n_0[1 - 2A)^{1/2} \quad \text{for} \quad x \geq b \quad (9.199)$$

outside the waveguide.

Several assumptions concerning the analysis must be made:

1. The graded-index portion of the waveguide has a radius b which is approximately given by

$$b \approx 100\lambda_0 \quad (9.200)$$

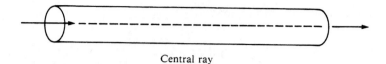

Central ray

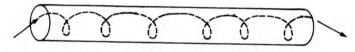

Sinusoidal-like path

Helical-like path

FIGURE 9.14 Various ray paths in a graded-index waveguide.

2. Variations in the index are small, so that modes can be considered transverse electro-magnetic modes in approximation.
3. The index variation is very small over distances of a wavelength, so that the waves may be approximated by plane waves.

Let Δt be the difference in travel time for two different modes m and m'.

$$\Delta t = [t_m^2 - t_{m'}^2]^{1/2} \tag{9.201}$$

The pulse width for a step-index (nongraded) waveguide can be shown to be

$$\Delta t_{(\text{step})} \approx \frac{LN_1 A}{2\sqrt{3}c} \tag{9.202}$$

where L = path length
N_0, N_1 = measure of group index
c = speed of light

Likewise, for a graded index (with $\alpha = 2$)

$$\Delta t_{(\alpha=2)} = \frac{LN_0 A}{c} \left(\frac{\sqrt{3} A}{24} \right) \tag{9.203}$$

creating the ratio:

$$\frac{\Delta t_{(\text{step})}}{\Delta t_{(\alpha=2)}} \simeq \frac{4}{A} \qquad (9.204)$$

A more precise evaluation yields the result of

$$\frac{\Delta t_{(\text{step})}}{\Delta t_{(\alpha=2)}} = \frac{10}{A} \qquad (9.205)$$

For real systems the graded-index constant $A = 0.01$ and $\Delta t_{(\text{step})}$ is typically 15 nsec/Km. This theoretically shows that a graded-index waveguide could reduce modal dispersion by a factor of 1000:

$$\Delta t_{(\alpha=2)} \approx 0.015 \text{ nsec/Km} \qquad (9.206)$$

Current levels of technology in graded-index waveguides have yielded dispersion in the region of

$$\Delta t \approx 1 \text{ nsec/Km} \quad (\text{LED source}) \qquad (9.207)$$

$$\Delta t \approx 0.08 \text{ nsec/Km} \quad (\text{GaAs laser source}) \qquad (9.208)$$

9.13 PLANAR WAVEGUIDES

John E. Midwinter (1979) has an excellent treatment of optical fiber waveguides and their applications. His work details the methods of manufacture and testing of various optical fiber systems. Some of the techniques discussed below for planar waveguides has its beginnings in the fiber-optics field.

Typical planar waveguides are of the step-index design and have a core with a slightly higher index than that of the surrounding material. This corresponds exactly to the work previously discussed for waveguides in general.

The thickness of the planar waveguide determines the number of modes that will propagate. Equation (9.161) may be more precisely derived and solved for the thickness b:

$$b = \left(\frac{m + \frac{1}{2}}{2}\right) \frac{\lambda_0}{\sqrt{n_1^2 - n_2^2}} \qquad (9.209)$$

where n_1 = index of guide region
  n_2 = index of substrate
  m = mode number

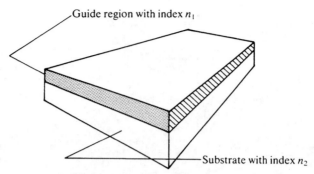

FIGURE 9.15 Planar waveguide.

Figure 9.15 shows a schematic of planar waveguide. Using typical optical glasses, equation (9.209) yields a thickness on the order of one or two wavelengths.

The index step is usually produced by exchanging some of the cations in a sodium borosilicate glass. The width of these waveguides is as small as possible. This allows for single-mode operation and thereby reduces modal dispersion. The exchange process is started by masking off the areas where the index is to be left unchanged. A metal salt such as silver iodine is painted over the surface of the planar waveguide blank. The blank is then heated to 300°C to 400°C where diffusion of the sodium out and the silver in takes place. After cooling the planar waveguide is ground and polished to allow coupling into optical systems (Ramaswamy et al., 1988).

9.14 LASER FABRICATION OF GEL-SILICA WAVEGUIDES

Recent work by Wang, Campbell, and Hench (1988) suggests that the optical transmission and index of refraction can be improved over currently available pure silica glasses. The range of useful wavelengths for gel silica is greatly improved over standard sodium borosilicate glass discussed above. Figure 9.16 shows the absorption for pure gel silica compared with that of a typical sodium borosilicate used in waveguides. The increased range is significant. The transmission for partially dense type VI gel silica (Hench et al., 1988) is shown in Figure 8.13, the transmission is still excellent as compared to commercial waveguide materials.

Miliou et al., (1988) have demonstrated that waveguiding regions can be produced by laser densification of sol-gel silica glass. For normal glasses there would be no densification effect. This would be true for laser heating or other heat treatment methods. With sol-gel silica partially dense type VI

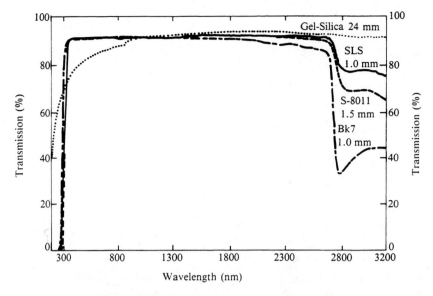

FIGURE 9.16 Partially dense gel-silica transmission.

silica can be used as a substrate. Upon laser heating, the region under the focussed beam is densified. The change in index reported above is $\Delta n \approx 0.06$ in a multimode waveguide region $40\mu m$ wide.

9.15 APPLICATION OF THE FRESNEL EQUATIONS

In order to determine the coefficients of absorption, the reflection, and transmission for a specific wavelength Fresnel's equations must be used.

For normally incident light on glass surface (see Figure 9.17):

$$I_0 - I_0R - I_\alpha - (I_0 - I_0R - I_\alpha)R = I_m \tag{9.210}$$

TABLE 9.2

Temp. (°C)	n	Trans. (1.3 μm)	Trans. (0.63 μm)	α(1.3 μm)	α(0.63 μm)	Thickness (cm)
1200	1.46	92.70	86.01	1.8×10^{-3}	0.03	2.40
900	1.43	88.01	86.50	0.10	0.13	0.60
850	1.42	83.50	85.50	0.27	0.21	0.45
775	1.38	81.50	75.50	0.34	0.51	0.45

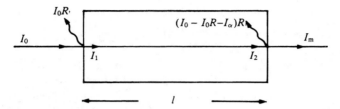

FIGURE 9.17 Transmission losses through a sample.

where I_0 = incident strength
I_α = internal absorbed strength
I_m = final strength
R = reflectance
$T = I_m/I_0$ external transmission
$A = I_\alpha/I_0$ internal absorption
$I_1 = I_0/I_0 R$
$I_2 = (I_0/I_0 R/ - I_\alpha)R$
I_2/I_1 = internal transmission = $\exp(-\alpha \cdot l)$
α = absorption coefficient
l = optical length

The solution for the external transmission

$$T = (1 - R)^2 \exp(-\alpha \cdot l) \tag{9.211}$$

For a known refractive index of the material, the reflectance is given by equation (9.170) for the angle of incidence of 90°:

$$R = [(n-1)/(n+1)]^2 \tag{9.212}$$

Also, the internal absorption coefficient is given by:

$$\alpha = \ln[(1 - r - A)/(1 - R)]/l \tag{9.213}$$

Substituting these into equation (9.86) and solving for A, we have:

$$A = [(1-R)^2 - T]/(1-R) \tag{9.214}$$

Using this method, the internal absorption coefficient of gel silica is 0.00276 cm^{-1} (or 0.012 dB/cm) at 1.3 μm (wavelength). The absorption coefficients and index of refraction are presented in Table 9.2 for different densification temperatures for sol-gel silica glass.

9.16 INFRARED WAVEGUIDES

The limits on performance of either fiber optics or planar waveguides based on silica are determined largely by the sum of absorption and scattering losses (see Chapter 8). As illustrated in Figure 9.18, the IR absorption loss for silica based optics intersects Rayleigh scattering losses at approximately 1.3 μm. This leads to a theoretical loss limit of $\sim$0.3 db/km for silica waveguides.

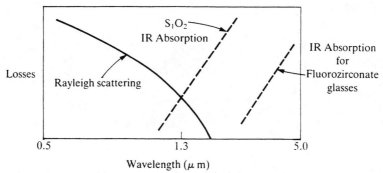

FIGURE 9.18 Losses in silica and fluorozirconate glasses.

For fluorozirconate glasses (Appedix A) losses as low as 0.01 db/km are possible at 3 μm.

9.17 WAVEGUIDE INTEGRATED OPTICS

An integrated optical phase shifter is shown in Figure 9.18. A strip waveguide is formed on an optically active substrate. Electrodes are overlaid on the substrate on both sides of the waveguide. The separation of the electrodes is D with a length L. The phase shift $\Delta\phi$ is produced when a voltage is applied to the surface electrodes. The phase shift is (see Chapter 10)

$$\Delta\phi = \frac{\pi}{\lambda} rn_1^3 V \frac{L}{D} \qquad (9.215)$$

where λ = wavelength
r = electro-optical coefficient

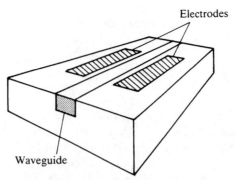

FIGURE 9.19 Optical phase shifter.

In these devices, if the ratio of

$$\frac{L}{D} \approx 1000 \tag{9.216}$$

then a phase shift of

$$\Delta\phi \approx \pi \text{ radians} \tag{9.217}$$

is possible for potentials near 1 V.

Using a similar device, an interferometric modulator can be produced. Figure 9.20 shows the modulator where one leg produces a phase shift after the guided wave is split. After the phase is shifted, the waves are mixed to form a modulated output.

Two types of optical switching devices are schematically shown in Figures 9.21 and 9.22. The first makes use of electro-static proximity coupling. If a voltage is applied across the electrodes, the electric field transfers energy to

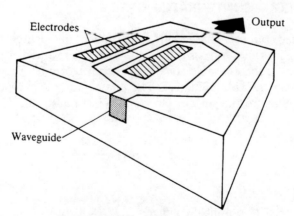

FIGURE 9.20 Optical modulator

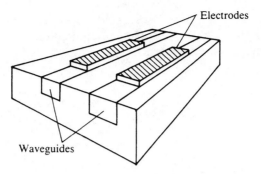

FIGURE 9.21 Optically coupled switch.

the other waveguide. The coupled light is now propagating out of the other waveguide. Thus, an optical switch has been created.

The second switch is based on Bragg diffraction through regions with a periodically changing index. The electrodes create local variations in the index of refraction when a potential is applied. Remove the potential and the incident ray is transmitted without switching.

As in Bragg diffraction from crystals by electrons, photons will diffract when they encounter a periodically changing index. The angle of diffraction is determined through the Bragg relation:

$$\sin \theta = \frac{m\lambda_0}{2(\Delta n)\, d} \qquad (9.218)$$

where Δn = change in index
 d = separation of electrodes
 m = integer of diffraction maxima
 λ_0 = photon wavelength

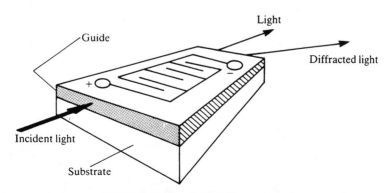

FIGURE 9.22 Optical Bragg switch.

432 CHAPTER 9: OPTICAL WAVEGUIDES

Coupling of waveguides and waveguide devices with electro-optic elements (Chapter 10), can produce many integrated optical functions that are analogous to integrated electronics.

PROBLEMS

9.1 Derive the wave equation (9.8) from Maxwell's equations and show that equations (9.3 and 9.4) represent solutions.

9.2 Verify that equation (9.89) represents a solution to equation (9.88).

9.3 Derive equations (9.45) and (9.46) showing that the guided wave is completely determined once E_{0z} and H_{0z} are known.

9.4 The index of refraction varies with radial distance from the core of a graded index waveguide. Using a parabolic graded index profile:
 (a) Calculate and plot the index as a function of radius, x. Assume that $n(core) = 1.4$ and $A = 0.01/(\text{sq.micron})$.
 (b) What is the diameter of this waveguide when the index of refraction is reduced to one?
 (c) How does this diameter compare to commercially available waveguides?

9.5 For the graded index waveguide in problem 9.4, calculate and plot the number of modes, m, as function of the radial distance, d, from the core.
 (a) For $b = 1$ to $10\ \lambda_0$: the wavelength of the guided wave.
 (b) Replot (a) for $\lambda_0 = 10.0$ microns.
 (c) Replot (a) for $\lambda_0 = 0.530$ microns.
 (d) For $b = 1$ and $m = 1, 2, 3,$ and 4, which of these mode can be excited for a 10.0 micron wavelength?
 (e) For $b = 1$ and $m = 1, 2, 3,$ and 4, which of these mode can be excited for a 0.530 micron wavelength?

9.6 Consider a range of step waveguides were the differences in indicies is from 0.01 to 0.1 where $n1 = 1.4$:
 (a) Calculate and plot the numerical apertures.
 (b) Calculate and plot the acceptance angles.
 (c) Calculate and plot the Brewster angles of internal reflection.

9.7 Using Table 5.1 calculate the index of refraction for the following list assuming the magnetic permeability is unity.

(a) SiO_2, (b) Al_2O_3, (c) Pyrex, (d) MgO, (e) NaCl, (f) BaO, and (g) diamond

9.8 (a) For the list in problem 9.7 calulate the reflectance that can be expected.

(b) Are "non reflective coated optics" coated with high index material or low index material? Why?

9.9 Derive equations (9.106) and (9.107) as complete solutions for a TM wave in a hollow waveguide.

READING LIST

Abraham, E., Seaton, C. T., and Smith, S. D., (1983). *Sci. Amer.*, **248**(2), 85.

Boyle, W. S., (1977). *Sci. Amer.*, **237**(2), 40.

Boyle, W. S., (1977). Light-wave communications. *Sci. Am.* **237**(2), 40.

Carson, D. and Lorrain, P., (1962). *Electromagnetic Fields and Waves*, Freeman, San Francisco, California, pp. 409.

Hench, L. L., Wang, S. H., and Nogues, J. L., (1988). *Multifunctional Materials*, **878**, (R. L. Gunshor, ed.), SPIE, Bellingham, Washington, pp. 76–85.

Jackson, J. D., (1967). *Classical Electrodynamics*, 6th ed., Wiley, New York, pp. 259.

Kraus, J. D., (1966). *Radio Astronomy*, McGraw-Hill, New York, pp. 131.

Midwinter, J. E., (1979). *Optical Fibers for Transmission*, Wiley, New York.

Miliou, A., Srivastava, R., Ramaswamy, V., Chia, T., West, J. K., (1988). USUK Conference, Pitlockery, Scotland.

Ramaswamy, V., Chia, T., Srivastava, R. and West, J. K., (1988). *Multifunctional Materials*. **878**, (R. L. Gunshor, ed.), SPIE, Bellingham, Washington, pp. 86–93.

Seippel, R. G., (1984). *Fiber Optics*, Reston Publ. Co., Reston, VA.

Snyder, A. W., and Love, J. D., (1983). *Optical Waveguide Theory*, Chapman and Hall, N.Y.

Wang, S. H., Campbell, C. and Hench, L. L., (1988). *Ultrastructrue Processing of Advanced Ceramics*," (J. D. Mackenzie and D. R. Ulrich, eds.) Wiley, New York, pp. 145–157.

Yariv, A., (1979). *Sci. Amer.*, **204**(1), 64.

10

ELECTRO-OPTICAL CERAMICS

10.1 INTRODUCTION

Electric and magnetic fields induce a change in the optical dielectric constants, ϵ_{ij}, or the refractive indices n_{ij}, of a material. Consequently, electric and magnetic fields affect light propagation through a material. Such induced optical effects are termed *electro-optical* when induced by an electric field, and *magneto-optical* when induced by a magnetic field. For example, a dc electric field can produce linear birefringence in a crystal, and a magnetic field can produce both linear and circular birefringences. This chapter is primarily concerned with electro-optical effects in ceramics.

It is the field-induced distortion of the electron distributions in a material that gives rise to electro-optical effects. A change in spatial distribution of electrons results in an anisotropic change in electronic polarizability and refractive index (see Chapter 8). The field-induced electron anisotropy can make isotropic crystals, such as GaAs, doubly refracting. It can also induce an additional optic axis into naturally doubly refracting crystals such as KDP (potassium dihydrogen phosphate, KH_2PO_4). Thus, the first step in understanding electro-optics is to understand the nature of birefringence.

10.2 BIREFRINGENCE

Optical materials fall into two classes:

1. Isotropic, e.g., opal, glass, diamond, garnet, spinel, fluorite. These are crystals in the isometric system or they are amorphous materials.

2. Anisotropic crystals which fall into four groups of crystal systems:
   A. Uniaxial (+); rhombic, hexagonal, tetragonal.
   B. Uniaxial (−); rhombic, hexagonal, tetragonal.
   C. Biaxial (+); orthorhombic, monoclinic, triclinic.
   D. Biaxial (−); orthorhombic, monoclinic, triclinic.

Light traveling through an isotropic substance has the same velocity in all

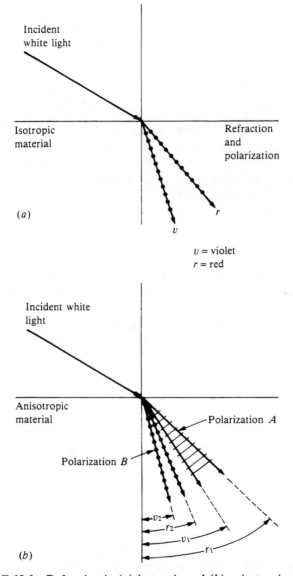

FIGURE 10.1 Refraction in (*a*) isotropic and (*b*) anisotropic materials.

directions. Consequently, the extent of refraction, *Snell's law* (Chapter 9), is independent of the direction of incidence of the light. There is also only one direction of polarization (Figure 10.1a). The wavelength components of white light each refract proportional to their dispersion, $\Delta n/\Delta \lambda$, but for each wavelength there is only one refracted ray exiting an isotropic material.

However, light traveling through anisotropic materials is doubly refracted, that is, the light beam is bent and is also broken into two beams, with each beam polarized (see Figure 10.1b). The extent of refraction depends upon the wavelength λ of the incident light, the indices of refraction, and the dispersion $dn/d\lambda$ (see Chapter 8).

10.3 OPTICAL PRINCIPAL AXIS

Uniaxial crystals (e.g., rhombic, tetragonal, and hexagonal crystals) do not exhibit double refraction if the light is observed in the direction of the principal (c-direction) crystal axis. By definition, the *direction of the optic axis is when the behavior of the incident light is isotropic*. In any other direction the material is anisotropic. The doubly refracting rays are polarized at right angles to each other.

One ray vibrates at right angles to the optic axis. This is called the *ordinary ray* (o). The other ray vibrates in a plane parallel to the optic axis, and is called the *extraordinary ray* (ϵ). The index of refraction of each ray is different.

For example, consider a calcite ($CaCO_3$) crystal. It is a rhombohedral crystal and therefore is uniaxial. The orientation of the ordinary and extraordinary rays of light leaving the calcite crystal is shown in Figure 10.2. A spot reviewed at any angle other than down the optic axis will appear as two spots. This is because the incident light is doubly refracted and the result is a transmitted double image. Each ray is oppositely polarized. The

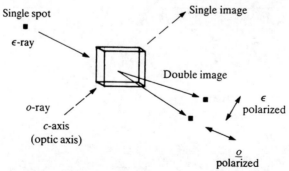

FIGURE 10.2 Calcite crystal showing double refraction of the extraordinary ray (ϵ).

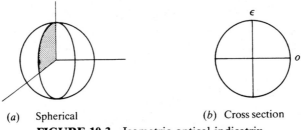

(a) Spherical (b) Cross section
FIGURE 10.3 Isometric optical indicatrix.

description of this anisotropic behavior of light is aided by use of a geometric construct called the *optical indicatrix*.

10.4 ISOMETRIC OPTICAL INDICATRIX

An optical indicatrix is a surface generated around the reference point with the indices of refraction of a material serving as the radii. The shape of the indicatrix surface thus generated illustrates the optical features of the material. Since the index is equivalent in all directions in an isometric material, the isometric indicatrix is a sphere (Figure 10.3). The diameter of the sphere will differ for various wavelengths depending upon the magnitude of dispersion of the material.

10.5 UNIAXIAL OPTICAL INDICATRIX

Since the indices of refraction differ in two directions for uniaxial crystals their optical indicatrix is a spheroid of revolution. Recall from Chapter 6 that the crystallographic features of tetragonal and hexagonal crystals (including the rhombohedral subgroup) are as shown in Table 10.1. Consequently, the index of refraction will be equal in the a and b directions for uniaxial crystals. The ray in these directions is termed the ordinary (o) ray. In contrast, the index in the c-direction, termed the extraordinary ray (ϵ), shows a range of values. The ordinary ray remains constant, as shown in Figure 10.4. When $\epsilon > o$, the spheroid is prolate and the crystal is optically positive (+), such as quartz (Figure 10.4a). When $\epsilon < o$, the spheroid is oblate and the crystal is negative (−), such as calcite (Figure 10.4b).

TABLE 10.1 Uniaxial Crystals

Tetragonal	Hexagonal (rhombic)
$a = b \neq c$	$a = b \neq c$
$\alpha = \beta = \gamma = 90°$	$\alpha = \beta = 90°; \gamma = 120°$

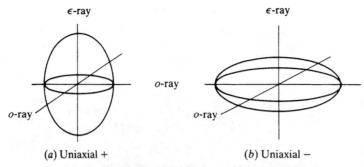

(a) Uniaxial + (b) Uniaxial −

FIGURE 10.4 Uniaxial optical indicatrix.

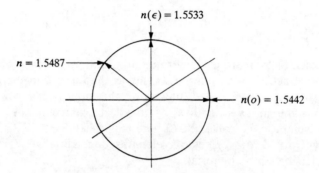

FIGURE 10.5 Quartz optical indicatrix.

The index of the extraordinary ray n_ϵ varies from n_ϵ to n_o in either case depending upon the angle of the incident ray in the crystal. For example, consider quartz in Figure 10.5. The index of the ordinary ray is 1.5442 and thus the prolate spheroid has a circular cross section of equal values of $n = 1.5442$ in all directions. However, the value of n_ϵ varies from 1.5533 to 1.5442 so that a ray with an angle of 45° from the two axes will have an index of $n_\epsilon = 1.5487$.

10.6 BIAXIAL OPTICAL INDICATRIX

Crystals with lower symmetry (Table 10.2), have three unequal refractive indices. Consequently, orthorhombic, monoclinic, and triclinic crystals have

TABLE 10.2 Biaxial Crystals

Orthorhombic	Monoclinic	Triclinic
$a \neq b \neq c$	$a \neq b \neq c$	$a \neq b \neq c$
$\angle a = \angle b = \angle c = 90°$	$\angle a = \angle b \neq \angle c$	$\angle a \neq \angle b \neq \angle c$
$\alpha = \beta = \gamma = 90°$	$\alpha = \beta = 90° \neq \gamma$	$\alpha \neq \beta \neq \gamma$

10.7 ELECTRIC FIELD DEPENDENCE OF INDEX OF REFRACTION

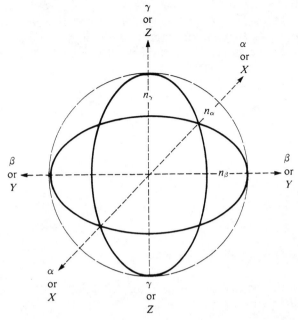

FIGURE 10.6 Notation for the biaxial optical indicatrix.

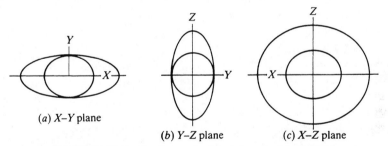

FIGURE 10.7 Sections of a biaxial optical indicatrix.

an optical indicatrix which is a triaxial ellipsoid. The origin is at the ellipsoid center, the coordinate axes are ellipsoid axes, and the coordinate planes are the principal planes.

The semi-axes of the ellipsoid are assigned refraction indices of n_α, n_β, n_γ. *Principal sections* are any combination in which a plane contains two axes. These sections are ellipses which have as major and minor semi-axes any pair of the values n_α, n_β, n_γ. This results in the optical indicatrix shown in Figure 10.6. Various cross sections of a biaxial optical indicatrix are shown in Figure 10.7.

10.7 ELECTRIC FIELD DEPENDENCE OF INDEX OF REFRACTION

When an electric field is applied to a material, the electron orbitals are polarized in the direction of the field (see Chapters 5 and 8). Consequently, the velocity of the light traveling through the material is lower in the direction of the field than perpendicular to it. The effect is to change the refractive index of the material such that it is higher in the direction of the field and lower at right angles to the field. The magnitude of the change in n with an applied electric field is given by equation (10.1). Higher-order terms have been dropped:

$$\Delta\left(\frac{1}{n^2}\right) = r\mathbf{E} + P\mathbf{E}^2 \qquad (10.1)$$

where r = linear electro-optical coefficient
       P = quadratic electro-optical coefficient
       $\mathbf{E}$ = electric field

10.8 POCKELS (LINEAR) EFFECT

The linear variation of n due to $r\mathbf{E}$ is termed the *Pockels effect*. The induced privileged directions are perpendicular to $\mathbf{E}$. Thus, when $\mathbf{E}$ is applied in the z-direction, an isotropic crystal will exhibit birefringence in the (x, y) directions, that is, an optically isotropic material becomes optically uniaxial. The electric field distorts the isometric indicatrix of Figure 10.3 to the form of the uniaxial indicatrix of Figure 10.4. The amount of distortion is given by the $r\mathbf{E}$ term.

Likewise, when an electric field is applied across an optically uniaxial material it will become biaxial, depending of course upon the orientation of the field with respect to the crystal axes. Thus, a material with a uniaxial indicatrix (Figure 10.4) will be polarized to produce a biaxial indicatrix such as Figure 10.6 with the magnitude of shift dependent on $r\mathbf{E}$.

10.9 KERR (QUADRATIC) EFFECT

Variations of n due to the $P\mathbf{E}^2$ term in equation (10.1) are termed the *Kerr effect*. (*Note:* This quadratic electro-optical Kerr effect should not be confused with the magneto-optical effect which is also called the Kerr effect.) The difference in refractive indices for light polarized parallel with, and perpendicular to the induced optic axis in a previously isotropic material is given by

$$\Delta n = n_\epsilon - n_o = K\lambda \mathbf{E}^2 \qquad (10.2)$$

where K is the electro-optical *Kerr constant* and λ is the vacuum wavelength of the incident light. Typical values of K are given in Table 10.3.

TABLE 10.3 Typical Values of Electro-optical Kerr Constant K at $\lambda = 589.3$ nm (20°C)

Material	$K (\times 10^{-12}/V^2)$
Water	5.2
Nitrobenzene	24
Silicate glasses	3–1.7

TABLE 10.4 Electro-optical Parameters

	$r(\times 10^{-12}/V)$	Polarizability $P(\times 10^{16} m^2/V^2)$	Index n_o	n_ϵ	Relative permittivity ϵ_r
KH$_2$PO$_4$(KDP)	10.6		1.51	1.47	42
AH$_2$PO$_4$(ADP)	8.5		1.52	1.48	12
CdTe	6.8		2.74		7.3
LiTaO$_3$	22–30.3		2.175	2.180	43
LiNbO$_3$	17–30.8		2.29	2.30	18
GaAs	1.6		3.6		11.5
ZnS	2.1		2.2		
Perovskites					
8.5La 65Zr 35Ti		38.6			
9La 65Zr 35Ti		3.8			
10La 65Zr 35Ti		0.8			

The magnitude of the linear and quadratic electro-optical coefficients depend upon the directional electronic polarizability of the material. Examples are given in Table 10.4 for several materials, along with their respective indices of refraction.

10.10 OPTICAL ACTIVITY

Some liquids and crystals affect the plane of polarization by rotation. The plane of the electric vector will change as the ray propagates through such a material. These materials are known as *optically active*. The rotation is a function of wavelength and thickness of the crystal. Quartz is an example. For yellow sodium light, the rotation is 21.7° for a 1 mm thick sample of quartz. However, for a sodium chloride crystal, the same conditions yield only a 3.67° rotation.

10.11 FARADAY EFFECT

When a beam of plane-polarized light (see Chapter 8) passes through a material subjected to a magnetic field, its plane of polarization is rotated by

TABLE 10.5 Values for Verdet Constant ($\lambda = 589.3$ nm)

Material	$V = (\text{rdm}^{-1}\,\text{T}^{-1}) = \left(\dfrac{\text{radians}}{\text{meter} \cdot \text{Tesla}}\right)$
Quartz (SiO$_2$)	4.0
Crown glass	6.4
Flint glass	23
Zinc sulfide	82

an amount proportional to the magnetic field component parallel to the direction of propagation. This is known as the *Faraday effect*. It is similar to the optical activity of crystals where the refractive index differs for right- and left-circularly polarized light. However, in the Faraday effect, the rotation of the plane of polarization is independent of the direction of propagation, whereas optical activity is related to the direction of propagation. The magnitude of rotation of the plane of polarization θ is

$$\theta = VBL = \frac{2\pi}{\lambda}(n_r - n_l)L \qquad (10.3)$$

where V = Verdet constant
  B = magnetic flux parallel to the direction of propagation
  L = path length in the material
  λ = wavelength of the light

and n_r and n_l are the refractive indices for right- and left-circularly polarized light, respectively. Typical values of Verdet constants are given in Table 10.5.

10.12 MATERIALS

Noncentro-symmetric ferroelectric ceramics, such as discussed in Chapter 6, are used in electro-optical applications. A major difference, however, is that electro-optical ceramics must be optically transparent, usually in visible wavelengths, whereas ferroelectric capacitors, transducers etc. can be opaque. There are a number of single-crystal ferroelectrics (Table 10.4), but they are susceptible to moisture, expensive, and difficult to manufacture with suitable homogeneity, orientation, and stoichiometry. In contrast, polycrystalline electro-optical ceramics are very stable chemically and can be routinely manufactured at reasonable cost.

Hot pressing of ferroelectric lead–zirconate–titanate (PZT®†) powders, developed by Haertling (1971), eliminates optical scattering due to porosity, grain boundaries, and internal refraction at domain walls. The optical

---

†Registered trademark of Vernitron, Inc.

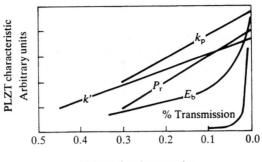

FIGURE 10.8 PLZT electro-optic ceramics characteristics as a function of porosity (after Haertling).

transmission of electro-optic ceramics is dramatically increased as the final volume fraction of porosity is removed and full density is achieved (Figure 10.8). The other important electrical characteristics are also improved as full density is reached (Figure 10.8), including: dielectric breakdown strength (E_b), remanent polarization (P_r), dielectric charging constant (k'), and planar coupling constant (k_p). Dielectric losses decrease considerably as porosity and grain boundary effects are eliminated.

Solid solution dopants of Ba, Sn, and La, with La being most effective, and an oxygen atmosphere during hot pressing are used to achieve full density and optical transparency of PLZT materials. The La-doped PLZT composition is

$$Pb_{1-x}La_x(Zr_yTi_{1-y})_{1-x/4}O_3 \qquad (10.4)$$

The La^{3+} ions replace Pb^{2+} ions in the A sites of the ABO_3 perovskite lattice (Figure 10.9). Electrical neutrality is maintained by vacancies on either A^{2+} sites or B^{4+} sites, depending in part on the atmospheric overpressures of PbO during densification.

The crystal phases of PLZT ceramics are primarily uniaxial negative tetragonal or rhombohedral structures depending upon Pb/Zr/La ratios (Figure 10.10). The boundary between the ferroelectric (FE), tetragonal, and rhombohedral phases is at a 65/35 ratio of $PbZrO_3$ to $PbTiO_3$. Addition of La results in an antiferroelectric (AFE) orthorhombic, a paraelectric (PE) cubic phase, or a mixture of phases, as illustrated in Figure 10.10.

Compositions of PLZT electro-optical ceramics located in the FE tetragonal phase region are useful for nonswitching, linear electro-optical effects such as Pockel cells, described in the following section. Figure 10.11 illustrates a typical light-output response of the tetragonal FE phase and its operational configuration. This linear birefringent effect exists for compositional ratios such as 12La–40Zr–60Ti (see Figure 10.10). The linear effect is intrinsic to the material after all the ferroelectric domains have been oriented by application of an electric field of sufficient strength to produce

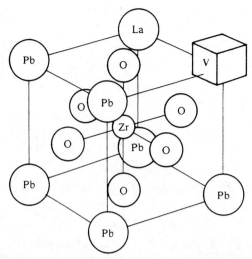

FIGURE 10.9 Perovskite ABO_3 crystal unit cell with La^{3+} substitution and a Pb^{2+} vacancy.

saturation remanent polarization (see Chapter 6). This process of domain orientation is termed *poling*. Once poled, the ferroelectric domains in the linear electro-optical material are not switched again. Thus, the uniaxial optical indicatrices of all the grains are oriented parallel. Unpoled, or virgin material will have the domains and indicatrices randomly distributed.

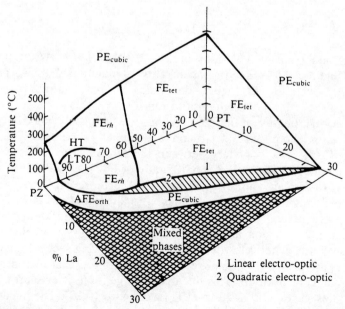

FIGURE 10.10 Room-temperature PZT® and PLZT phase diagrams (after Haertling).

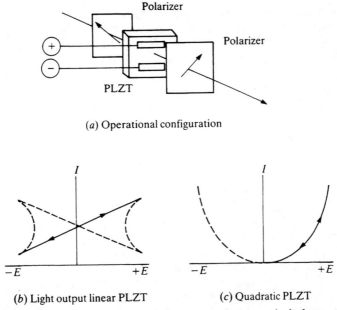

FIGURE 10.11 PLZT operation as an electro-optical element.

Optical information is extracted by applying an electric field that causes linear changes in the birefringence without polarization reversal or domain switching.

This effect can be used over many wavelengths, since polycrystalline plates of the 12–40–60 PLZT composition have reasonably good optical transmission from the blue visible, 500 nm (Figure 10.12) out to 6.5 μm in the IR decreasing to zero at 12 μm. Since the 12–40–60 composition is ferroelectric, it has lower transmission than the 9–65–35 composition, due to the presence of domain walls that scatter light.

The quadratic (Kerr) effect is the most widely used electro-optical response of electro-optical ceramics. This effect is quite large, and is due to the electric field polarizing an optically isotropic cubic phase into an

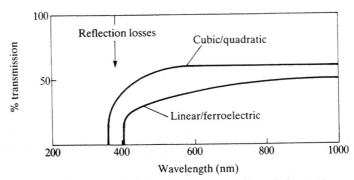

FIGURE 10.12 PLZT optical transmission characteristics.

optically uniaxial ferroelectric rhombohedral or tetragonal phase. When the field is removed the polar phase relaxes back to its cubic state. The PLZT composition 9La–35Zr–65Ti exhibits this behavior, as do other compositions in the shaded boundary between the paraelectric cubic phase field and the ferroelectric fields (see Figure 10.10).

When the unpolarized cubic material is put in the path of an incoming polarized beam of light (Figure 10.11a), there is no retardation of the beam since the material is optically isotropic. This is the OFF state. When an electric field is applied, polarization and birefringence is induced in the material and optical retardation is observed between crossed polarizers, the ON state (Figure 10.11c). When the field is turned off the material relaxes back to its OFF state within 1–100 μsec. Turn-on times are about the same. ON/OFF ratios as high as 5000 to 1 are possible.

Figure 10.13 illustrates the phenomenon of optical phase retardation in an electro-optical ceramic. When linearly polarized monochromatic light enters the ceramic in the ON state, the light is resolved into two perpendicular components c_1 and c_2, whose vibration directions are determined by the uniaxial optical indicatrix of the material. The propagation velocity of each component of light will be different because of the different refractive indices, n_o and n_ϵ (see Figure 10.4). This difference in velocity results in a phase shift, termed *retardation*. The magnitude of retardation Γ is a function of both $\Delta n = (n_o - n_\epsilon)$ and the path length L which is the thickness of the electro-optical plate; (i.e.,)

$$\Gamma = \Delta n L \tag{10.5}$$

Full half-wave retardation of c_1 relative to c_2 is achieved when sufficient voltage is applied (Figure 10.11). This results in a 90° net rotation of the vibration direction of the linearly polarized light. Thus, the polarized light will be transmitted through the second polarizer when it is in the crossed

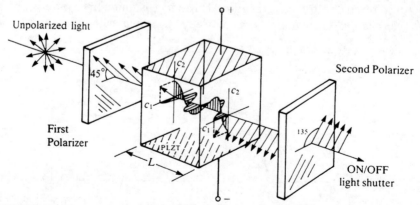

FIGURE 10.13 Optical phase retardation in an activated electro-optical ceramic (after Haertling, 1986).

condition or blocked if it is in the parallel position. Consequently, switching the Kerr-effect electro-optical ceramic from zero retardation to half-wave retardation results in an ON/OFF light shutter. For discussion of the use of the Kerr effect to produce electrically induced interference colors, etc., refer to Haertling (1971).

Since there are no permanent domain walls in the 9–65–35 cubic phase PLZT ceramics, their optical transmission extends to the violet region of the EM spectrum (Figure 10.12).

10.13 MEMORY EFFECT

Figures 10.14a and b shows the operational configuration and the typical light output I for a memory PLZT electro-optical ceramic. This effect is used for image storage devices such as a "memory oscilloscope." The effect is caused by light scattering from the domains within the material. The domain orientation is electrically altered with increasing electric field, E_+. As the electric field is reduced the image remains. Even as E is reduced to zero there remains a finite value of I. The light is preferentially scattered along the polar direction of the domains and the light transmitted can be controlled by the applied electric field. The image can be erased by flooding the plate with light while a voltage is applied in the positive saturation direction.

A summary of the electro-optical effects and their uses is given in Table 10.6.

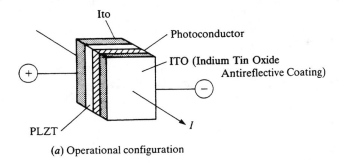

(a) Operational configuration

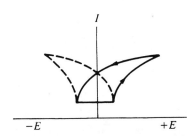

(b) Light output from memory PLZT

FIGURE 10.14 PLZT memory effect of electro-optical ceramic.

TABLE 10.6 Electro-optic Devices

Electro-optical effect	Devices
Birefringence	Shutters
Depolarization	Modulators
	Color filters
	Goggles
	Displays
	Memories
Scattering	Image storage
	Modulators
	Displays
Diffraction	Image storage
Photoferroelectric	Holographic recording
Photoelastic	Optical waveguides modulators

10.14 PARAMETRIC AMPLIFICATION AND OSCILLATION

The photon model of electromagnetic radiation is useful in discussing a nonlinear phenomenon known as *parametric amplification* and *oscillation*. Power from an incident photon with an angular frequency ω_3 is transferred to two photons of ω_2 and ω_1 where

$$\omega_3 \rightleftarrows \omega_1 + \omega_2 \tag{10.6}$$

This "transfer" can be applied to a nonlinear material like lithium niobate, where the two smaller frequencies can increase above the noise. ω_1 is known as the *signal frequency* while ω_2 is known as the *idler frequency*.

If only ω_3 is fixed then the idler and signal frequencies are free to range over an infinite set of values. This effect is known as *parametric amplification*.

The photons must conserve angular momentum in the transfer of equation (10.6). Therefore, the indices must match exactly according to the relationship:

$$\frac{h}{2\pi}\frac{n'}{\lambda_1} + \frac{h}{2\pi}\frac{n''}{\lambda_2} = \frac{h}{2\pi}\frac{n'''}{\lambda_3} \tag{10.7}$$

where n = index of refraction
 λ_i = wavelength
 h = Planck's constant

This usually requires that the process should occur within an optical cavity with highly reflective mirrors for ω_1 and ω_2, but not ω_3.

This effect is known as parametric oscillation. Tuning of the oscillator is accomplished by varying the index of refraction. This has been demonstrated by varying the temperature of the crystal. Coherent wavelengths from the IR to the UV are selected and transmitted through the cavity. Efficiencies for a parametric oscillator are on the order of a few percent. However, the photons are coherent as well as monochromatic.

10.15 FLUORESCENCE

Optical filters are generally subtractive. Portions of the spectrum are removed from the light as the light causes electrons to be excited into higher energy states. Luminescence on the other hand is additive. Electrons decay into lower energy levels and release photons. These photons have a specific energy or color, if they are in the visible portion of the **EM** spectrum.

The term *fluorescence* in general refers to a process of electronic absorption immediately followed by luminescence. The term *phosphorescence* is used to describe the same process, except where there is a significant time delay between absorption and emission. This results in a phenomenon known as *afterglow*.

Fluorescence involves at least two excited states and one ground state. A photon excites an electron into the higher of the two excited states, E_2. Then a nonradiative decay occurs into a lower excited state, E_1. From E_1 to the ground state, E_0, a decay occurs with the emission of a photon. By definition

$$E_2 > E_1 \tag{10.8}$$

Therefore the photon that was absorbed has an energy

$$h\nu_2 = E_2 - E_0 \tag{10.9}$$

and the emitted photon (fluorescence) has an energy

$$h\nu_1 = E_1 - E_0 \tag{10.10}$$

This yields the result characteristic of all fluorescence, that the wavelength of the fluorescence is longer (frequency less) than that of the absorbed photon:

$$\nu_2 > \nu_1 \tag{10.11}$$

The lifetime of such transitions are of the order of 10 nsec to a few hundred nanoseconds (see McKeever, 1985, for other processes).

Phosphorescence is conceptually identical with fluorescence except that the intermediate state E_1 is metastable. A metastable state is usually a very well-defined energy level with a long lifetime. The state is governed by the Heisenberg uncertainty principle which requires the product of the change in energy and the change in time to be a constant (see Chapter 2).

$$\Delta E \cdot \Delta t \approx h \qquad (10.12)$$

Phosphorescence typically has a lifetime of seconds (Δt). This means that

$$\Delta E \approx \frac{h}{\Delta t} \qquad (10.13)$$

Thus, for a long lifetime, the bandwidth of the metastable state ΔE is extremely narrow. This also indicates that photons emitted during this process are very much the same wavelength or color.

Most rare-earth elements have fluorescence bands in the visible spectrum; some possess several. In cases where there are several emission bands, the specific emission may be controlled by adjusting the color of the incident light.

Luminescence and fluorescence both require metastable states to trap the holes for a short time, to create the lifetime of the fluorescence. Figure 10.15 shows the sequence of events to create a metastable state by hole trapping. First a photon is absorbed and the resulting hole is immediately trapped in an acceptor state, E_a. The electron may spend some time in a donor state, E_d. Finally, the electrons recombine with the trapped hole. The probability (inverse lifetime $= 1/\Delta t$) for the recombination depends on the depth of the donor state and the temperature:

$$\frac{1}{\Delta t} = Q \exp\left[\frac{E_c - E_d}{kT}\right] \qquad (10.14)$$

where

$$Q \approx 1 \times 10^{+8} \text{ sec}^{-1} \qquad (10.15)$$

for typical luminescent materials.

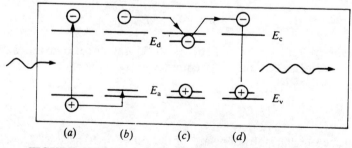

FIGURE 10.15 Luminescence caused by hole trapping.

10.16 LIGHT EMITTING DIODES (LED)

From Chapter 3, we have seen that the forward bias current of a diode is an exponential function of the applied potential. This forward current creates an excess of charge carriers as they cross the p–n interface. Diffusion moves them away from the interface but they must recombine across the band gap eventually. Thus

$$h\nu_g = E_c - E_v \tag{10.16}$$

and radiation is emitted of a frequency ν_g when the recombination occurs.

This occurs in the IR region of the EM spectrum for silicon diodes but the radiation is immediately reabsorbed and no light is emitted. In the case of GaAs, however, the band-gap transitions are in the visible and the photon conduction is high. Thus for a GaAs band gap

$$E_g = 1.44 \text{ eV} \tag{10.17}$$

recombination yields a wavelength of emission near

$$\lambda_g = 0.86 \text{ }\mu\text{m} \tag{10.18}$$

The intensity, i, of the emitted radiation is proportional to the number of minority charge carriers injected through the interface or the current I:

$$i = \alpha I \tag{10.19}$$

Therefore, substituting the diode equation (3.109), we obtain:

$$i = \alpha I_0 \exp\left[\left(\frac{eV}{kT}\right) - 1\right] \tag{10.20}$$

where V = potential
e = electronic charge
I_0 = saturation current
T = temperature

The intensity of the emitted radiation increases exponentially with the forward bias voltage across the LED.

The efficiency of this process is termed *quantum efficiency* (η) and is the ratio of the number of emitted photons to the number of injected electrons:

$$\eta = \frac{N_{ph}}{N_{e^-}} \tag{10.21}$$

The quantum efficiency is relatively small because recombination of electrons and holes can occur in different ways. Recombination through interband transitions produce visible photons while recombination through impurity centers and exciton recombinations do not produce visible photons. The quantum efficiency for silicon-doped GaAs is approximately 10%. Other GaAs-doped systems are less than 1%. However, because of the diode equation, small increases in forward bias potential V generate sufficient photon intensity for the LED devices to be useful.

10.17 BLUE LIGHT EMMITTING DIODES

SiC optical transmission from the absorption edge at 400 nm to 50 μm is shown in Figure 10.16. The undoped SiC crystals (curve 10.16b) are nearly fully transparent in the visible and near-infrared region. The strong absorption band in the 10-13 μm region corresponds to the fundamental vibration mode. The absorption observed in the 4-5 μm and 5-10 μm regions are attributed to combination bands of lattice vibrations. For Al-doped crystals (p-type) the optical transmission behavior (curve 10.16a) is quite different. In the visible region the transmission is considerably less; in the near-infrared the transmission is almost zero due to strong impurity absorption. In the far-infrared the crystal is again transparent, as for the undoped samples. The aluminum concentration must be adjusted so that there is a good trade-off between transparency (requiring lower Al concentration) and

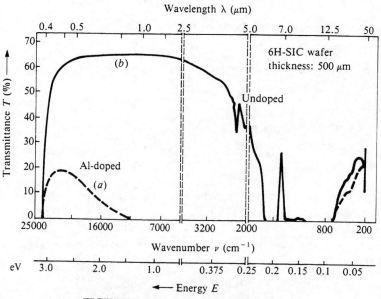

FIGURE 10.16 Silicon carbide spectrum.

conductivity (requiring higher Al concentration). A value of 2×10^{18} acceptors/cm^3 is found to be nearly optimal.

Next we turn to p–n junctions made from 6H polytype SiC crystals using a mesa preparation technique (see Van Vleit et al., 1988).

The emitted light of the p–n junction devices was found to come mainly from the n-region which is relatively low doped. The injection efficiency of holes from the Al-doped p-region is high. The light is collected through the relatively transparent p-layer and substrate. The optical radiation stems from pair transitions between the aluminum acceptors and nitrogen donors in the compensated n-region. Typical emission spectra of the device are shown in Figure 10.17. The spectra show that at higher currents the peak of the LED shifts to the blue (470 nm).

Uncapsulated devices of a mesa structure yield an external quantum efficiency of up to 2×10^{-4}. This is only about one order of magnitude less than for GaP or GaAsP LEDs. In Figure 10.18, the emission spectra of the various commercially available LEDs are compared (see Van Vleit et al. in the reading list). There is an important difference in light-emission mechanism for the devices shown in Figure 10.18. Silicon carbide, in contrast to GaAs or GaP, is an indirect-band-gap material. Thus the assistance of phonon processes in the emission is necessary. This requires normal temperature operation. The data shown in Figure 10.18 are for room temperature, 300°K, operation. The external quantum efficiency can be written as

$$\eta_{\text{ext}} = \eta_{\text{inj}} \times \eta_{\text{light}} \times \eta_{\text{opt}} \tag{10.22}$$

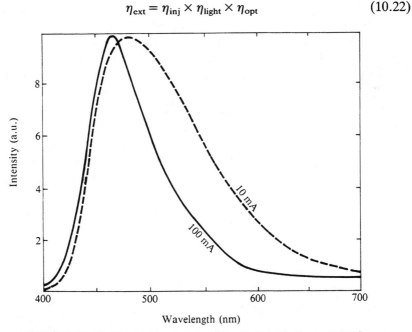

FIGURE 10.17 Typical emission spectra [after Hoffman (1982)].

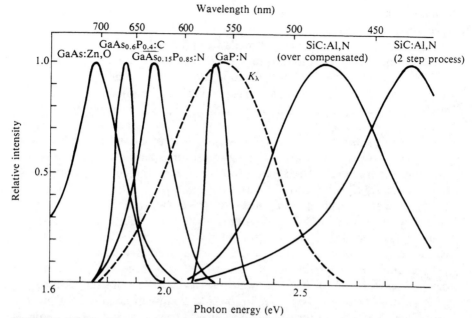

FIGURE 10.18 Photon energy (eV) for different LEDs (after van Vleit, et al., 1988.

The efficiency, η_{inj}, is close to unity for the aluminum concentration described above. The quantum efficiency η_{opt} depends on the optical transmission, as well as on the refractive index. Wide-band semiconductors generally have smaller refractive indices; for SiC, $n = 2.7$ is the accepted value. This is better than for GaP ($n = 3.3$) or GaAs ($n = 3.6$). The main limiting mechanism, therefore is η_{light} or the internal quantum efficiency, for which

$$\eta_{light} = \frac{1/\tau_r}{1/\tau_r + 1/\tau_{nr}} \qquad (10.23)$$

where τ_r is the lifetime of the radiative process, while τ_{nr} is the lifetime of all nonradiative processes. The former should be as short as possible and the latter as long as possible. Unfortunately, so far the nonradiative processes have a lifetime which is short (1–17 nsec). A study of these processes leading to an improvement in lifetime would therefore have great practical value. Noise studies are very useful in order to determine lifetimes (trapping times, detrapping times, recombination times) in single crystal material (see Van Vleit et al., 1988, for a detailed analysis of noise characteristics in SiC).

10.18 AVALANCHE PHOTODIODES (APDS)

Silicon carbide 6H polytype crystals can also be used for avalanche photodiodes. Epitaxially grown mesa-shaped p–n junctions are prepared

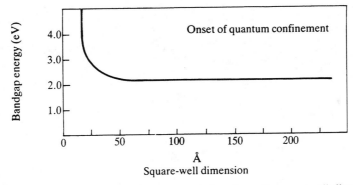

FIGURE 10.19 Quantum confinement as a function of square-well dimension.

with p-layer doping between 10^{16} and 2×10^{18} cm^{-3}. The breakdown field of the diode behaves as

$$\mathbf{E}_m \propto N_a^{1/\beta} \quad (10.24)$$

where $\beta \approx 5$, with fields exceeding 3×10^6 V/cm. The breakdown occurs for higher voltages than in corresponding silicon diodes. This is due to the band-gap dependence of breakdown voltage V_B:

$$V_B = 60(E_G/1.1)^{3/2}(N_A/10^{16})^{-3/4} \text{ V} \quad (10.25)$$

where E_G is in electron volts and the acceptor doping N_A is in cubic centimeters. Applying this relationship to 6H SiC, with $E_G = 2.9$ eV and $N_A = 10^{17}$ cm^{-3}, yields $V_B = 46$ V, in fair agreement with experimental data. The mean free path for optical phonon emission is smaller than in silicon, about 35 Å. Noise measurements in the SiC APDs are lower than in silicon or GaAs which is a major advantage for this type of device (see Van Vleit et al., 1988). Appropriate SiC polytypes with a superlattice structure in the Brillouin zone, giving rise to band discontinuities, could do even better.

10.19 EXCITONS

A simple model of an exciton can be based upon the *Bohr model* of atomic hydrogen. The electron is considered to be in an orbital about an electron hole. In Chapter 2 we showed how the quantum energies are obtained for each orbital by assuming the quantization of the orbital angular momentum. With the integer n representing the primary quantum number, the energy of each orbital was shown to be

$$E_n = \frac{mZ^2 e^4}{8n^2 h^2 \epsilon_0^2} \quad (10.26)$$

CHAPTER 10: ELECTRO-OPTICAL CERAMICS

For an exciton in its ground state,

$$n = 1 \tag{10.27}$$

$$Z = 1 \tag{10.28}$$

$$\epsilon_0 \rightarrow \epsilon_r \epsilon_0 \tag{10.29}$$

Thus

$$E_{ex} = \frac{me^4}{8h^2 \epsilon_r^2 \epsilon_0^2} \tag{10.30}$$

If we let the mass m_e become the reduced mass, m_r^*:

$$\frac{1}{m_r^*} = \frac{1}{m_e^*} + \frac{1}{m_h^*} \tag{10.31}$$

where m_e^* = effective electron mass
m_h^* = effective hole mass

The reduced mass is usually expressed in units of the electron mass m. Therefore we can express the binding energy as follows:

$$E_{ex} = \left(\frac{e^4 m}{8h^2 \epsilon_0^2}\right)\left(\frac{m_r^*}{m}\right)\left(\frac{1}{\epsilon_r}\right)^2 \tag{10.32}$$

or

$$E_{ex} = A\left(\frac{m_r^*}{m}\right)\left(\frac{1}{\epsilon_r}\right)^2 \tag{10.33}$$

Evaluating the proportionality constant

$$A = \frac{(1.6 \times 10^{-19} \text{ C})^4 (9.11 \times 10^{-31} \text{ kg})}{8(6.63 \times 10^{-34} \text{ J sec})^2 (8.854 \times 10^{-12} \text{ F/m})^2} \tag{10.34}$$

$$= (2.1657 \times 10^{-3}) \frac{\times 10^{-107}}{1 \times 10^{-92}} \tag{10.35}$$

$$= 2.1657 \times 10^{-18} \text{ J} \times \frac{1 \text{ eV}}{1.602 \times 10^{-19} \text{ J}} \tag{10.36}$$

$$= 13.518 \text{ eV} \tag{10.37}$$

Therefore the exciton binding energy is usually written as

$$E_{ex} = 13.5 \left(\frac{m_r^*}{m}\right)\left(\frac{1}{\epsilon_r}\right)^2 \text{ eV} \tag{10.38}$$

TABLE 10.7 Optical Properties of Semiconductors

	Band gap E_g(eV)	Relative permittivity ϵ_r	Effective mass m_e^*	m_h^*	Refractive index n
Si	1.12	11.8	0.12m	0.23	3.49
Ge	0.74	16.0	0.26m	0.38m	4.0
GaAs	1.4	10.9	0.068m	0.56m	3.6
CdS	2.42	6.0	0.19m	0.74m	2.50
CdTe	1.45	7.3	0.32m	—	2.74
α-SiC	2.9	9.7	0.60m	1.00,m	2.85
CdSe	1.74	—	0.13m	0.45m	—
PbS	0.35	15.3	0.32m	1.38m	3.91

In the case of GaAs, we can evaluate the exciton binding energy by using (see Table 10.7)

$$m_e^* = 0.068m$$
$$m_h^* = 0.56m \qquad (10.39)$$
$$\epsilon_r = 10.9$$

Evaluating the reduced mass,

$$\frac{1}{m_r^*} = \frac{1}{0.068m} + \frac{1}{0.56m} \qquad (10.40)$$

$$m_r^* = 0.0607m \qquad (10.41)$$

Substituting these values into equation (10.38), we obtain the theoretical exciton binding energy in GaAs:

$$E_{ex}(\text{GaAs}) = 0.0069 \text{ eV} \qquad (10.42)$$

Typical values for donor or acceptor levels in GaAs would be

$$E_c - E_d \approx 0.006 \qquad (10.43)$$
$$E_a - E_v \approx 0.02 \qquad (10.44)$$

Thus the excitons are extremely close to the conduction band edges.

10.20 QUANTUM CONFINEMENT

As discussed in Chapter 2, the electrons in conduction bands act much like free electrons. The crystal boundary is a potential wall but the electrons do not in general experience its effects. The conduction band is a continuum of allowed electron energies. If excitons form, because of hole trapping, then their energies are very close to the conduction edge. Small thermal energies, e.g., phonons, can cause the exciton to dissociate. Quantum confinement allows greater band gap energies to form and larger binding energies for excitons.

If the dimensions of the crystal are of the order of magnitude of the de Broglie wavelength of the particle, then quantum confinement effects can be observed. Instead of the continuum of energies available in the conduction band, discrete quantum energy levels dominate in quantum confinement conditions. In this situation, the solution to the energy eigenvalues can be modeled by a finite square-well potential as done in Chapter 2. The energy levels separate and this increases the binding energy of the excitons, making them more useful. Figure 10.19 indicates schematically the onset of quantum confinement as the dimensions of the crystal are reduced. Pure CdS crystals have been shown to exhibit quantum confinement effects by Potter and Simmons (1988). Their work indicates that the translational mass of the exciton is a more accurate model than either the reduced mass or the effective mass used in the Nair model.

Figure 10.20 shows the results of the Potter and Simmons experiments and calculations. The lowest excited-state energy is compared with the three models and plotted as a function of the inverse average square crystallite radius. Their experimental values fall very close to the translational mass model. The effective mass m_{eff} used is

$$m_{\text{eff}} = 0.30m \qquad (10.45)$$

and the translational mass M is

$$0.9m \leq M \leq 1.22m \qquad (10.46)$$

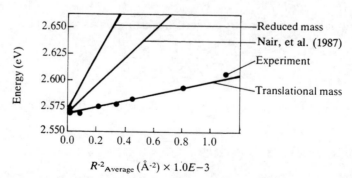

FIGURE 10.20 Lowest excited-state energy of the exciton (after Potter and Simmons, 1988).

We can study luminescence in CdS by using the simple Bohr model (equation 10.38) for the binding energy of the exciton and a simplified recombination energy, $h\nu$:

$$h\nu \approx E_c - E_{ex} \qquad (10.47)$$

The reduced mass, m_r^*, is defined in equation (10.31) as

$$\frac{1}{m_r^*} = \frac{1}{m_e^*} + \frac{1}{m_h^*} \qquad (10.48)$$

For CdS this becomes:

$$m_r^* = 0.15m \qquad (10.49)$$

Using this reduced mass of the exciton (ignoring quantum confinement effects), we can calculate the expected photoluminescent transition for the recombination of an exciton in CdS by using equations (10.38) and (10.47).

The binding energy for the Bohr model becomes:

$$E_{ex}(m_r^*) = 0.056 \text{ eV} \qquad (10.50)$$

Subtracting this energy from the band gap gives an approximation of the recombination energy or the photoluminescent photon energy:

$$h\nu(m_r^*) = 2.36 \text{ eV} \qquad (10.51)$$

Converting this energy to the wavelength of the transition:

$$\lambda(m_r^*) = 511 \text{ nm} \qquad (10.52)$$

The photoluminescence peak observed by Potter and Simmons at 9°K is at 490 nm. The agreement with this simple Bohr model is good and shows the concepts of exciton transition. More accurate calculations must include the exact band gap and exciton Hamiltonian in a crystal field in order to determine the precise recombination energy. The observed band gap is uncertain by ±10 nm and remains a problem in theoretical calculations.

10.21 PHOTOCONDUCTIVE SEMICONDUCTORS

Both cadmium sulfide (CdS) and cadmium selenide (CdSe) are used as visible radiation sensors. Light meters for cameras and optical sensors for parts sensing in automation rely on these inexpensive semiconductors. The typical response time is approximately 50 ms.

Lead sulfide is sensitive to radiation in the near-infrared. Its response is good for 1–3.4 µm wavelengths. A typical response time of 200 µsec, coupled with a high initial impedance of approximately 1 Mohm make PbS a good commercial detector.

10.22 NONLINEAR OPTICS

Lasers have provided the power densities necessary to study and commercialize *nonlinear optical* (NLO) effects in optical materials. Nearly all optical materials exhibit some NLO response. The magnitude of the effect depends on the susceptibility of the material, χ.

The terms arise from the expression of the electric polarization, **P**, or dipole moment per unit volume of a dielectric:

$$\mathbf{P} = \epsilon_0(\chi\mathbf{E} + \chi_2\mathbf{E}^2 + \chi_3\mathbf{E}^3 + \cdots) \tag{10.53}$$

The equation for the polarization is a polynomial and thus if χ_2 is large, the nonlinear effect is large, resulting in the term *NLO material*.

For an electro-magnetic wave in a NLO material, the polarization increases more rapidly with increasing electric vector. A second harmonic is generated as a direct result of the $\mathbf{E}^2$ term in the polarization:

$$\mathbf{E} = \mathbf{E}_0 \sin \omega t \tag{10.54}$$

where $\mathbf{E}_0$ = maximum value of electric field
ω = frequency
t = time

Then

$$\mathbf{E}^2 = \mathbf{E}_0^2(\sin \omega t)^2 \tag{10.55}$$

Recalling the identity

$$\sin x = [\tfrac{1}{2}(1 - \cos 2x)]^{1/2} \tag{10.56}$$

By squaring both sides, we obtain

$$(\sin x)^2 = \tfrac{1}{2}(1 - \cos 2x) \tag{10.57}$$

let $x = \omega t$ and substitute into the equation for $\mathbf{E}^2$:

$$\mathbf{E}^2 = \tfrac{1}{2}\mathbf{E}_0^2(1 - \cos 2\omega t) \tag{10.58}$$

Thus, a harmonic is generated with frequency 2ω if X_2 or $\mathbf{E}$ is large enough. Electric fields of 10^6 V/m are usually required for the NLO effect to be observed.

The intensity of the second harmonic is significantly reduced, but with anisotropic media such as quartz, ADP crystals and KDP crystals (see Table 10.4) several percent frequency conversion is possible. Uses include shifting laser output to higher frequencies and optical switching.

PROBLEMS

10.1 (a) Plot the ranges of Faraday rotation for a crystal whose $n(\text{right}) = 1.515$ and $n(\text{left}) = 1.596$ for visible light for a length of 0.01 mm.

(b) Plot the ranges of Faraday rotation as in (a) for a wavelength of 600 nm with the path length varying between zero and 100 mm.

10.2 For a specific luminescent material there exists donor states 0.5 eV below the conduction edge.

(a) What would be the expected lifetime of the luminescence at room temperature?

(b) At 100°C?

(c) For 77°K? (liquid nitrogen temperatures)?

(d) Where must the donor level be in order for the lifetime to exceed 9 nsec?

10.3 Calculate the binding energy for a Bohr exciton: (a) In silicon, (b) in GaAs, and (c) in Germanium.

10.4 Calculate the photoluminescence wavelengths for Bohr excitons: (a) In Silicon, (b) in GaAs, (c) in Germanium, and (d) compare these values with the literature.

10.5 Consider a Bragg switching device as described in Chapter 9. Calculate diffraction angle θ for the first Bragg maxima for yellow (500 nm) light if: $\Delta n = 0.1$ and

(a) The electrode spacing $d = 1$ micron.

(b) The electrode spacing $d = 2$ microns.

(c) The electrode spacing $d = 3$ microns.

(d) Which of the three spacings makes a valid Bragg switch?

(e) What is the minimum separation allowable for UV (200 nm) light?

(f) Will larger or small Δn help matters?

(g) If 10 electrodes are necessary to produce this device, how does it compare to the current size of a transistor in a VLIC (very large integrated circuit)?

(h) Comment on the size of an optical computer assembled from Bragg optical switches.

10.6 For diodes made from the following materials calculate the breakdown voltage if the $N(\text{acceptors}) = 1 \times 10^{17}\text{ cm}^{-3}$: **(a)** For Silicon, **(b)** for GaAs, **(c)** for Germanium, **(d)** for SiC, **(e)** discuss the importance of breakdown voltage in the above semiconductor materials, **(f)** compare these calculated values with the experimental values, and **(g)** plot the breakdown voltage for $1E_{12} < N < 1E_{17}$ for each of these materials.

10.7 The intensity of a LED increases exponentially with forward bias voltage. At room temperature:

  (a) Plot the change in output intensity, i, for a change in bias of $0.01 < eV < 0.1$.

  (b) Typical LEDs junctions the p–n active region is usually no more than 10 microns and the electrode separation approximately 100 microns. What voltage bias must be applied to generate 0.1 eV in the active region?

READING LIST

Elliot, S. R., (1984). *Physics of Amorphous Materials*. Longman, Essex, England.

Gustafson, T. K., and Smith, P. W. (1988). *Photonic Switching*, Springer-Verlag, Berlin, Heidelberg, N.Y.

Haertling, G., (1971). *J. Amer. Ceramic Soc.*, **54**, 303.

Haertling, G., (1986). *Ceramic Materials for Electronics* (R. C. Buchanan, ed.), Marcel Dekker, Inc., N.Y. pp. 139–225.

Haus, H. A., (1984). *Waves and Fields in Optoelectronics*, Prentice-Hall, Englewood Cliffs, N.J.

Hoffman, L., Ziegler, G., Theis, D., and Weyrich, C., (1982). *J. Appl. Phys.*, **53**, 6952.

Hummel, R. E., (1985). "Electronic Properties of Materials." Springer-Verlag, Berlin.

Laud, B. B., (1987). *Lasers and Non-linear Optics*, Wiley, N.Y.

McKeever, S. W. S., (1985). *Thermal Luminescence of Solids*, Cambridge University Press, UK.

Nair, S. V., Sinha, S., and Rustagi, K. C., (1987). *Phys. Rev. B*, **35**, 4098.

Potter, B. G., Jr., and Simmons, J. H., (1988). *Phys. Rev. B*, **37**(18), 838.

Schubert, M., and Wilhelmi, B., (1986). *Nonlinear Optics and Quantum Electronics*, Wiley, N.Y.

Shen, Y. R., (1971). *Physics of Electronic Ceramics*, Part B, (L. L. Hench and D. B. Dove, eds.), Dekker, New York, pp. 987.

Van Vliet, C. M., Bosman, G. and Hench, L. L., (1988). *Annu. Rev. Mater. Sci.*, **18**, 381–421.

Wilson, J. and Hawkes, J. F. B., (1983). *Optoelectronics: An Introduction*, Prentice-Hall, Englewood Cliffs, New Jersey.

Ziegler, G., Lanig, P., Theis, D., and Weyrich, C., (1983). *IEEE Trans. Electr. Devices*, **ED-30**, 277.

11

GLASS AND CRYSTALLINE LASERS

11.1 INTRODUCTION

Laser light is different from ordinary light in several ways. The name *laser* comes from the phrase: Light Amplification by Stimulated Emission of Radiation. Lasers are truly optical oscillators (see *parametric oscillators*: Chapter 10) with a feedback mechanism created by the mirrors at the ends of the optical cavity. Part of the energy of the emission is returned to the cavity and part is allowed to escape to form a beam of light. The laser light that is allowed to escape is extremely coherent and monochromatic.

Laser emissions can either arise from electronic transitions or from molecular transitions. Figure 11.1 shows the ranges in energy, wavelength, frequency, and type of transitions associated with lasers. As indicated, lasers can be designed and made to operate from the ultraviolet through the microwave regions of the electromagnetic spectrum.

The first laser (optical maser) was invented by Theodore H. Maiman. It was the result of a population inversion between energy levels of Cr^{3+} ions in a ruby crystal when pumped by a xenon flashlamp. He submitted his findings to *Physical Review Letters*; they were rejected. He finally held a news conference in New York City on July 7, 1960 to announce that he had successfully built an operative laser. Earlier work by Townes, Prokhorov and Basov earned them the Nobel Prize in Physics in 1964 for the *maser*: Microwave Amplification by Stimulated Emission of Radiation.

Light from a laser appears differently from normal light when it is reflected from a surface; objects appear to sparkle. This sparkling, or *speckle*, makes it difficult for the observer to focus on any object so illuminated. The cause is interference of the reflected light. Normal light

464 CHAPTER 11: GLASS AND CRYSTALLINE LASERS

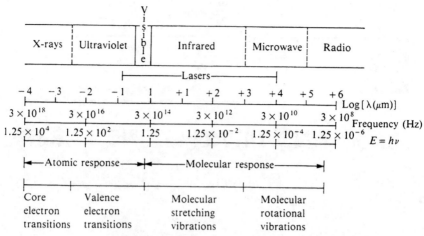

FIGURE 11.1 Laser transitions.

does not interfere, because it is not coherent (it is not in phase). The speckle pattern is due to destructive and constructive interference of the reflected waves, which creates dark and bright spots, respectively.

Laser light is also extremely monochromatic. This is due to the fact that the emitted photons all have very nearly the same energy. This energy arises from electron transitions from a metastable state E_m to the ground state E_g.

$$E_m - E_g = h\nu \tag{11.1}$$

where h = Planck's constant
ν = frequency

The transition energy is very nearly the same for all of the emitted photons. Thus, the frequency, the wavelength, or the color of the light is fixed into a very sharp distribution, that is, it is monochromatic.

11.2 OPTICAL CAVITIES: FEEDBACK AND LASERS

Energy stored in an electric field per unit volume is

$$U_E = \tfrac{1}{2} \epsilon \mathbf{E}^2 \tag{11.2}$$

Also, the energy stored in a magnetic field per unit volume is

$$U_B = \frac{1}{2} \frac{\mathbf{B}^2}{\mu} \tag{11.3}$$

11.2 OPTICAL CAVITIES: FEEDBACK AND LASERS

where the variables are the same as defined in Chapter 9. If an optical cavity is resonating, then there will be an integer number q of half wavelengths within the cavity:

$$q = \frac{d}{(\lambda/2)} \tag{11.4}$$

where d = length of the cavity
λ = wavelength
q = longitudinal cavity modes

The total energy in the optical cavity is

$$U = U_B + U_E \tag{11.5}$$

The total energy is not constant

$$\frac{dU}{dt} = \text{Energy added} - \text{Energy lost} = U_A - U_L \tag{11.6}$$

Thus,

$$\frac{dU}{dt} - U_A + U_L = 0 \tag{11.7}$$

We can define the coefficients α and β where

$$\frac{dU}{dt} - \beta U + \alpha U = 0 \tag{11.8}$$

If we let the energy fluctuate as

$$U = U_0 \exp[-(\alpha - \beta)t] \tag{11.9}$$

then

$$\frac{dU}{dt} = U_0(-\alpha + \beta)\exp[-(\alpha - \beta)t] \tag{11.10}$$

By substituting back into (11.8) we obtain

$$(-\alpha + \beta)U_0 \exp[-(\alpha + \beta)t] - \beta U_0 \exp[-(\alpha + \beta)t]$$
$$+ \alpha U_0 \exp[-(\alpha + \beta)t] = 0 \tag{11.11}$$

leaving

$$0 = 0 \tag{11.12}$$

Therefore the solution

$$U = U_0 \exp[-(\alpha - \beta)t] \quad (11.13)$$

satisfies the differential equation for the energy of the optical cavity. The coefficient α is the total loss coefficient which is a function of: (1) scattering losses, (2) absorption losses, (3) isotropic losses, and (4) reflection losses. The coefficient β is basically the energy of the stimulated photon emissions plus the energy reflected back into the cavity (optical feedback):

$$\beta = \text{stimulated emission and feedback} \quad (11.14)$$

$$\beta = Nh\nu + RU_0 \quad (11.15)$$

where R = reflected energy fraction
N = number of emitted photons
h = Planck's constant
ν = frequency

If the losses are greater than the sources of energy,

$$\alpha > \beta \quad (11.16)$$

then the energy

$$U = U_0 \exp[-(\alpha - \beta)t] \quad (11.17)$$

is an exponential decay function, and the oscillations die away with time. However, if

$$\alpha \approx \beta \quad (11.18)$$

then

$$U \approx U_0 \quad (11.19)$$

with no exponential decay, and laser operation is possible. The fraction of the energy lost per unit cycle,

$$\frac{\Delta U}{U_0} \approx \frac{\alpha U_0}{U_0} \quad (11.20)$$

is a measure of the quality (or Q) of an oscillator. The Q of an oscillator is defined as:

$$Q \approx 2\pi \left(\frac{U_0}{\Delta U}\right) \quad (11.21)$$

$$Q \approx \frac{2\pi}{\alpha} \quad (11.22)$$

From this we can see that a "high-Q" oscillator has s small energy loss. Conversely, a low-Q oscillator has a larger energy loss. We can determine the approximate conditions for sustained laser operation by looking at the conditions required to achieve equation (11.19):

$$\alpha = \beta \quad (11.23)$$

or

$$\alpha = Nh\nu + RU_0 \quad (11.24)$$

For typical solid-state lasers

$$R \approx 0.9 \quad (11.25)$$

and

$$Nh\nu \approx 0.1 U_0 \quad (11.26)$$

therefore

$$\alpha \approx 1 \quad (11.27)$$

or

$$Q \approx 2\pi \quad (11.28)$$

$$Q \approx 6.28 \quad (11.29)$$

in order for the laser cavity to sustain operation.

11.3 METASTABLE STATES AND STIMULATED EMISSION

The key to the photon amplification process is the existence of metastable energy states. For example, in a ruby laser the metastable states are due to substitution of about 0.05% of Cr^{3+} ions for Al^{3+} ions in the Al_2O_3 crystal lattice. These states are unique. Under normal conditions, an electron can be excited into a higher state. We have all seen neon signs with their characteristic glow. This glow is just the emission of photons which occurs after the neon valence electrons are excited into a higher energy state. The excited electrons quickly decay back to their original level, emitting a photon. This process occurs randomly and there are always considerably more electrons in the ground state than in excited states. Population inversion has not occurred. Because the process of decay is also random, stimulated emission also has not occurred.

In the case of a ruby crystal laser, there is a significant difference in optical behavior. The Cr^{3+} ions in the Al_2O_3 crystal lattice have strong absorption peaks in the green and blue part of the visible spectrum (see Figure 11.2). In a nonlasing material, the transitions corresponding to these absorptions would decay and the crystal would glow green and blue. But

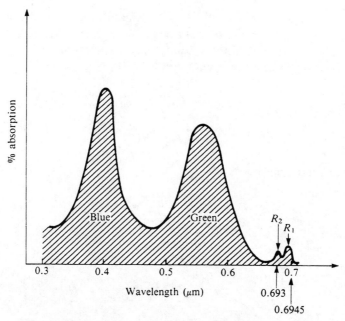

FIGURE 11.2 Ruby laser absorption Cr^{3+} in Al_2O_3 crystal.

this does not occur in a laser. There is an energy level below these absorption levels that accepts electron transitions. The energy of this transition is low and the lattice accepts a phonon; no photons are emitted. This increases the thermal energy of the lattice and plays an important role in overall efficiency. This intermediate state is called a *metastable state* because electrons "like" being there more than they like the original absorption level. Thus, they do not decay immediately.

This creates an unusual situation. The population of electrons in the metastable state can exceed that of the ground state, creating a *population inversion*. A population inversion must occur for there to be a sufficient number of excited electrons to be available for energy input into the optical cavity. Referring back to equations (11.24) and (11.26), $Nh\nu$ must be large enough for the energy source to be equal to the energy losses. Note also, that if the metastable state did not exist, there could be no population inversion, and laser operation would be impossible.

11.4 THREE-LEVEL LASER (Cr–Al$_2$O$_3$)

Figure 11.3 shows the energy relationships between the Cr-doped Al_2O_3 three-level laser. Blue and green light are absorbed by electron transitions to excited states. They immediately decay to the metastable state by release of a phonon. In the metastable state, electrons "wait" for a photon of nearly

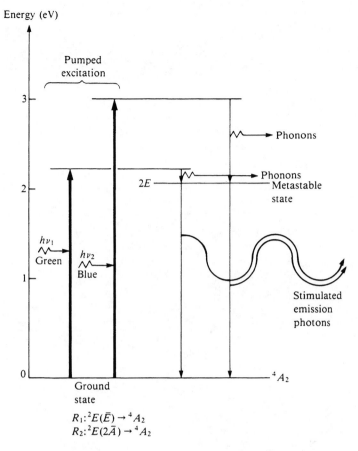

FIGURE 11.3 Metastable state: C^{3+} in Al_2O_3.

identical energy. They then undergo a transition back to the ground state with their released photons exactly in phase. This is called *stimulated emission* and is the cause of the extremely coherent nature of laser light.

Figure 11.4 illustrates the initiation of the stimulated emissions and the resulting plane wave that forms near the end of the optical cavity. This illustration also shows why only a small percentage of the light is available for resonance within the cavity. Most of the light is lost because it radiates isotopically from the emitting atoms. Only a small fraction near the optical axis is reflected back into the cavity for the feedback mechanism.

There is a shift in the transition line due to temperature. The fluorescence peak R is empirically related to temperature by

$$\lambda = 6943.25 + 0.068(T - 20) \tag{11.30}$$

where T is in degrees Celsius.

The general properties of ruby lasers are summarized in Table 11.1.

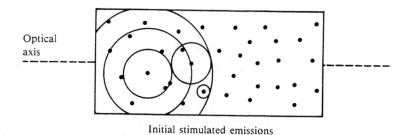

FIGURE 11.4 Stimulated emission.

TABLE 11.1 Properties of Ruby Lasers

Ion	Host	Wavelength (μm)	Comments
Cr^{3+}	Sapphire	0.6943	Pink ruby
Cr^{3+}	Sapphire	0.7009	Red ruby
Cr^{3+}	Sapphire	0.7041	Red ruby

11.5 FOUR-LEVEL LASERS (Nd–YAG, Nd–Glass)

We have just seen that a three-level laser is one for which the ground state is the lower level. A four-level laser has an excited state as the lower level. Many fewer electrons must be excited into the upper lasing level to achieve population inversions in the four-level laser, since there are only a few electrons in the lower level. Consequently, the pumping intensity of light required to reach lasing threshold is ~1000 times less for a four-level laser than for a three-level laser. Because of the large amount of pumping energy, three-level lasers have to be operated in a pulsed mode, whereas four-level

11.5 FOUR-LEVEL LASERS (Nd–YAG, Nd–Glass)

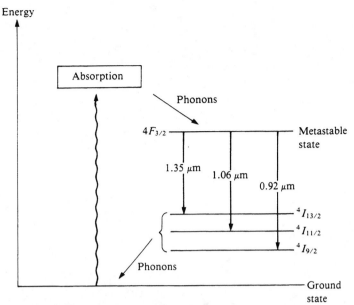

FIGURE 11.5 Energy level diagram for Nd–YAG laser.

lasers can be operated in a continuous mode, or at much higher power in a pulsed mode (see Table 11.2).

The Nd–YAG laser is an important commercial four-level laser. Neodymium is the substitutional impurity which provides the metastable states in the host, yttrium–aluminum–garnet ($Y_3Al_5O_{12}$) crystal. About 0.5–2% Nd is used. The energy level diagram for the Nd–YAG laser is shown in Figure 11.5. There are three widely spaced lasing transitions. However, the one at 1.064 μm is dominant, and operation at 1.35 μm is also possible (see Figure 11.5). Use of Ho instead of Nd in YAG results in lasing at 2.10 μm.

As the temperature of the laser is raised from 77°K to room temperature, the ground state of the Nd–YAG laser broadens. This enables it to absorb a greater percentage of the pumping radiation. Consequently, the lasing threshold of the Nd–YAG laser decreases with increasing temperature, another important advantage over the three-level ruby laser.

The high gain and good thermal conductivity of the crystalline Nd–YAG laser makes it applicable for continuous wave oscillators and amplifiers as well as high-power pulsed lasers. Characteristics of Nd–YAG lasers in pulsed mode are summarized in Table 11.3 and in continuous-wave (CW) mode in Table 11.4.

The world's most powerful lasers are the four-level Nd–glass lasers. Because of their lower cross sections, glass lasers store energy well and consequently make good short-pulse lasers and amplifiers. For example, a Nd–glass Shiva laser at Lawrence Livermore Laboratory produced >10 KJ in a 0.8 nsec pulse and 20 TW in a 0.06 nsec pulse at a 20 cm aperture.

TABLE 11.2 Classes, Types, and Representative Examples of Laser Sources

Class	Type (characteristic)	Representative example	Nominal operating wavelength (nm)	Method(s) of excitation
Gas	Atomic, neutral (electronic transition)	Neon–Helium (Ne–He)	633	Glow discharge
	Atomic, ionic (electronic transition)	Argon (Ar$^+$)	488	Arc discharge
	Molecule, neutral (electronic transition)	Krypton fluoride (KrF)	248	Glow discharge: e-beam
	Molecule, neutral (vibrational transition)	Carbon dioxide (CO_2)	10,600	Glow discharge; gasdynamic flow
	Molecule, neutral (rotational transition)	Methyl fluoride (CH_3F)	496,000	Laser pumping
	Molecule, ionic (electronic transition)	Nitrogen ion (N_2^+)	420	E-beam
Liquid	Organic solvent (dye-chromophor)	Rhodamine dye (Rh6G)	580–610	Flashlamp; laser pumping
	Organic solvent (rare-earth chelate)	Europium–TTF	612	Flashlamp
	Inorganic solvent (trivalent rare-earth ion)	Neodymimum–$POCl_4$	1060	Flashlamp
Solid	Insulator, crystal (impurity)	Neodymium–YAG	1064	Flashlamp, arc lamp
	Insulator, crystal (stoichiometric)	Neodymium–UP(NdP_5O_{14})	1052	Flashlamp
	Insulator, amorphous (impurity)	Neodymium–glass	1061	Flashlamp
	Semiconductor (p–n junction)	GaAs	820	Injection current
	Semiconductor (electron–hole plasma)	GaAs	890	E-beam, laser pumping

11.5 FOUR-LEVEL LASERS (Nd–YAG, Nd–Glass)

TABLE 11.3 Performance of Solid-State Pulsed Lasers

Parameter	Unit	Nd–YAG (xenon-flashlamp excited)	Nd–glass (xenon-flashlamp excited)
Gain medium density	ions/cm^3	1.5, 1	3, 2
Wavelength	nm	1064	1061
Laser cross section	cm^{-2}	7	2.8
Radiative lifetime	μsec	2.6	4.1
Decay lifetime	μsec	2.3	3.7
Gain bandwidth	nm	0.5	26
Homogeneous saturation fluorescence	J/cm^{-2}	0.6	~5
Decay lifetime	μsec	<1	<1
Inversion density	cm^{-3}	4	3
Small-signal gain coefficient	cm^{-1}	0.3	8
Medium excitation energy density	J/cm^{-3}	0.15	0.6
Output energy density	J/cm^{-3}	5	2
Laser dimensions (diameter, length)	cm, cm	0.6, 7.5	0.6, 8.3
Excitation peak power	W	4	9
Output pulse energy	J	0.1	1.0
Output pulse length	sec	2	1
Output pulse power	W	5	1
Efficiency	%	1.5	3.7

An important feature of glass lasers is that the glass host matrix composition can be varied over enormous ranges. Table 11.5 summarizes the composition range of glass network formers used for a glass laser hosts. Within a particular type of glass laser host, it is possible to vary the laser properties systematically by changing the amount and species of network-modifying ions in the composition. Commercial laser glasses are silicate, phosphate, or fluorophosphate. A typical silicate composition (in mole%) is: 60SiO$_2$, 27.5Li$_2$O, 10.0CaO, 2.5Al$_2$O$_3$. A typical phosphate glass laser host is (in mole%): 59P$_2$O$_5$, 8BaO, 25K$_2$O, 5Al$_2$O$_3$, 3SiO$_2$. A typical fluorophosphate glass laser host is (in mole%): 4Al(PO$_3$)$_3$, 36AlF$_3$, 10MgF$_2$, 30CaF$_2$, 10SrF$_2$, 10BaF$_2$.

Neodymium, in the range of a few mole%, is the active ion used in commercial glass lasers because of its high-efficiency, room temperature operation, and convenient 1.06 μm wavelength. The Nd ion provides four-level laser operation with a high radiative efficiency (0.9) of the $^4F_{3/2}$ state and a long radiative lifetime (150 to >800 nsec). The absorption bands of Nd–glass lasers are also well-matched with the spectra of xenon

TABLE 11.4 Performance of Solid-State Continuous-Wave (CW) Lasers

Parameter	Unit	Nd–YAG (krypton-arc-lamp excited)	GaAs (dc-injection excited)
Gain medium composition		Nd–YAG	p–Ln–GaAs
Gain medium density	ions/cc	1.5, 2	2, 3, 3
Wavelength	nm	1064	810
Laser cross section	cm^{-2}	7	~6
Radiative lifetime	μsec	2.6	~1
Decay lifetime	μsec	2.3	~1
Gain bandwidth	nm	0.5	10
Type, gain saturation		Homogeneous	Homogeneous
Homogeneous saturation flux	W cm^{-2}	2.3	~2
Decay lifetime	μsec	<1	<1
Inversion density	cm^{-3}	6	1
Small-signal gain coefficient	cm^{-1}	5	40
Pump power density	W cm^{-3}	150	7
Output power density	W cm^{-3}	95	5
Laser dimensions (diameter, length)	cm, cm	0.6, 10	5, 7, 29
Excitation current/voltage	A/V	90/125	1.0/1.7
Excitation current density	A cm^{-2}	140	4.5
Excitation power	KW	11.2	1.7
Output power	W	300	120
Efficiency	%	2.6	7

flashlamps used for pumping, which increases laser efficiency. Figure 11.6a illustrates the absorption spectrum, the laser transitions, and laser wavelengths of a Nd–glass laser. The dominant transition for the Nd–glass laser is $^4F_{3/2} \rightarrow {}^4I_{11/2}$ which yields a 1.06 μm laser output (Figure 11.6a).

The transition for an Er–glass laser is $^4I_{13/2} \rightarrow {}^4I_{15/2}$ which results in a 1.54 μm line (Figure 11.6b).

TABLE 11.5 Laser Host Compositions

Compound	Glass laser host network former	Concentration range of network former (cationic %)	Active ion(s)
Silicates	SiO$_2$	35–65	Nd, Gd, Ho, Er, Tm, Yb
Germanates	GeO$_2$	50–60	Nd
Tellurites	TeO$_2$	65–80	Nd
Borates	B$_2$O$_3$	50–80	Nd, Tb, Yb
Phosphates	P$_2$O$_5$	45–70	Nd, Er
Fluoroberyllates	BeF$_2$	45–60	Nd
Fluorozirconates	ZrF$_4$	50–60	Nd

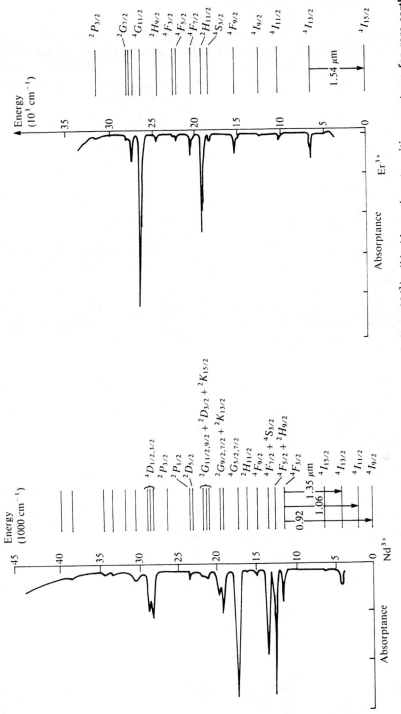

FIGURE 11.6 (*a*) Absorption, transition spectrum for rare-earth ions: Nd^{3+}. (*b*) Absorption, transition spectrum for rare-earth ions: Er^{3+}.

11.6 EMISSION CROSS SECTION

In general, the spectroscopic properties of the lasing ion dominates the laser characteristics. With these characteristics, the laser performance may be predicted. The gain coefficient, g, is a product of the ion-excited state density and the emission cross section of the laser transition. The density of the metastable state depends on the pumping sources, intensities of the absorption bands, and the lifetime of the fluorescence state.

The most critical parameter is the emission cross section σ:

$$g = \sigma N \qquad (11.31)$$

where g = gain coefficient
σ = emission cross section
N = excited-state population

The gain coefficient is related to the total gain β in equation (11.13) as

$$\beta = g + \text{feedback} \qquad (11.32)$$

If the lifetime of the emission state is short, then N is small and the gain is small. This is a linear relationship. The emission cross section increases as the fourth power of the index of refraction:

$$\sigma(\lambda_p) \approx \frac{\lambda_p}{\Delta\lambda_{\text{eff}}} \frac{(n^2 + 2)^2}{n} S \qquad (11.33)$$

where S = line strength
n = index of refraction
λ_p = wavelength of emission peak
$\Delta\lambda_{\text{eff}}$ = effective emission line width

Therefore, small changes in the host homogeneity cause a significant change in emission cross section and thereby the gain of the laser. The compositional dependence of the emission cross section for Nd is shown in Table 11.6.

Nonradiative transitions (phonons) can occur that reduce the population of the metastable state, N. If this rate of phonon decay is large, then the quantum efficiency of the laser is small. A low radiative quantum efficiency requires that pumping be done in short pulses. The phonon decay rate can be approximated by

$$W_{\text{nr}} = W_0 \exp[-\gamma \Delta E / kT] \qquad (11.34)$$

where W_0 = coefficient of nonradiative decay
γ = host glass coefficient
ΔE = energy gap between neighboring states

TABLE 11.6 Emission Cross-Section for Nd

$\sigma(\times 10^{-20}\ cm^2)$	Host composition
1.8–4.8	Borate
1.8–4.7	Phosphate
1.0–3.6	Silicate
1.6–3.5	Germanate
3.0–5.1	Tellurite
2.2–4.3	Fluorophosphate

Thus phonon decay increases with temperature. As [OH$^-$] increases in the host glass, the quantum efficiency decreases. This is particularly true of [OH$^-$] in phosphate Nd^{3+} glass lasers. This effect is measured by the fluorescence lifetime τ in seconds. For example, the lifetime of the $^4F_{3/2}$ state of Nd^{3+} is given by

$$\frac{1}{\tau} = 3.3 \times 10^3 + 2.2 \times 10^2 \alpha \qquad (11.35)$$

where α is the absorption coefficient at 2.85 μm (OH$^-$) in units of cm^{-1}.

11.7 LASER MODES AND FOCUSING

Laser communications, industrial applications, and so on, require a very specific distribution of the intensity of the laser beam. The focusing, dispersion, and general operation are all related to the modes exiting from the laser. The output modes for a He–Ne laser are shown in Figure 11.7.

We have discussed in Chapter 9 the various TEM, transverse electric and magnetic modes found in waveguides. Figures 9.1 and 9.2 show these modes pictorially. The indices are usually written

$$\text{TEM}_{mnq} \qquad (11.36)$$

where m = transverse mode number in x-axis

n = transverse mode number in y-axis

q = longitudinal mode number or z-axis standing waves

The longitudinal mode number is defined by equation (11.4).

For many applications, the TEM$_{00}$ mode is desirable. It has no side lobes and has a Gaussian distribution of intensity about the optical axis.

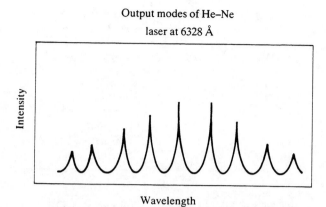

FIGURE 11.7 Output modes of He–He laser at 6328 Å.

There are cases where a "blunt tool" is required for a laser application. The beam may be spread into a square four-point tool by selecting the TEM_{11} mode. Likewise, a square nine-point tool could be selected by using the TEM_{22} mode (see Chapter 9).

Because of the feedback by reflection along the optical axis, the output laser beam diverges very little. Figure 11.8 shows a simple lens of focal length F. With this arrangement, the depth of focus b and the minimum

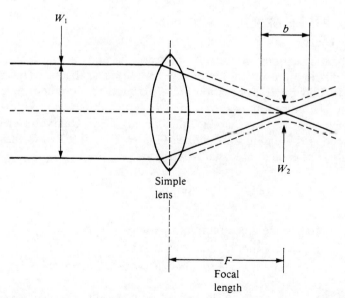

FIGURE 11.8 Laser focusing. b: depth of focus; W_1: beam diameter; W_2: minimum focus diameter.

diameter of the focused beam W_2 can be approximated by:

$$W_2 = \frac{8\lambda}{\pi}\left(\frac{F}{W_1}\right) \qquad (11.37)$$

and the depth of focus by:

$$b = \frac{8\lambda}{\pi}\left(\frac{F}{W_1}\right)^2 \qquad (11.38)$$

where λ = wavelength
F = focal length
W_1 = initial beam diameter
W_2 = focused beam diameter
b = depth of focus

11.8 ETALONS: FABRY–PEROT INTERFEROMETER

The modes that are excited in optical cavities are illustrated in Figure 11.7. These modes represent the fine structure of an emission line. However, these modes also represent different wavelengths, whereas single-mode operation is desirable for most applications. The mode selection is often done with an etalon.

Figure 11.9 shows a typical schematic for the position of the etalon in a laser cavity. The tilted etalon reduces the Q for all modes except the one of interest.

The original device invented by C. Fabry and A. Perot was designed to establish a standard for the length of a meter by counting the number of standing waves in an optical cavity for red cadmium light. The importance of the device, along with its simplicity, soon led to improvements and various other uses. These applications include, but are not limited to: solar spectrum analysis, Zeeman effect, viscosity of air, electrostatic voltmeter, fine structure of atomic emission, and resonant cavity of a laser.

Equation (11.4) represents the longitudinal modes allowed for a laser cavity. A Fabry–Perot interferometer is used to select various longitudinal modes of resonance due to the separation between two end mirrors.

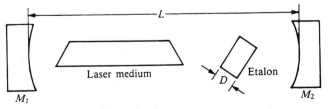

FIGURE 11.9 The "etalon" method of single-moding a laser.

Usually a second etalon is placed in the beam path to improve longitudinal mode selectivity of the system. So in effect, this becomes a narrow-band-pass filter. These etalons can be positioned in such a way as to permit only one longitudinal mode to lase. These devices also allow tuning of a laser to a specific mode and frequency. Common methods of tuning include tilting the etalon and varying the temperature of the etalon which changes its dimension due to thermal expansion (see Chapter 1).

The phase difference between a reflected and transmitted ray can be determined from the difference in path length.

$$\frac{\text{phase difference}}{2\pi} = \frac{\text{path difference}}{\lambda} \tag{11.39}$$

If we let

$$\delta \equiv \text{phase difference} \tag{11.40}$$

and then we can determine the path difference from Snell's law (equation 9.49), and geometry.

$$n_1 \sin \theta = n_2 \sin \phi \tag{11.41}$$

The refracted ray path length for ray AB is different from the reflected ray path AC (see Figure 11.10).

$$AC - AB = \frac{2d}{\cos \phi} \tag{11.42}$$

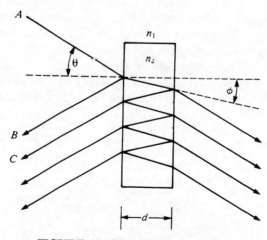

FIGURE 11.10 Parallel-face etalon.

11.8 ETALONS: FABRY–PEROT INTERFEROMETER

If we let $n_1 = 1$ and $n_2 = n$, then equation (11.41) becomes

$$\sin \phi = \frac{\sin \theta}{n} \tag{11.43}$$

using the identity

$$\cos^2 \phi + \sin^2 \phi = 1 \tag{11.44}$$

then

$$\cos^2 \phi + \left(\frac{\sin \theta}{n}\right)^2 = 1 \tag{11.45}$$

or

$$\cos \phi = \left[1 - \left(\frac{\sin \theta}{n}\right)^2\right]^{1/2} \tag{11.46}$$

then

$$AC - AB = \frac{2d}{\left[1 - \left(\frac{\sin \theta}{n}\right)^2\right]^{1/2}} \tag{11.47}$$

and equation (11.39) becomes

$$\frac{\delta}{2\pi} = \frac{2d}{\lambda \left[1 - \left(\frac{\sin \theta}{n}\right)^2\right]^{1/2}} \tag{11.48}$$

$$\delta = \frac{4\pi d}{\lambda} \left[1 - \left(\frac{\sin \theta}{n}\right)^2\right]^{-1/2} \tag{11.49}$$

$$\delta = \frac{4\pi d}{\lambda} \left[1 - \frac{1}{n^2} \sin^2 \theta\right]^{-1/2} \tag{11.50}$$

Remembering from equation (11.44), we obtain

$$\sin^2 \theta = 1 - \cos^2 \theta \tag{11.51}$$

then

$$\delta = \frac{4\pi d}{\lambda} \left[1 - \frac{1}{n^2}(1 - \cos^2 \theta)\right]^{-1/2} \tag{11.52}$$

$$= \frac{4\pi d}{\lambda} \left[1 - \left(\frac{1 - \cos^2 \theta}{n^2}\right)\right]^{-1/2} \tag{11.53}$$

Multiply by $\dfrac{n}{n}$:

$$\delta = \frac{4\pi dn}{\lambda}\left[\frac{1}{n^2 - n^2\left(\dfrac{1 - \cos^2\theta}{n^2}\right)}\right]^{1/2} \qquad (11.54)$$

$$\delta = \frac{4\pi dn}{\lambda}\left[\frac{1}{n^2 - 1 + \cos^2\theta}\right]^{1/2} \qquad (11.55)$$

Multiply by $\dfrac{\cos\theta}{\cos\theta}$:

$$\delta = \frac{4\pi dn \cos\theta}{\lambda}\left[\frac{1}{n^2\cos^2\theta - \cos^2\theta + \cos^4\theta}\right]^{1/2} \qquad (11.56)$$

For small θ and for the index of silica glasses, the factor

$$[n^2\cos^2\theta - \cos^2\theta + \cos^4\theta]^{-1/2} \approx 1 \qquad (11.57)$$

Finally then the phase difference can be approximated by;

$$\delta = \frac{4\pi dn \cos\theta}{\lambda} \qquad (11.58)$$

where δ = phase difference
 θ = incident angle with respect to the normal
 n = index of refraction of the etalon
 λ = wavelength
 d = thickness of the etalon

This relation (11.59) is known as the first of *Airy's formulae* for multiple-beam fringes of equal inclination.

The other two Airy formulae represent the reflected intensity I_R and the transmitted intensity I_T. They can be derived from Fresnel's equations developed in Chapter 9 (see equations 9.57–9.60).

If we let

$$u_1 \approx u_2 \approx 1 \qquad (11.59)$$

and let

$$k = \frac{n}{\lambda} \qquad (11.60)$$

11.8 ETALONS: FABRY–PEROT INTERFEROMETER

then the intensity of the reflected ray,

$$I_R = \left(\frac{\mathbf{E}_R}{\mathbf{E}_0}\right)^2 \tag{11.61}$$

and that of the transmitted ray,

$$I_T = \left(\frac{\mathbf{E}_T}{\mathbf{E}_0}\right)^2 \tag{11.62}$$

are normalized to the incident electric field E_0.

For reflection and transmission at the interface between two dielectrics, Fresnel's equations become:

$$\left(\frac{\mathbf{E}_R}{\mathbf{E}_0}\right)_N = \frac{\left(\dfrac{n_1}{n_2}\right)\cos\theta - \cos\phi}{\left(\dfrac{n_1}{n_2}\right)\cos\theta + \cos\phi} \tag{11.63}$$

and

$$\left(\frac{\mathbf{E}_T}{\mathbf{E}_0}\right)_N = \frac{2\left(\dfrac{n_1}{n_2}\right)\cos\theta}{\left(\dfrac{n_1}{n_2}\right)\cos\theta + \cos\phi} \tag{11.64}$$

A similar relation for parallel electric fields can also be derived. The coefficient of reflection R is defined as

$$R = \left(\frac{\mathbf{E}_R}{\mathbf{E}_0}\right)^2 \tag{11.65}$$

Likewise, the coefficient of transmission T is defined as

$$T = \frac{n_2}{n_1}\frac{\mathbf{E}_T^2 \cos\phi}{\mathbf{E}_0^2 \cos\theta} \tag{11.66}$$

Note that $R + T = 1$ in all cases. A typical value of R for silica glass is

$$R = 0.1 \tag{11.67}$$

Thus, a typical value of T for silica glass is

$$T = 0.9 \tag{11.68}$$

Thus, Airy's other two formulae for etalons become

$$I_T = \frac{T}{(1-R)^2 + 4R \sin^2\left(\frac{\delta}{2}\right)} \tag{11.69}$$

$$I_R = \frac{4R \sin^2\left(\frac{\delta}{2}\right)}{(1-R)^2 + 4R \sin^2\left(\frac{\delta}{2}\right)} \tag{11.70}$$

With recent technology in nonreflective coatings and carefully controlled surface finishes and flatnesses, transmission intensities (I_T) of more than 10% are possible.

The resolving power of an etalon can be calculated from

$$\text{Resolving power} \equiv \frac{\lambda}{\Delta\lambda} \tag{11.71}$$

Using equation (11.4),

$$q = \frac{2d}{\lambda} \tag{11.72}$$

with

$$\Delta\lambda = 2\pi^2 \, dn \left(\frac{4R}{(1-R)^2}\right)^{1/2} \tag{11.73}$$

or

$$\frac{\lambda}{\Delta\lambda} = \frac{1}{q\pi^2 n} \left(\frac{(1-R)^2}{4R}\right)^{1/2} \tag{11.74}$$

The inverse of the resolving power divided by the phase change is used as a measure of the quality of an etalon, called *finesse*

If we define:

$$\frac{\Delta\lambda}{\delta\lambda} \equiv \text{finesse} \tag{11.75}$$

this yields

$$\frac{\Delta\lambda}{\delta\lambda} = \frac{\pi}{2}\left(\frac{4R}{(1-R)^2}\right)^{1/2} \tag{11.76}$$

Values of finesse as high as 10,000 are now possible for the best etalons with coated nonreflective optics.

11.9 BREWSTER ANGLE WINDOWS

These devices are simply high-quality silica plano–plano lenses with parallel faces. These windows (or prisms) are placed in the laser beam path to take advantage of the polarization that occurs upon reflection near or at the critical angle, called Brewster's angle. In Chapter 9, equation (9.159) was developed for the angle of incidence for total reflection:

$$\frac{n_2}{n_1} = \tan \theta_B \qquad (11.77)$$

Only the normal component of the electric field is reflected. A second relation is that the angle between the incident and refracted rays is 90°:

$$\theta_B + \phi_B = \frac{\pi}{2} \qquad (11.78)$$

As the beam is refracted through a Brewster angle window, only parallel polarized light is allowed to remain in the laser cavity. The normal component is reflected away.

11.10 Q-SWITCHING

This term relates back to our discussion on the quality of the optical cavity. If feedback is changed by increasing transmission, then the Q is lowered. Likewise, if reflection is increased (increased feedback) then Q increases.

By increasing Q for a short time, then reducing it, large high-energy pulses can be generated from the laser cavity. An almost total population inversion can be achieved with this technique. When Q is changed, a giant pulse of photons will occur as electrons drop down to their ground states almost in unison.

Practical application of this technique is called *Q-switching*. Many control schemes to achieve Q-switching are possible, including bleachable absorbers, rotating prisms and mirrors, mechanical choppers, ultrasonic cells, and electro-optic shutters, as discussed in Chapter 10.

11.11 GaAs LASERS

Optical information can be processed much more rapidly and with less error if mode dispersion is reduced. Single-mode lasers have no theoretical mode dispersion and therefore are of great interest. They can transmit at a higher pulse rate and over longer distances than any other laser. The dimensions of the laser optical cavity determine the number of modes that are allowed. Using equation (9.160) for a single-mode waveguide,

$$\cos \theta_i = \frac{\lambda_0}{2b} \tag{11.79}$$

If we let θ_i approach zero, then

$$b \cong \frac{\lambda_0}{2} \tag{11.80}$$

and the width of the waveguide is half the wavelength for single-mode operation. A more precise relation for the width, b, of the waveguide is

$$b = \left(\frac{3}{2}\lambda_0\right)\left(\frac{1}{\sqrt{n_1^2 - n_2^2}}\right) \tag{11.81}$$

where n_1 = index of the waveguide
n_2 = index of the surrounding material

This requires the manufacturing of very small structures. These widths are on the order of micrometers for proper operation.

Single-mode lasers are made by etching a step structure on GaAs semiconductors and regrowing an insulating semiconductor around it. This surrounded structure is called a *buried heterostructure* (BH). If, on the other hand, the active lasing area is deposited onto a concave semiconductor substrate, then it is called a *constricted double heterostructure* (CDH). Figure 11.11 shows a buried (BH) structure fabricated from GaAs, a type III–V semiconductor.

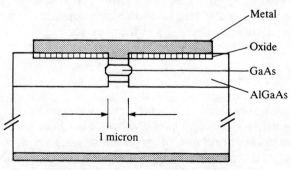

FIGURE 11.11 CDH GaAs laser.

11.12 QUANTUM WELL LASERS

Quantum well lasers are made from heterostructures where the optical cavity is very small. The thickness of the crystal must be less than the de Broglie wavelength. This dimension is of the order of 500 Å for electrons in GaAs crystals.

This effect is caused by electrons bound in an infinite square-well potential (see Chapter 2). The energy levels of the band structure become discrete instead of continuous. This confinement creates a lower ground state in which there is a larger density of states. Therefore, it is easier to create a population inversion, that is, less energy is required (see Chapter 10).

11.13 LIMITATIONS IN LASER POWER

Heat absorption and the resulting increase in stress and temperature limit the total power of a laser system. The greatest limitation is due to heat lensing caused by a local increase in temperature, which leads to a slight bulge in the optics at the beam. This effect is directly proportional to the coefficient of thermal expansion (CTE), and inversely proportional to the thermal conductivity of the laser host material. Heat lensing defocuses the beam and increases the total loss coefficient α.

Another effect that increases the total loss coefficient is strain birefringence (see Chapter 10), where the local index of refraction is a function of the local strain within a sample. The higher the thermal gradients near the beam, the greater is the change in index of refraction. This again defocuses the laser optics, which increases α.

As we have seen, the longitudinal modes within the laser cavity are directly proportional to the cavity length. The thermal expansion of the cavity can change the mode if the CTE is too large. This leads to deresonance and eventually lasing stops. Thus, there is a limit to the power that can be effectively emitted for a given mode. For this reason, glass lasers are generally pulsed to allow the glass to remain thermally stable.

Finally, the index of refraction can also change with temperature and light intensity. These nonlinear effects (see Chapter 10) also defocus the optical cavity, lenses, etalons, and prisms (see Weber, 1981 or Laud, 1987).

To summarize, the effective power that can be sustained by a laser system is a function of CTE, strain birefringence, nonlinear refractive index, and thermal effects on index of refraction. Thus,

$$\text{laser power efficiency} = f\left(\text{CTE}, \frac{dn}{dt}, \frac{dn}{dI}, \frac{dn}{dT}\right) \quad (11.82)$$

11.14 FLUORESCENCE LIFETIMES

The lifetime τ of the fluorescence of a lasing medium is related to the probabilities of electronic radiative and nonradiative transitions:

$$\frac{1}{\tau} = P_{rad} + P_{nonrad} \tag{11.83}$$

These probabilities are also related to the quantum yield or efficiency, η, developed in Chapter 10:

$$\eta = \frac{P_{rad}}{P_{rad} + P_{nonrad}} \tag{11.84}$$

or

$$\eta = (P_{rad})\tau \tag{11.85}$$

From equation (11.18) where ($\alpha \approx \beta$), another quantity can be defined, the *threshold of stimulated emission*, $\Delta N/V$. Stimulated emission requires that an upper energy level N_2 is more populated than the lower level N_1; V is the volume of the resonator cavity:

$$\frac{\Delta N}{V} = \frac{N_2 - N_1}{V} \tag{11.86}$$

For lasing to occur

$$\frac{\Delta N}{V} > \frac{2\Delta\nu\tau}{\lambda^2}\left(\frac{1-\rho}{1} + \alpha\right) \tag{11.87}$$

where $\Delta\nu$ = linewidth of spontaneous emission
  λ = wavelength for stimulated emission
  ρ = reflectance of the front surface of the resonator, length l and volume V
  α = absorption factor inside the resonator

If $\Delta N/V$ is less than the right-hand side of equation (11.87), then laser emission cannot occur. Therefore the threshold intensity ϕ_p of the pumping radiation has a minimum value defined by

$$\phi_p = h\nu_p \frac{\Delta N}{\tau\eta} \tag{11.88}$$

where $h\nu_p$ is the mean energy of the pumping radiation.

11.14 FLUORESCENCE LIFETIMES

TABLE 11.7 Extrema of Output Parameters of Laser Devices or Systems

Parameter	Value	Laser medium
Peak power	2×10^{43} W (collimated)	Nd–glass
Peak power density	10^{18} W/cm^2 focused)	Nd–glass
Pulse energy	$>10^4$ J	CO_2, Nd–glass
Average power	10^6 W	CO_2
Pulse duration	3×10^{-13} sec continuous wave (CW)	Rh6G dye; various gases, liquids, solids
Wavelength	60 nm–385 μm	Many required
Efficiency (nonlaser-pumped)	70%	CO_2
Beam quality	Diffraction limited	Various gases, liquids, solids
Spectral line width	20 Hz (for 10^{-1} sec)	Neon–helium
Spatial coherence	10 m	Ruby

TABLE 11.8 Laser Emission Wavelengths

Laser	Wavelength (nm)	Laser	Wavelength (nm)
Ar_2	125	HgCl	560
Kr_2	146	Cu	578
F_2	165	GaSe	585
H_2	180	Ne	630
Xe	186	CdSe	670
ArCl	188	Ruby (Cr–Al$_2$O$_3$)	695
ArF	197	F	710
KrCl	221	Pb	725
KrF	248	CdTe	770
Cu$^+$	250	GaAs	830
Cl_2	258	InP	900
Ag$^+$	260	Nd–YAG	945
XeBr	281	Yb–glass	1020
Br	290	Nd–glass	1050
XeCl	309	Nd–YAG	1064
Au$^+$	312	Ne	1150
Cd$^+$	325	Nd–YAG	1300
ZnS	330	GaSb	1530
I_2	341	Er–glass	1560
Kr^{2+}	348	Xe	2050
XeF	350	Ho–YAG	2100
Ar^{2+}	352	InAs	3050
ZnO	368	HF	2600–3300
Pb	404	Ne	3380
Kr$_2$F	408	Xe	3500
N$_2^+$	420	HCl	3650–4000
ZnSe	460	DF	3800–4400
Ar$^+$	488	PbS	4300
CdS	490	HBr	4000–4600
HgBr	501	CO	4900–6800
Cu	510	DCl	5000–5600
Xe$_2$Cl	511	DBR and NO	5800–6300
Ar$^+$	518	OCS	8200–8400
ZnTe	530	PbSe	8600
XeO	540	CO_2	9050–11,000
ArO	560	N_2O	10,050–11,100
KrO	560	CS_2	11,200–11,800

In order to understand more fully the problem of creating the population inversion as required for $\Delta N/V$, we need to evaluate the lifetime in terms of the Heisenberg uncertainty principle (see Chapter 2):

$$\Delta E \cdot \tau \approx h \tag{11.89}$$

The lifetime τ of the fluorescence is bound by this principle. If the energy level is broad (ΔE large) then pumping is easy. However, Heisenberg's uncertainty principle then requires τ to be small. With a small τ, ϕ_p becomes impractically large for lasing to occur (equation 11.88).

If we produce a lasing media with τ large, then by the same uncertainty, ΔE becomes small and population inversion becomes practically impossible. Therefore, a balance between large ΔE (or $\Delta \nu$) and a large τ must be found for a practical laser to be produced.

11.15 PARAMETERS OF VARIOUS LASERS

Table 11.2 catalogs the spectrum of laser devices. GaAs, molecular, solid-state, and semiconductor lasers are summarized. Table 11.7 shows the extrema for various laser systems. For example Nd–glass lasers yield the highest power, but the CO_2 laser has the highest efficiency. Active laser ions, crystals, and molecules are listed in Table 11.8 in order of emission

TABLE 11.9 Laser Pulse Width

Form	Technique	Log pulse width range (sec)
Continuous wave	Excitation is continuous; resonator Q is held constant at some moderate value	∞
Pulsed	Excitation is pulsed; resonator Q is held constant at some moderate value	-6 to -3
Q-switched	Excitation is continuous or pulsed; resonator Q is switched from a very low value to a moderate value	-8 to -6
Cavity-dumped	Excitation is continuous or pulsed; resonator Q is switched from a very high value to a low value	-7 to -5
Mode-locked	Excitation is continuous or pulsed; phase or loss of the resonator modes are modulated at a rate related to the resonator transit time	-12 to -9

wavelength. The shortest wavelength laser listed is Argon gas at 125 nm (UV) to CS_2 at 11,200 nm (IR).

The methods of pulsing laser systems are shown in Table 11.9. There are continuous-wave (CW) lasers with moderate power, to mode-locked with a pulse width as short as 10^{-12} sec at extreme power outputs.

Tables 11.3 and 11.4 compare the detailed operational parameters of Nd–glass lasers and GaAs lasers to those of Nd–YAG lasers.

PROBLEMS

11.1 If the beam diameter of a given laser is 1 mm, what would be the minimum focus diameter obtainable from a lens of focal length of 100 mm?
   (a) For a CO_2 laser?
   (b) For a InAs laser?
   (c) For a GaAs laser?
   (c) For a Nd–YAG?
   (d) For Argon ion laser?
   (e) For an Argon gas laser?

11.2 Calculate the depth of focus in millimeters for the lasers listed in problem 11.1.

11.3 Select the lens material for each of the five lasers listed in problem 10.1.

11.4 Develop the second of Airy's formula for the transmitted intensity of an etalon. (Equation 11.69.)

11.5 Develop the third of Airy's formula for the reflected intensity of an etalon. (Equation 11.70.)

11.6 The threshold intensity I_p of a laser is given in equation (11.88). Develop it from the definitions of the fluorescence lifetime and the quantum efficiency.

11.7 In a GaAs laser cavity, the surrounding material, AlGaAs, has a lower index of refraction. What is the width of the planar waveguide cavity for this device to operate:
   (a) In a single mode?
   (b) How many modes can be excited if the cavity width is 1.1 microns?

READING LIST

Abraham, E. et al., (1983). *Sci. Amer.*, **248**(2), 85.

Anderson, D. Z., (1986). *Sci. Amer.*, **254**(4), 94.

Bertolotti, M., (1983). *Masers and Lasers: A Historical Approach*, (Adam Hilger, ed.), IOP Publishing, Ltd., Bristol and Philadelphia.

Craxton, R. S. et al., (1986). *Sci. Amer.*, **255**(2), 52.

Halliday, D. and Resnick, R., (1962). *Physics*, Wiley, New York.

Hecht, J., (1986). *The Laser Guidebook*, McGraw-Hill, N.Y.

Higgins, T., (1987). "Etalons," *Lasers and Optronics*, High Tech Publications, Inc., May.

Hilger, A., (1986). *Laser Damage in Optical Materials*, IOP Publishing, Ltd., Bristol and Boston.

Hummel, R. E., (1985). *Electronic Properties of Materials*, Springer-Verlag, Berlin, Chapter: 13.

Jones, K. A., (1987). *Introduction to Optical Electronics*, Harper and Row, New York.

Kingery, W. D., Bowen, H. K., and Uhlmann, D. R., (1976). *Introduction to Ceramics*, 2nd ed., Wiley, New York.

Laud, B. B., (1987). *Lasers and Non-Linear Optics*, Wiley, N.Y.

Milonni, P. W., and Ebberly, J. H., (1988). *Lasers*, Wiley, N.Y.

Schawlons, A. L., (1968). *Sci. Amer.*, **219**(3), 120.

Weber, M. J., (1981). *Handbook of Laser Science and Technology, Vol. I: Lasers and Masers*, CRC Press, Boca Raton, Florida.

12

CERAMIC SUPERCONDUCTORS

12.1 HIGH-TEMPERATURE SUPERCONDUCTING CERAMICS

In April 1986, Bednorz and Muller announced the discovery of 40°K superconductivity of Ba–La–Cu–O. On March 2, 1987, Paul Chu and colleagues at the University of Houston reported that superconductivity above 90°K was discovered in a ceramic material (Wu et al., 1987). The material was a Y–Ba–Cu–O perovskite similar to that schematically shown in Figure 12.1.

The following quotation from the *New York Times Magazine* summarizes the competiton experienced during the spring of 1987 just after Chu's announcement.

> "...The secrecy, the petulance, the jockeying for science's top prize, the raw displays of ego and ambition—all these have risen nakedly to the surface in recent months. For researchers, the Nobel Prize is certain, but the precise names it will honor are not. For industry, a patent battle likely to burn through the next decade will hinge on the events of this year.
>
> Still, when these conflicts recede from memory, a story will remain of scientific discovery in its purest form. The heroes will be a few obsessive physicists driven to understand the strange, shimmering, electronic qualities of crystalline matter and who chose a path that their colleagues either scorned or overlooked. They blended intuition with experiment, mixing weeks and months of patient trial-and-error with an occasionally uncanny insight into structures too small to see... ."

The superconducting compositions are variations of a perovskite structure that are slightly oxygen deficient. Figure 12.2 shows the orthorhombic

494 CHAPTER 12: CERAMIC SUPERCONDUCTORS

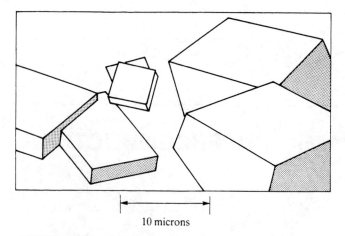

10 microns

FIGURE 12.1 Schematic of single crystals of $YBa_2Cu_3O_x$.

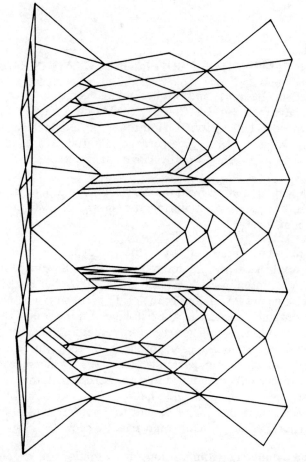

FIGURE 12.2 Schematic of the Y–Ba–Cu–O crystal structure.

TABLE 12.1 Superconducting Data for some Selected Compounds

Chemical formula	T_c^{onset} (°K)	$T_c^{R=0}$ (°K)	Cu–O sheets	Reading list
$YBa_2Cu_3O_7$	125	83	1	Hirabayashi et al. (1987)
$Y_{0.4}Ba_{0.6}CuO_{2.22}$	91.7	88	1	Syono et al. (1987)
$YBa_2Cu_3O_{6.9}$	93.5	91	1	Murphy et al. (1987)
$YBa_2Cu_3O_{6.8}$	94	92	1	Haucket et al. (1987)
$YBaCu_2O_{6-y}$	93	92.1	1	Semba et al. (1987)
$La_{1.85}Ba_{0.15}CuO_{4-y}$	32	25	1	Takagi et al. (1987)
$LaBa_2Cu_3O_{6.6}$	77	60	1	Murphy et al. (1987)
$LaBa_2Cu_3O_{6+x}$	91	75	1	Hor et al. (1987)
$La_{1.85}Sr_{0.15}CuO_4$	41	24.5	1	Yoshizaki et al. (1987)
$La_{1.4}Sr_{0.6}CuO_{3.1}$	54.0	30.5	1	Ihara et al. (1987)
$YBa_2Cu_3F_2O_y$	—	155	1	Ovshinsky et al. (1987)
$Bi_{1.8}Pb_{0.2}Ca_2Sr_2Cu_3O_{10}$	—	105	2	Politis (1988)
$Tl_2-Ca-Ba_2-Cu_3-O_x$	125	107	2	Sleight (1988)
$Bi-Ca-Sr-Cu-O$	110	—	2	Syono et al. (1988)
$Bi_2-CaSr_2Cu_2O_{9-\Delta}$	120	—	2	Sleight (1988)
$Tl_2Ca_2Ba_2Cu_3O_x$	—	118	3	Sleight (1988)

structure of a 1–2–3 compound $YBa_2Cu_3O_{7-\delta}$, determined by neutron diffraction at the McMaster reactor by John Greedan and co-workers. The designation *1–2–3 compound* refers to the molar ratio of rare earth (1) to alkaline earth (2) to copper (3) in the crystal.

Table 12.1 shows the superconductivity transition temperature T_c, and chemical formulas for various high-temperature superconductors. The highest reported T_c is approximately 155°K for fluorine-doped 1–2–3 compounds.

12.2 BACKGROUND

The first superconductor was discovered in 1911 by Kamerlingh Onnes in the Netherlands. In a span of about 0.05°K near 4°K, the resistance of mercury goes to a very low value, perhaps zero. If a current is established in a superconducting circuit, it can persist for weeks without significant decay. Figure 12.3 shows the abrupt drop in resistivity near 4°K for mercury.

The temperature dependence of copper is shown for comparison with the superconducting phenomenon. Copper metal shows no decrease to zero resistivity at very low temperature. Instead, copper, as with many other metals, shows a nearly linear reduction in resistivity ρ as temperature is decreased:

$$\rho = \rho_0[1 + \alpha(T - T_0)] \qquad (12.1)$$

where α is the linear temperature coefficient of resistivity.

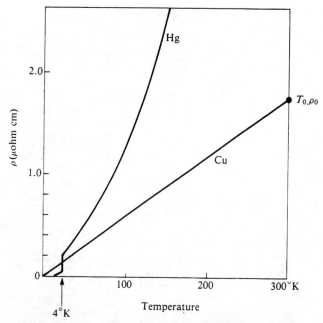

FIGURE 12.3 Discovery of superconductivity in Hg.

Careful measurements of the superconducting phase transition yield the lowest value of electrical resistivity to be

$$\text{Minimum } \rho_s = 4 \times 10^{-23} \text{ ohm cm}$$

while the lowest value measured for a nonsuperconducting metal is

$$\text{Minimum } \rho_m = 1 \times 10^{-13} \text{ ohm cm}$$

12.3 MEISSNER EFFECT

Very low resistivity is a necessary but not a sufficient condition for a material to be a superconductor. After the transition into the superconducting phase, the material will expel a magnetic field. In other words, a superconducting material becomes diamagnetic. This effect is known as the *Meissner effect*, after W. Meissner and R. Ochsenfeld who discovered the effect in Germany in 1933.

From electro-dynamics, the magnetic field inside the material is given by:

$$\mathbf{B}_{int} = \mu_0 \mathbf{H} + \mu_0 \mathbf{M} \tag{12.2}$$

$$= \mathbf{B}_{ext} + \mu_0 \mathbf{M} \tag{12.3}$$

where **M** is the magnetic moment per unit volume. If the field inside the superconductor is truly zero, then equation (12.3) becomes

$$0 = \mathbf{B}_{ext} + \mu_0 \mathbf{M} \qquad (12.4)$$

Solving for **M**, we have

$$\mathbf{M} = -\frac{\mathbf{B}_{ext}}{\mu_0} \qquad (12.5)$$

Thus the material becomes a perfect diamagnet (see Chapter 7). However it is an effect caused by electrons in circular motion near the surface of the material. It is not a structural diamagnetic effect at the atomic level. The electron currents simply generate an equal and opposite magnetic field as shown in Figure 12.4. This is possible because of the zero resistivity. This diamagnetic effect can be destroyed by externally applied magnetic fields that are too high. With high magnetic fields, the material will undergo a transition from the superconducting phase back to the normal phase.

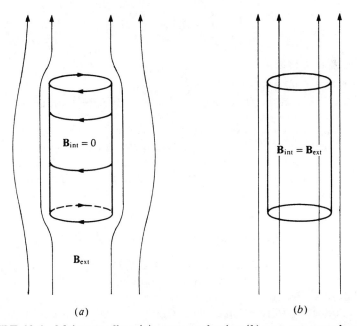

FIGURE 12.4 Meissner effect (*a*) superconducting (*b*) non-superconducting.

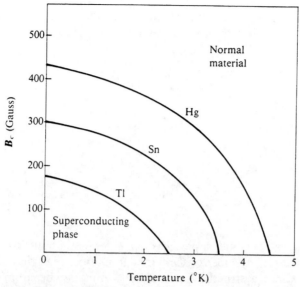

FIGURE 12.5 Critical magnetic fields of metallic superconductors.

12.4 THE CRITICAL FIELD

The critical magnetic field, $\mathbf{B}_c$, that destroys the superconducting effect obeys a parabolic law of the form:

$$\mathbf{B}_c = \mathbf{B}_0\left[1 - \left(\frac{T}{T_c}\right)^2\right] \tag{12.6}$$

where $\mathbf{B}_0$ = constant
T = temperature
T_c = critical temperature for the onset of superconductivity

Figure 12.5 shows the phase boundaries between normal conductivity and superconductivity for several metals. In general, the higher T_c, the higher $\mathbf{B}_c$. However, for recent developments in ceramic high-T_c superconductors, the value of the critical field seems to be much lower than earlier values for metallic superconductors.

12.5 EARLY THEORIES ON SUPERCONDUCTIVITY

Zero resistance is difficult to explain at absolute zero, let alone above that temperature. Early empirical theories included the idea of two conduction mechanisms. These were a normal conducting fluid of electrons plus a superconducting fluid of electrons. There was no theoretical basis for these ideas, it was simply a model used to describe the effect.

12.6 THE BCS THEORY

The behavior of electrons in the superconducting phase was first described adequately in 1957. This theory of superconductors is based on the pairing of electrons as the temperature of the material is lowered. It was formulated by John Bardeen, Leon H. Cooper, and J. Robert Schrieffer of the University of Illinois. They were awarded the Nobel Prize for this work and it is now called the BCS theory. Quantum-mechanical calculations indicate that at temperatures below the superconducting transition temperature T_c, it is energetically favorable to form a more ordered state. This new state is formed from electrons combining into what is known as *Cooper pairs*.

The formation of these electron pairs can best be understood by considering conduction mechanisms in metals. The repulsive Coulomb force between any two particles of the same charge is small compared with the overall potential of the lattice. As a result, the net interaction of electrons in a metal is attractive and arises from lattice vibrations or phonons that accompany the moving electrons. As the electrons pass near the positive lattice ions, phonons propagate as a result of the mutual electrical attraction.

12.7 ELECTRON PAIRS

The questions fundamental to understanding the BCS theory are: (1) How do the two like charges on the electrons in a Cooper pair not repel each other? (2) How does this pair cause the electrical resistance to go to zero? This result appears contradictory to the accepted principle that electrons repel each other and that Ohm's law is never violated.

First remember that superconductivity requires that the conducting material must be cold. At temperatures near absolute zero, atoms, electrons, and molecules tend to be near their quantum-mechanical ground state (Chapter 2). They are not at a state of zero energy, but at their lowest energy. Only in the unique condition of lower energy of a lattice can electron pairing take place. The coupling or attraction of electron pairs is very weak, and normal temperatures cause thermal motion so large that any attraction is totally destroyed.

Let us consider a state or temperature just before the transition into the superconducting state. As an electrical potential is applied to the material under test (there must be either an applied potential or an external magnetic field for the superconducting effect to be seen), electrons begin to move parallel to the applied field in accordance with Gauss's law. Because of this motion, the lattice responds to this moving electron by expanding around it and letting it through. This "footprint in sand" remains for a period of time; although short, it is finite. Another nearby electron moving under the same applied potential "sees" the footprint left by the first

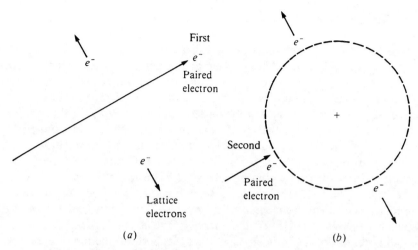

FIGURE 12.6 Highly ordered electron structure. (*a*) Lattice electrons displace. (*b*) Second electron is attracted by positive region in lattice.

electron. The footprint is a generally positive area and the second electron is attracted to it, again by Gauss's law. Physics is not being violated. At normal temperatures these footprints would be totally washed away by the tide of thermal motion of the lattice. However, at low temperatures they persist and cause electrons to form pairs (see Figure 12.6).

These footprints generally form a wave motion, or *phonon*, as the lattice responds to the moving electrons. The electrons tend to line up and follow each other, staying in the continuously moving and positively charged footprints. In fact, the electrons actually give up energy to stay lock-stepped together. They energetically fall into these regions and their electrical resistance in a classical sense approaches zero. As the temperature of the lattice falls below T_c, more and more paired electrons can exist, thus creating the superconducting phase.

This concept of superconductivity generally holds. The issue today is what is the causes of the attraction between the electron pairs and between the pairs and the lattice. The high T_c of the ceramic superconductors and the relatively large thermal energy should preclude a simple phonon coupling mechanism.

Quantum calculations confirm that the free energy of the lattice can be lowered if electrons pair. A classical model of this pairing is shown in Figure 12.7. The coupling constant λ arises from the positively charged phonon attracting the second electron. We have also seen in Chapter 7 that the classical energy of a coupled electron pair is less than the energy of the two free electrons (equation (7.59)).

The energy eigenvalue, ϵ, for the Cooper pair can be determined

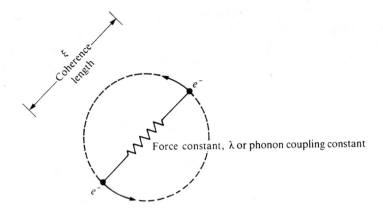

FIGURE 12.7 Cooper pair model.

by integrating $\psi^*\langle E\rangle\psi\,dE$ from twice the Fermi level, $2E_f$, to twice the phonon energy, $2h\nu_{max}$, where ν_{max} is the Debye frequency. The integral can be assumed to have the following form by using equations (2.11) and (11.34):

$$\epsilon = \int_{2E_f}^{2E_f + 2h\nu_{max}} W\exp[-E/kT]\,dE \quad (12.7)$$

Thus,

$$\epsilon = -WkT\exp[-E/kT]_{2E_f}^{2E_f + 2h\nu_{max}} \quad (12.8)$$

The eigenvalue can be defined as $2(E_f - \Delta)$ where Δ is the binding energy of one of the electrons. Using a series expansion of $\exp(-ax) \cong 1 - ax + \cdots$:

$$2E_f - 2\Delta \cong -WkT + 2Wh\nu_D \quad (12.9)$$

Therefore, the Cooper Pair binding energy is

$$2\Delta \cong WkT \quad (12.10)$$

if the Fermi level is close to the phonon energy.

The electrons most probably continuously exchange places between different Cooper pairs. The paired electrons form a very ordered structure. The population of Cooper pairs starts to increase as the temperature is lowered near T_c. As soon as enough pairs have formed, superconduction is possible. At absolute zero all the conduction electrons should be paired.

A Cooper pair in this model functions as a quantum particle. It has a little over twice the mass of two electrons. It can only exist in a lattice where conduction electrons are present. It has an effective diameter called its *coherence length*, ξ, where

$$100 \text{ nm} < \xi < 1000 \text{ nm} \tag{12.11}$$

The formation mechanism for Cooper pairs implies that all pairs are in the same state of motion. Either their centers of mass are at rest or if the superconductor is carrying a current, they all move with the same velocity in the direction of the current, that is, the coupled structure moves.

The resistance of normal conduction electrons is different from that of superconducting Cooper pairs primarily because of the differences in their de Broglie wavelengths.

For a normal conductor, the electrons can be described as a superposition of many plane wave de Broglie wavepackets. Their wavelengths are basically short (from equation (2.16)):

$$\lambda_e = \frac{h}{p_e} \tag{12.12}$$

At these lower temperatures, the difference in momentum in the conduction band is small so

$$\Delta p_e = \text{small} \tag{12.13}$$

Therefore, the superposition creates relatively small wavepackets and they scatter within the lattice easily. This creates electrical resistance.

For the Cooper pairs, the de Broglie wavelength is:

$$\lambda_c = \frac{h}{p_c} \tag{12.14}$$

Under an applied potential V, the energy of the Cooper pair becomes

$$V = \tfrac{1}{2}mv^2 = \tfrac{1}{2}(2m_e)v^2 \tag{12.15}$$

$$V = m_e v^2 \tag{12.16}$$

$$= \frac{(m_e v)^2}{m_e} = \frac{p_c^2}{m_e} \tag{12.17}$$

Therefore

$$p_c = \sqrt{m_e V} \qquad (12.18)$$

for an electron under the same potential

$$p_c = \sqrt{2m_e V} \qquad (12.19)$$

Substituting into equation (12.12) and (12.14)

$$\lambda_c = h \frac{1}{m_e V} \qquad (12.20)$$

$$\lambda_e = h \frac{1}{2m_e V} \qquad (12.21)$$

Therefore

$$\lambda_c > \lambda_e \qquad (12.22)$$

and a single Cooper pair is less likely to scatter from the lattice than the conduction electrons. If the Cooper pair phonon and its coherence length ξ are considered in the analysis, the effective de Broglie wavelength is even larger. This implies even less scattering by the lattice.

Finally, the lattice plays a major role in keeping the Cooper pairs moving at very nearly the same velocity. Figure 12.8 shows the phonon frequency v_D. Cooper pairs moving under an applied potential tend to move into low-energy spaces (low-amplitude vibrations). They pick spots where v_{e^-} is nearly zero and all the energy is transferred to the opposite electron. The opposite electron is easily moved because of its higher energy. The lattice is the source of these phonons and they tend to be very regular. The Cooper pairs move much less randomly than normal conduction electrons and their velocities can thereby be much lower to yield the same conduction current. This lower velocity again increases the effective de Broglie wavelength making scattering even less likely. Consequently, the pairs form a structure, or Cooper pair lattice. This in turn makes scattering even less likely.

So the combination of higher effective mass of the Cooper pair, a form of resonance with the lattice phonons, higher currents at lower velocities, and very little scattering by the lattice reduces the resistivity in the superconducting phase to zero. This results in the superconducting electrons forming a highly ordered structure just above the Fermi surface with no scattering.

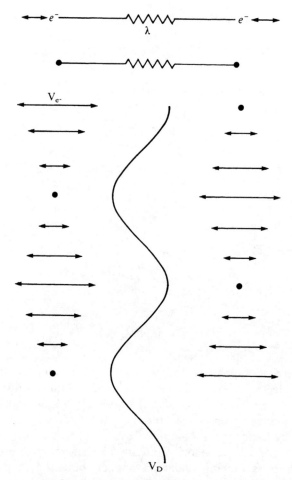

FIGURE 12.8 Cooper pair oscillations.

12.8 ENERGY OF THE COOPER PAIR

The electron pair is formed only if it is energetically favorable to do so. The energy gap between the electrons in a Cooper pair is denoted by 2Δ. Therefore, the energy required to excite either electron is Δ. The BCS theory predicts that this energy gap is (from equation (12.10))

$$2\Delta = 3.5\,kT \qquad (12.23)$$

where T = temperature (°K)
 k = Boltzmann's constant

FTIR (Fourier transform infrared) spectroscopy on high-T_c superconducting ceramics indicate that Cooper pairs of this energy are present (Tanner, 1987).

The conditions for electron pair attractions require that the difference in energy of the initial and final states for each electron is less than the phonon energy. This condition is satisfied when all the electron energies are just above the Fermi level, E_f.

12.9 THE ISOTOPE EFFECT

Before the BCS theory was formalized, experimental results on certain superconductors indicated that the lattice structure played an important role. It was shown that ordinary (white) tin is a superconductor but the other allotropic form (gray tin) did not show any superconducting effect. Likewise, bismuth changed into a superconductor at extremely high pressure. Therefore, superconductivity appears to be a property of the structure of the atomic lattice and not of the atom itself.

If the lattice "force constant" plays a major role in superconductivity, changes in mass of the lattice atoms should change its effective frequency analogous to equation (12.10). *Increasing the atomic mass of the lattice should reduce the lattice frequency* and *thereby reduce the effective coupling between Cooper pairs*. This simple analysis predicts that T_c should decrease as atomic masses increase.

Calculations indicate that T_c should be related to atomic mass M in the first-order form as follows:

$$\log T_c = -\alpha \log M + \beta \qquad (12.24)$$

where $\alpha = \frac{1}{2}$.

By exchanging isotopes of Hg and measuring T_c, Reynolds, Serin, and Hesbitt, verified this relationship in 1951. Their data showed

$$\alpha = 0.504 \qquad (12.25)$$

For Hg, the isotope effect demonstrated clearly that superconductivity is intimately related to phonons (or lattice vibrations).

The story is not over yet, however. Recent studies with some high-T_c superconducting ceramics show a clear absence of the isotope effect. In 1987, Batlogg and his fellow workers at AT&T Bell Laboratories did extremely careful studies on two 90°K superconductors, $Ba_2YCu_3O_7$ and $Ba_7EuCu_3O_7$.

They exchanged O^{16} for O^{18} within the lattices by processing test and control samples in precisely identical conditions. Care was taken to remove differences in handling and environmental conditions. They successfully replaced 90% of the O^{16} with O^{18}.

By exchanging this much O^{16} with O^{18}, the *Raman spectroscopy* of the lattice vibration is *predicted* to shift by $\Delta\omega_p/\omega_p$:

$$\frac{\Delta\omega_p}{\omega_p} \approx 4.3 \pm 0.4\% \qquad (12.26)$$

The average *observed* shift, $\Delta\omega_0/\omega_0$, taken from several experiments was

$$\frac{\Delta\omega_0}{\omega_0} \approx -4.2 \pm 0.5\% \qquad (12.27)$$

These shifts correspond to an increase in mass of

$$\frac{\Delta M}{M} = 9.3 \pm 1\% \qquad (12.28)$$

Substitution of this mass change into equation (12.24) yields:

$$\log \Delta T_c = -\tfrac{1}{2} \log \Delta M \qquad (12.29)$$

$$\log \Delta T_c \cong -\tfrac{1}{2} \log (0.093) \qquad (12.30)$$

$$\log \Delta T_c \cong 0.5157 \qquad (12.31)$$

$$\Delta T_c \cong 3.3°K \qquad (12.32)$$

This change in T_c can be more accurately calculated to be

$$\Delta T_c = 3.7 \pm 0.5°K \qquad (12.33)$$

This was calculated from the full isotope effect and if the phonon frequencies account for the superconducting electron pairs.

However, the *observed* shift in T_c for both resistive effects and Meissner effects in the O^{18} substituted material were a few tenths of a degree Kelvin at best.

The slope α of the isotope effect determined for these data was:

$$\alpha = -\frac{\partial(\log T_c)}{\partial(\log M)} = \frac{1}{2}(1-\zeta) \cong 0 \qquad (12.34)$$

At about the same time, a research group from University of California at Berkeley led by Bourne (1987) reported very similar results for $YBa_2Cu_3O_{7-\delta}$, i.e.,

$$\alpha = 0.0 \pm 0.027 \qquad (12.35)$$

for this superconducting ceramic ($T_c = 90°K$) with a substitution of O^{18} for O^{16} of 90%. In addition, reduced isotope effects have previously been observed in some metals, such as Ru and Zr.

This was a significant result. Cooper pairs had been observed in FTIR spectroscopy, but the phonon coupling mechanism appears to be missing. Therefore, the coupling mechanism has to be re-evaluated. In order to do this we have to review the concept of the *Debye Temperature*, θ_D.

12.10 DEBYE THEORY OF SPECIFIC HEAT

The introduction of a quantity called the *Debye temperature*, θ_p, is important. It arises from the quantization of lattice vibrations. In a like manner to that of the quantization of light, phonons have an energy $h\nu$. For low temperatures the heat capacity, C_v, is given by:

$$C_v = \text{constant} \left(\frac{T}{\theta_D}\right)^3 \tag{12.36}$$

where T = temperature
θ_D = Debye temperature

(see Kittel (1986) for the derivation). The Debye temperature is the ratio of the maximum energy of the phonon, that is, the maximum frequency $h\nu_{max}$, to Boltzmann's constant k. This ratio

$$\theta_D \equiv \frac{h\nu_{max}}{k} \tag{12.37}$$

has units of temperature and is named after Debye. if we substitute this definition for θ_D into the equation for the heat capacity C_v, then

$$C_v = \text{constant} \left(\frac{kT}{h\nu_{max}}\right)^3 \tag{12.38}$$

As the temperature is increased, the thermal energy approaches the quantum energy of the phonon and C_r becomes a constant. This leads to the classical value of

$$C_v = 5.96 \frac{\text{cal}}{\text{g atom°C}} \tag{12.39}$$

At low temperatures, the quantum energy dominates and C_v approaches zero. This simple analysis matches the observational evidence well and represents a conceptional link to the analysis of phonon vibrations to the BCS theory of superconducting electrons.

12.11 BCS PHONON–ELECTRON COUPLING

The Debye temperature is a factor in the famous BCS equation for the value of the critical temperature T_c:

$$T_c = 1.14 \theta_D \exp\left(-\frac{1}{\lambda}\right) \qquad (12.40)$$

The parameter λ represents the phonon–electron coupling constant. This relationship shows a strong dependence on the elastic properties of the lattice. The Debye temperature is related to the velocity of sound in the material, and the coupling constant is related to the freedom of the lattice to respond to passing electrons.

The highest T_c available from a phonon mechanism can be estimated if both phonon and Coulomb interactions are included in λ. Increasing the phonon coupling does not increase T_c very much. On the other hand, increasing the Coulomb interaction will increase T_c. Using realistic values for a strong Coulomb interaction the maximum T_c is

$$T_c(\text{phonons}) \leq 40°K \qquad (12.41)$$

This clearly is below the values of ceramic high-T_c superconductors. Thus, an electronic mechanism of attraction has to be introduced for the high-T_c materials. If the coupling phonons are replaced by excitons, then T_c values as high as 300°K can be predicted:

$$T_c(\text{excitons}) \leq 300°K \qquad (12.42)$$

Excitons disturb the lattice to a greater extent than a single electron. The increased disturbance creates a greater electron–electron coupling.

12.12 TYPE I SUPERCONDUCTORS

There are generally two classes of superconductors. The division of these classes is based on the rate of transition from the normal state to the superconducting state. Single elements that show superconducting properties are typically type I. The transition is very rapid with respect to temperature or magnetic field. There exists a nearly perfect diamagnetic effect during the superconducting phase and as the external magnetic field or the temperature is increased slightly above the phase boundary, the

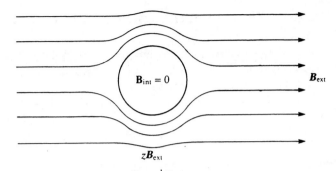

FIGURE 12.9 Intermediate type I superconducting state.

material returns to the normal state with no residual diamagnetic effect:

$$\mathbf{B}_{int} = \mathbf{B}_{ext} \quad (12.43)$$

For type II superconductors, the transition is gradual. As the magnetic field or the temperature is increased above the phase boundary, the material retains much of its superconducting effect.

A type I superconducting phase change is also sensitive to sample geometry. If the external magnetic field $\mathbf{B}_{ext}$ is applied parallel with a long thin axis of a type I sample, then the transition is sharp. If $\mathbf{B}_{ext}$ is perpendicular to the long axis of a thin sample, then the lines of force are split around the sample (see Figure 12.9). The field strength over the edge of the samples is at least twice the uniform $\mathbf{B}_{ext}$.

Therefore as $\mathbf{B}_{ext}$ approaches the critical value $\mathbf{B}_c$, local areas of the material will experience fields near $\mathbf{B}_c$:

$$\mathbf{B}_{ext} \approx \tfrac{1}{2}\mathbf{B}_c \quad (12.44)$$

This is called the *onset of the intermediate state*. The transition to the normal state is not completed, however, until:

$$\mathbf{B}_{ext} = \mathbf{B}_c \quad (12.45)$$

This analysis indicates that the field near the edge of the material reaches B_c and a portion of the material transforms to the normal state while the inside remains superconducting. These structures have been investigated using several techniques. Field-sensitive probes, magneto-optical effects, and magnetic powder patterns all show a complex structure throughout the intermediate state.

The intermediate state is dependent upon the geometry of the sample, the direction and strength of the external magnetic field. It is completely different from the state of type II superconductors.

12.13 MAGNETIC FIELD PENETRATION DEPTH

The diamagnetic response to an external magnetic field is the result of induced electric currents from superconducting electrons. These currents must exist within a certain depth of the superconductor. If these electric currents existed only in an infinitely thin region at the surface of the material, then B_c would be reached for any B_{ext}, no matter how small. Therefore, there is a finite penetration depth λ_p of these superconducting currents.

The penetration depth cannot be derived from Maxwell's equations alone. Additional equations known as the *London equations* govern these superconducting magnetic properties:

$$\nabla \mathbf{J}_s = -\left(\frac{n_s e^2}{m_e}\right)\mathbf{B} \tag{12.46}$$

and

$$\frac{\partial J_s}{\partial t} = \left(\frac{n_s e^2}{m_e}\right)\mathbf{E} \tag{12.47}$$

where n_s = number density of superconducting (paired) electrons
  m_e = mass of e^-
  $\mathbf{E}$ = electric field
  $\mathbf{J}$ = current density of superconducting current
  $\mathbf{B}$ = magnetic induction

If we postulate quasi-static conditions so that

$$\frac{\partial}{\partial t}\mathbf{J}_s = 0 \tag{12.48}$$

then

$$\mathbf{E} = 0 \tag{12.49}$$

Therefore there should not be any contribution to the current from nonsuperconducting electrons; so the only current left is the superconducting current.

Using Maxwell's first equation:

$$\nabla \times \mathbf{H} = \mathbf{J} + \frac{\partial \mathbf{D}}{\partial t} \tag{12.50}$$

$$\left(\frac{1}{\mu}\right)\nabla \times \mathbf{B} = \mathbf{J} + \epsilon \frac{\partial}{\partial t}\mathbf{E} \tag{12.51}$$

12.13 MAGNETIC FIELD PENETRATION DEPTH

Using equation (12.49) the second term on the right side becomes zero and equation (12.51) becomes

$$\frac{1}{\mu} \nabla \times \mathbf{B} = \mathbf{J}_s \tag{12.52}$$

or

$$\nabla \times \mathbf{B} = \mu \mathbf{J}_s \tag{12.53}$$

If we take the curl of equation (12.53), it becomes

$$\nabla \times \nabla \times \mathbf{B} = \mu \nabla \mathbf{J}_s \tag{12.54}$$

By substituting the London equation (12.46) into the right-hand side, we have

$$\nabla \times \nabla \times \mathbf{B} = \mu \left[-\left(\frac{n_s e^2}{m_e} \right) \right] \mathbf{B} \tag{12.55}$$

Using the vector identity:

$$\mathbf{A} \times \mathbf{B} \times \mathbf{C} = \mathbf{B}(\mathbf{A} \cdot \mathbf{C}) - \mathbf{C}(\mathbf{A} \cdot \mathbf{B}) \tag{12.56}$$

and substituting the left-hand side of equation (12.55), it becomes

$$\nabla \times \nabla \times \mathbf{B} = \nabla(\nabla \cdot \mathbf{B}) - (\nabla \cdot \nabla)\mathbf{B} \tag{12.57}$$

because the dot product is a scalar. If we now use Maxwell's fourth equation

$$\nabla \cdot \mathbf{B} = 0 \tag{12.58}$$

that requires that there be no magnetic monopoles, equation (12.57) becomes

$$\nabla \times \nabla \times \mathbf{B} = -\nabla^2 \mathbf{B} \tag{12.59}$$

and substituting this into equation (12.53) we have

$$\nabla^2 \mathbf{B} = \mu \left(\frac{n_s e^2}{m_e} \right) \mathbf{B} \tag{12.60}$$

Likewise, by taking the curl of equation (12.46) it can be shown that

$$\nabla^2 \mathbf{J}_s = \mu \left(\frac{n_s e^2}{m_e} \right) \mathbf{J}_s \tag{12.61}$$

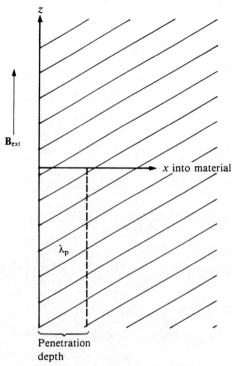

FIGURE 12.10 Semi-infinite blank.

The solution to these equations is a simple exponential if B_{ext} is parallel to the surface of a semi-infinite blank, as shown in Figure 12.10:

$$B_z = B_{ext} \exp\left[-x\left(\frac{\mu n_s e^2}{m_e}\right)^{1/2}\right] \quad (12.62)$$

By defining the penetration depth of the magnetic field as λ_p:

$$\lambda_p \equiv \frac{m_e}{\mu n_s e^2} \quad (12.63)$$

Then the solution has the form

$$B_z = B_{ext} \exp\left[-\frac{x}{\lambda_p}\right] \quad (12.64)$$

The values of λ_p can be estimated by assuming one electron per atom as n_s

and letting $\mu = \mu_0$ and

$$n_s \cong 1 \times 10^{28}/\text{m}^3$$
$$\mu_0 = 1.26 \times 10^{-6} \text{ H/m}$$
$$e = 1.602 \times 10^{-19} \text{ C}$$
$$m_e = 9.1 \times 10^{-31} \text{ Kg}$$

Substituting:

$$\lambda_p = \left[\frac{9.1 \times 10^{-31}}{(1.26 \times 10^{-6})(10^{28})(1.602 \times 10^{-19})^2}\right]^{1/2} \quad (12.65)$$

$$\lambda_p = 5.3 \times 10^{-8} \text{ m} \quad (12.66)$$

As the number of superconducting electron pairs decrease, the penetration depth increases. By experimentally measuring the penetration depth as a function of temperature, λ_p has the empirical form (see Figure 12.10)

$$\lambda_p = \lambda_{0p}\left[1 - \left(\frac{T}{T_c}\right)^4\right]^{-1/2} \quad (12.67)$$

By setting the right-hand side of equation (12.67) equal to the right-hand side of equation (12.63), we see that number of superconducting electrons, n_s, changes rapidly near T_c:

$$n_s \propto \left[1 - \left(\frac{T}{T_c}\right)^4\right] \quad (12.68)$$

For a type I superconductor, then, this semi-empirical analysis explains the very rapid change from the normal state to the superconducting state. As n_s increases, the conductivity increases. At the transition temperature T_c, the number of superconducting electrons is nearly zero and the material has a normal resistance.

12.14 TYPE II SUPERCONDUCTORS

Type II superconductors are of practical interest because of the gradual transition. This gives engineers a larger window of acceptable performance near T_c. Also, the new high-T_c superconducting ceramics are of this type.

In the intermediate state of the type I superconductors, the boundary between the normal resistive phase and the superconducting phase appears to be a relatively simple surface. This would indicate a positive free energy at this surface. On the other hand, the interface between the superconduct-

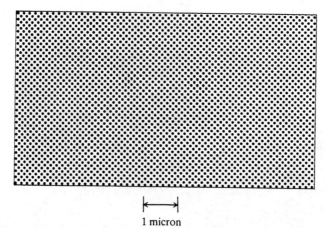

1 micron

FIGURE 12.11 Schematic of type II superconductor "mixed state". Dark spots indicate fluxoid quanta (see Essmann and Träuble, 1971.)

ing phase and the normal phase in type II materials is a filamentary structure such as indicated in Figure 12.11. This extremely high surface area indicates that the surface area has been maximized at the interface. Therefore the free energy is negative.

This difference in surface energy can be analyzed using the two parameters:

$$\xi = \text{coherence length (see Figure 12.7)} \tag{12.69}$$

and

$$\lambda_p = \text{penetration depth} \tag{12.70}$$

If the ratio

$$\frac{\xi}{\lambda_p} > \sqrt{2} \tag{12.71}$$

as shown by the Ginzburg–Landau theory (Ginzburg, 1982), then the surface energy will be positive and we have a type I superconductor. The energy removed by electron pairing is relatively small, because the coherence length is large compared with the penetration depth. Cooper pairs simply do not have enough room to form.

However, the opposite is true for type II superconductors where

$$\frac{\xi}{\lambda_p} < \sqrt{2} \tag{12.72}$$

This implies that the surface energy be negative. The electron pair can more easily form within the penetration depth, reducing the surface energy.

12.15 STRUCTURE OF HIGH-T_c SUPERCONDUCTING CERAMICS

Thus, the surface area of the interface tries to maximize itself, and the filamentary structure results.

Therefore, the relative size of λ_p and ξ determines the sign of the free surface energy and the type I or II character of the superconductor.

Finally, the filamentary structure seems closely related to the lattice structure. Work by Essmann and Träuble (1971) indicates that each filament of magnetic flux carries one *fluxoid quantum*, defined as

$$\phi_0 \equiv \frac{h}{2e} \qquad (12.73)$$

Essmann obtained SEM micrographs of these quantized filaments by sprinkling iron filings over the surface of the material while it was in the mixed state. After moving the sample into an SEM, micrographs showed the residual iron powder captured over the filaments (see Figure 12.11). By counting the filaments and comparing these counts with the applied magnetic field, $\mathbf{B}_{ext}$, he confirmed the value of the fluxoid quantum ϕ_0, within experimental error:

$$\phi = \text{total flux} \qquad (12.74)$$

$$\phi = \int \mathbf{B} \cdot d\mathbf{s} \qquad (12.75)$$

over a surface **s** was found to be

$$\phi = n\phi_0 \qquad (12.76)$$

where

$$\phi_0 = \frac{h}{2e} \qquad (12.77)$$

or

$$\phi_0 = 2.07 \times 10^{-15} \text{ Wb} \qquad (12.78)$$

The experimental verification of equation (12.77) demonstrates that the superconducting electrons are *paired*. The quantized magnetic flux is a function of twice the electron charge, i.e., ϕ_0, the fluxoid quantum.

The designation for the initial penetration into the type II superconductor is $\mathbf{B}_{c1}$. The second field strength $\mathbf{B}_{c2}$ is the upper magnetic field that totally destroys the superconducting phase.

12.15 STRUCTURE OF HIGH-T_c SUPERCONDUCTING CERAMICS

The key to achieving high-T_c superconductivity appears to be related to the crystal structure of the material. The following is a partial list of the characteristics of the high-T_c ceramic superconductors:

1. Cu–O chains are not required but two-dimensional sheets are necessary (Syono et al., 1988).
2. They have a tetragonal to orthohombic phase transition (Gu et al., 1988; Rickett et al., 1988).
3. Increasing T_c tends to be correlated with an increasing number of Cu–O sheets (Tanner, 1987), see Table 12.1.
4. Cu–O defects are required (Crabtree et al., 1987; Terakura et al., 1987; Takagi et al., 1987; Chen et al., 1987; Chen et al., 1987).
5. Rare-earth elements are not required (Collin et al., 1988).
6. Reduced Cu–Cu lattice parameter, a, tends to correlate with increasing T_c (Sarikaya et al., 1988; Chen et al., 1987; Mitchell et al., 1988; Meng et al., 1988; Crabtree et al., 1987).
7. Decreasing unit cell volume is correlated with increasing T_c (Chen et al., 1987).
8. They appear to have an antiferromagnetic phase (Ramakrishna et al., 1988; Venturini et al., 1988; Takada, 1988; Rickett et al., 1988; Sinha, 1988) (see Figure 12.12)
9. Electron pairing is required (Yin, 1988; Tanner, 1987, and Allen, 1988)

12.16 HIGH-T_c ELECTRON PAIRS AND COUPLING MECHANISMS

There seems to be a gap between theory and practice. Phonons are probably not the primary coupling mechanism for the electron pair coupling in the new ceramic superconductors. A strong coupling mechanism (exciton) was proposed by Little in 1964 and expanded by Ginzburg in 1982.

A few infrared studies have confirmed the presence of Cooper pairs in the new high-T_c superconductors (Tanner, 1987). This only adds to the confusion. Their binding energies seem to be very close to that predicted by the BCS theory, that is,

$$2\Delta = 3.5kT \qquad (12.79)$$

Other mechanisms for high-T_c superconductivity have been proposed which differ markedly from the BCS approach. One such mechanism can be considered as holes trapped in an antiferromagnetic region. This leads to an effective attractive interaction between the polarons and to pairing into bosons which then Bose-condense into a superfluid state.

Bose–Einstein particles are indistinguishable and may occupy any allowed state (see Kittel). At absolute zero, all particles would then fill the lowest state, i.e., Bose condensation. At temperatures above absolute zero, the state occupation density decreases exponentially with increasing energy. They are also called *symmetric particles*.

Another possible mechanism includes the boundary charge trapping of the Cooper pairs along grain boundaries. Others include resonating valence bond (RVB) states which trap holes that act like charged bosons. The RVB states should then condense into a charged superfluid at a temperature which depends on their density and effective mass. Other proposed coupling mechanisms include plasmons, bipolarons, spin coupling on copper ions, etc. (see Halley, 1988).

12.17 RESONATING VALENCE BONDS

A strong coupling mechanism that has promise is the *resonating valence bond* (RVB). The RVB models the local magnetic moments on the copper sites and their interaction through virtual hopping through the oxygen sites. Conduction occurs by holes created by doping the otherwise insulating ceramic (Liu, 1988).

There is a striking correlation between the antiferromagnetic phase and the onset of superconductivity (Sinha, 1988) (see Figure 12.12). The RVB model may give the required strong interactions with neighboring electronic states to cause superconductivity.

12.18 SPIN BAG MODEL

In the spin bag model (Liu, 1988) magnetic electrons are in band states which cause antiferromagnetic spin correlation. This does yield a strong

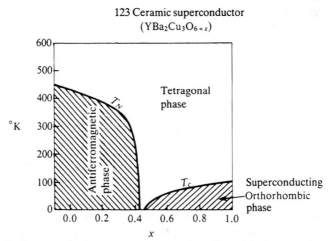

FIGURE 12.12 Antiferromagnetic phase in superconductors: T_N: Neel state temperature. T_c: critical superconducting temperature (after Sinha, 1988).

pairing interaction for a wide range of temperatures. However, if the *coherence length* becomes short for the spin correlation, then pairing is less likely. Attempts are being made to show that the correlation length is sensitive to the dopant level, i.e., the number of holes.

12.19 PHASE DIAGRAMS OF HIGH-T_c SUPERCONDUCTING CERAMICS

Phase equilibrium diagrams of the complex oxide systems that exhibit superconductivity are beginning to be established. The diagrams make it possible to identify, for each system, what are (if more than one) the phases responsible for superconductivity. The Y–Ba–Cu–O system has received the most attention because of $T_c > 90°K$.

The copper–oxygen planes are important to the high-T_c superconducting phase. The oxygen can be removed easily by raising the material to 500°C in a vacuum. The reduced number of oxygens (increased δ) prevents any superconducting phase. The process is reversible; by raising the material to 500°C in the presence of oxygen, the structure is restored. Likewise, the high-T_c phase is also restored.

This indicates the importance of the Cu–O defects. The defect sites are of great interest and are the focus of the antiferromagnetic nature of 1–2–3 compounds.

Various substitutions for Cu, Fe, Be, Cr, Ag, and other metals have been investigated. Small concentrations appear to have little or no effect, but larger substitution tends to destroy the superconducting phase.

Figure 12.13 shows a phase diagram of part of the system $YO_{1.5}$–BaO–CuO as developed in March 1987. Figure 12.14 shows the same system as it appears in another article in July 1987. Comparing these two figures, one can see that progress was made, in the sense that more phases and its corresponding phase field were identified. It is interesting to note that a straight line joining $Y_2Cu_2O_5$ and $BaCuO_2$, which is present in Figure 12.13 disappears in the newer phase diagram. In Figure 12.15 is shown another phase diagram for this system. This diagram looks very similar to the one shown in Figure 12.14, but one can see that there are apparent disagreements in several phase fields. One more phase diagram for this system is shown in Figure 12.16. Again, this diagram appears to be different in many ways of the previous ones. (It is interesting to note that Figures 12.14–12.16, all come from the same publication—same issue.)

Recent scanning Auger electron microscopy by Cota et al. shows that two phases are present for Y–Ba–Cu–O ceramic superconductors. Onset of superconductivity occurred at 105.6°K with $T_c = 97.9°K$. One of the phases corresponded to the standard compositon of 1–2–3 compounds while the other was Ba enriched. The first showed a platelet shape, while the Ba-enriched phase was granular in appearance.

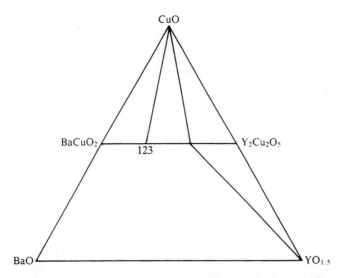

FIGURE 12.13 Partial Y–Ba–Cu–O phase diagram (after Muromachi et al., 1987).

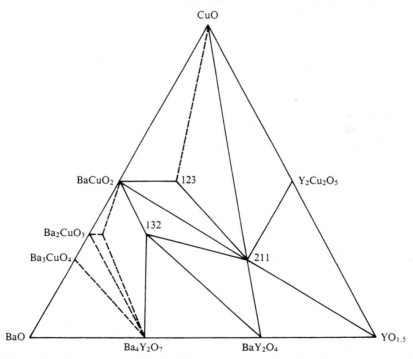

FIGURE 12.14 Compatibility regions in the pseudo-ternary Y_2O_3–BaO–CuO phase diagram, determined at 1223°K (after Frase and Clark, 1987).

519

520 CHAPTER 12: CERAMIC SUPERCONDUCTORS

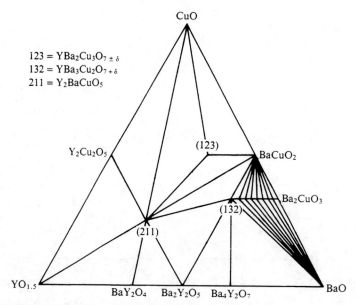

FIGURE 12.15 Subsolidus phase diagram for the Y_2O_3–BaO–CuO system in air at 1223°K (after Wang et al., 1987).

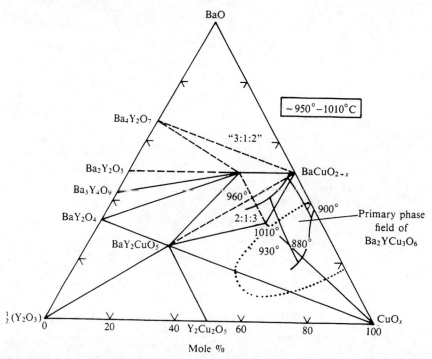

FIGURE 12.16 Preliminary melting data in air for part of the system $\frac{1}{2}Y_2O_3$–BaO–CuO_x superimposed on the phase diagram for this system at 1223–1283°K (after Roth et al., 1987).

12.20 CRYSTALLINE STRUCTURE

The Ba enrichment suggests a further depletion of oxygen, or probably an increase in oxygen vacancies. This phase generally covered the surface of the standard composition platelets.

12.20 CRYSTALLINE STRUCTURE

In 1964, it was suggested by Ginzburg that "sandwich" type structures could favor an electronic mechanism of superconductivity over lower T_c phonon mechanisms. Such systems should contain a metallic film surrounded on both sides by a dielectric or semiconducting film, which would secure the exciton mechanism of superconductivity for the electrons of the central metallic film. Indeed, these layered structures seem to be a characteristic of the new superconductors with high T_c.

Since a superconducting phase in the La–Ba–Cu–O system was found to be of the K_2NiF_4 type, the study of the La_2CuO_4 structure has attracted

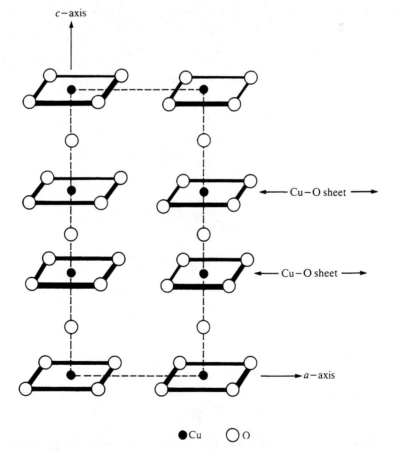

FIGURE 12.17 The Cu–O sublattice of 123 ceramic superconductors (after Hirabayashi et al., 1987).

great attention. The structure of La_2CuO_4 is orthorhombic at low temperatures and its pure form has been referred to as either superconducting or nonsuperconducting. The crystalline structure of La_2CuO_4 can be described as an alternating layering of perovskite $LaCuO_4$ and rock-salt LaO. It is interesting to note that besides the orthorhombic form, k phases with a tetragonal symmetry also can be superconductng. Examples of some specific compounds that have been identified as superconductors in this system include: $(La_{1-x}Ba_x)_2CuO_{4-y}$ (Fujita, et al., 1987; Onoda et al., 1987), $La_{1.85}Ba_{0.15}CuO_4$ (Jorgensen et al., 1987; Kirschner et al., 1987) and $LaBaCu_2O_{5+x}$.

In the Y–Ba–Cu–O system, compounds that have been identified as superconducting include: $YBa_2Cu_3O_{7-x}$, $Y_{1-x}Ba_xCuO_y$ $(0.6 < x < 0.7)$ and $Y_{0.4}Ba_{0.6}CuO_{2.22}$. The sublattice of Cu–O of the $YBa_2Cu_3O_{7-x}$ phase is shown in Figure 12.17. This type of structure can be described as an orthorhombic-layered perovskite with ordered oxygen vacancies along its c-axis. The unit cell of the 1–2–3 compound is shown in Figure 12.18. An orthorhombic structure has been found at room temperature, and the tetragonal one was found to be stable above 1030°K.

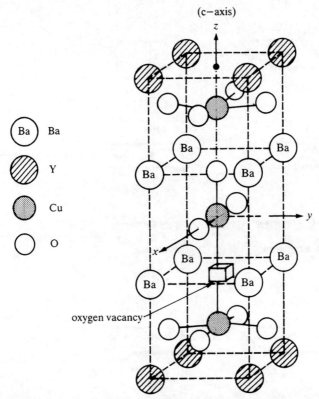

FIGURE 12.18 A unit cell of $YBa_2Cu_3O_7$ (after Stavola et al., 1987).

The problem of determining the critical aspect of crystal structure responsible for superconductive behavior of the 1–2–3 oxides, or other high-T_c oxides, is far from being solved. In fact, several polymorphic structures are possible. This is not hard to believe, especially if one considers the complexity of those type of structures and the fact that as one varies the processing conditions (particularly the pressure of oxygen in the atmosphere), the oxygen content in the final structure will vary. As a consequence of this, the vacancy concentration also will vary, as well as the distortion of the structure, interatomic spacings along the cell axes, and so on.

12.21 DEFECT CHEMISTRY OF SUPERCONDUCTING CERAMICS

Even though the phases and crystalline structures of the oxide superconductors are not well known, the distortion of the structure associated with the presence of oxygen vacancies seems to be a key factor in obtaining high T_c values. By varying the oxygen content (and related oxygen deficiency) and the amount of doping (content of Ba, Sr, or Ca, for example), the lattice parameters vary, with an accompanying variation in T_c. Figure 12.19 shows how T_c varies as the Ba and Sr content varies in $YBa_{2-x}Sr_xCu_3O_{7-\delta}$.

Despite the apparent need for oxygen nonstoichiometry to attain the superconducting state, when the oxygen deficiency increases too much it

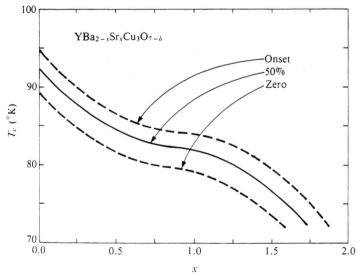

FIGURE 12.19 Superconducting transition temperature of $YBa_{2-x}Sr_xCu_3O_{7-\delta}$ (after Crabtree et al., 1987).

524 CHAPTER 12: CERAMIC SUPERCONDUCTORS

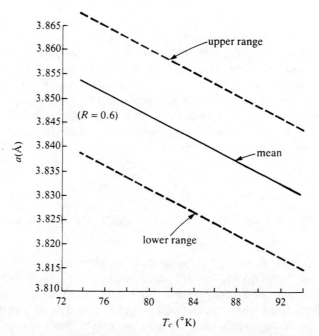

FIGURE 12.20 Tc dependence on the shortest lattice parameter (a) (i.e.,) the Cu–Cu spacing (after Chen et al., 1987).

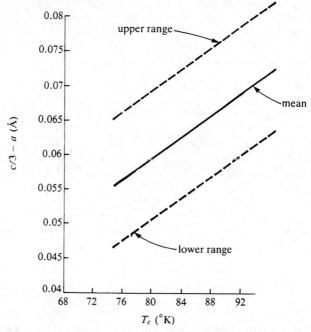

FIGURE 12.21 Tc dependence on the orthorhombic distortion ($c/3 - a$) (after Chen et al., 1987).

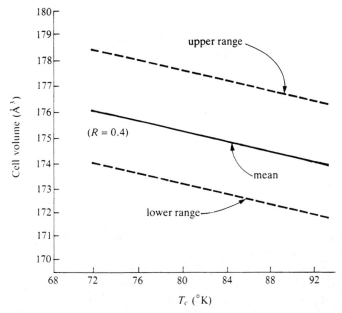

FIGURE 12.22 T_c dependence on the unit cell volume (after Chen et al., 1987).

tends to lead to the disappearance of superconductivity. A small (crystalline) cell volume, a close Cu–oxygen-vacancy–Cu spacing, and a large orthorhombic distortion all favor a high T_c. Figures 12.20, 12.21, and 12.22 respectively, show the influences of the lattice parameter a, the orthorhombic distortion $c/3 - a$, and the unit cell volume on T_c. It should be noted, however, the correlations are not good and the data have a high degree of scatter.

If silver is introduced into a Y–Ba–Cu–O superconductor, it seems to reduce the porosity of the ceramic. At the same time, T_c was not reduced but environmental stability was improved, a major problem with many of the oxide superconductors.

12.22 CONCLUSIONS

The need for electronic pairing to cause the superconducting phase transition seems to be widely accepted. The primary issue is the strength of the coupling. Theoretical and experimental evidence indicates that multiple coupling mechanisms may be present in the oxide superconductors. Several terms may be operating simultaneously to produce the bonding forces for the electronic pairs. Phonons may be the interaction mechanism below 40°K. However, excitons may operate when phonons are destroyed by thermal energies. Other electron–lattice interactions may phase in then out as the temperature increases and the lattice adjusts.

CHAPTER 12: CERAMIC SUPERCONDUCTORS

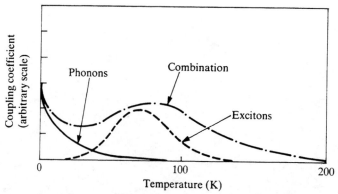

FIGURE 12.23 Linear combination of coupling coefficients.

This multiterm approach is not new to physics. The structure of the nucleus is dominated by the short-range and complex weak and strong nuclear forces. The Coulomb repulsion still exists. The positively charged protons do not like to be together at atomic distances, but at nuclear distances the attractive force dominates.

In much the same manner, phonon coupling may dominate at low temperatures, whereas exciton and resonating valence bonds may dominate at higher temperatures. Figure 12.23 represents a schematic of the effective electron coupling when a linear combination of the various mechanisms may account for the total interaction.

PROBLEMS

12.1 The differential equations for the magnetic field, **B**, and the current density, $\mathbf{J}_s$, have the same form. (see equations 12.50–12.66)

(a) Write down a solution for the current density, $\mathbf{J}_s$, for type I superconductors.

(b) Plot the current density as a function of x in units of the external current density, $\mathbf{J}_{ext}$, for $0.1 < x < 9$ Å and the penetration depth of 5.3 Å.

(c) Compare this with the interatomic separation for mercury.

12.2 At what wavelength and wavenumber would you look to see absorption of IR for Cooper paired electrons at the onset of the superconducting phase?

(a) Mercury

(b) 1–2–3 Y–Ba–Cu–O

(c) 1–2–3 La–Ba–Cu–O

(d) AlNb$_3$.

(e) Hypothetical room temperature ceramic superconductor.

12.3 The coherence length in the BCS phonon coupling theory is inversely related to the coupling constant. Calculate the coupling constant for the following materials:
  (a) Hg: where the Debye temperature = 87°K and T_c = 4.16°K
  (b) Aluminum.
  (c) Copper.
  (d) Nb.
  (e) V.
  (f) 1–2–3: Y–B–Cu–O and assume that the Debye temperature = 300°K:
  (g) On the average, how many times larger is the coupling constant for the 1–2–3 compound than for the others?
  (h) If the coherence length on the elemental superconductors is 100 Angstroms, what would it be for the 1–2–3 compound?
  (i) Compare this result with the literature.

12.4 The heat capacities at the critical temperature, T_c, of the superconductors should be related to the Debye temperature.
  (a) Determine the heat capacities of the six materials in problem 12.3. (Hint: use *CRC Handbook of Chemistry and Physics* and use Cu–O for the 1–2–3 material)
  (b) Determine the proportionality constant in the Debye relation using Hg.
  (c) Using this new relationship, calculate the Debye temperature of the list of materials including the 1–2–3 high T_c sperconductor.
  (d) Discuss the consequences associated with a Debye temperature as calculated for the 1–2–3 compound compared to 300°K.

12.5 The velocity of sound in a material is an indication of the rigidity of the material. Likewise, the coupling constant is also an indication of the rigidity.
  (a) Determine the velocity of sound for the materials listed in problem 12.3. (Hint: Vel of sound = (Young's modulus/ square root of the density)
  (b) Calculate the phonon frequency from the Debye temperature for this list of materials.
  (c) Calculate the phonon wavelengths for these materials.
  (d) Compare the phonon wavelengths with the minimum lattice separation for these materials. Identify all assumptions made during the analysis. Does the ratio of the phonon wavelength to the lattice separation, (λ/a), increase as the critical temperature increases?

12.6 The critical magnetic field is an important feature of superconductors. Low critical field are interesting but the materials would have little practical use. From the literature:

(a) Determine the critical field strengths for the superconducting materials listed in problem 12.3.

(b) Determine the current density that can be maintained in a magnetic field for these materials.

(c) Compare these values with those needed to produce practical devices.

READING LIST

Akimitsu, J., Ekino, T., Sawa, H., Tomimoto, K., Nakamichi, T., Oshiro, M., Matsubara, Y., Fujiki, H., and Kitamura, N. (1987). *Jpn. J. Appl. Phys.* **26**, Part 2, L449.

Allen, P. B. (1988). *MRS Proc., Ext. Abstr. High Tech. Supercond. 2nd, 1988* p. 3.

Anderson, P. W., Baskaran, G., Zou, Z., and Hsu, T. (1987). *Phys. Rev. Lett.* **58**, 2790.

Bardeen, J., Cooper, L. N., and Schrieffer, J. R. (1957). *Phys. Rev.* **106**, 162.

Batlogg, B. et al. (1987). *Phys. Rev. Lett.* **58**(22), 2333.

Beardsley, T. M. (1987). *Sci. Am.* **257**(4), 32.

Bednorz, J. and Muller, K. A. (1986). *Z. Phys. B* **64**, 189.

Behrman, E. C., Amarakoon, V. R. W., Axelson, S. R., Bhargava, A., Brooks, K. G., Burdick, V. L., Carson, S. W., Corah, N. L., Cordaro, J. F., Cormack, A. N., DiCarlo, D. G., Dwivedi, A., Fishman, G. S., Friel, J., Hanagan, M. J., Hexemer, R. L., Heuberger, M., Hong, K.-S. Hsu, J.-Y., Johnson, P. F., LaCourse, W. C., LaGraff, J. R., Lakshminarasimha, M., Laughner, J. W., Longobardo, A. V., Malone, P. F., McCluskey, P. H., McPherson, D. M., Mroz, T. J., Rabidoux, C. W., Reed, J. S., Sainamthip, P., Sanchex, S. C., Sheckler, C. A., Schulze, W. A., Seth, V. K., Shelby, J. E., Shieh, S. H. M., Simmins, J. J., Simpson, J. C., Snyder, R. L., Swiler, D., Taylor, J. A. T., Udaykumar, R., Varshneya, A. K., Vitch, S. M., and Votava, W. E. (1987). *Adv. Ceram. Mater.* **2**, Spec. Issue, 539.

Bianconi, A., Catellano, A. C., De Santis, M., Politis, C., Marcell, A., Mobilio, S., and Savoia, A. (1987). *Z. Phys. B* **67**, 307.

Bonn, D. A., Greedan, J. E., Stager, C. V., Timusk, T., Doss, M. G., Herr, S. L., Kamaras, K., and Tanner, D. B. (1987). *Phys. Rev. Lett.* **58**, 2249.

Bourne, L. C. et al. (1987). *Phys. Rev. Lett.* **58**(22), 2337.

Burton, R. L., and Bartholomew, C. S. (1988). *MRS Proc., Ext. Abstr. High-Tech. Supercond., 2nd, 1988* p. 77.

Cai, X., Joynt, R., and Larbalestier, D. C., (1987). *Phys. Rev. Lett.* **58**, 2798.

Cava, R. J., Batlogg, B., van Dover, R. B., Murphy, D. W., Sunshine, S., Siegrist, T., Remeika, J. P., Rietman, E. A., Zahurak, S., and Spinosa, G. P. (1987). *Phys. Rev. Lett.* **58**, 1676.

Chen, I.-W., Keating, S., Keating, C. Y., Wu, X., Xu, J., Reys-Morel, P. E., and Tien, T. Y. (1987). *Adv. Ceram. Mater.* **2**, Spec. Issue, 457.

Clark, D. R. (1987). *Adv. Ceram. Mater.* **2**, Spec. Issue, 273.

Collin, G., Audier, A. C., Albouy, P. A., Senoussi, S., Oussena, M., Comes, R., Konczykowski, M., Rullier-Albenque, F., Petitgrand, D., and Schweiss, P. (1988). *MRS Proc., Ext. Abstr. High-Temp. Supercond., 2nd, 1988.*

Cooper, L. N. (1956). *Phys. Rev.* **104**, 1189.

Cota, L., Morales de la Garza, L., Hirata, G., Martínez, L., Orozco, E., Carrillo, E., Mendoza, A., Albarrán, J. L., Fuentes-Maya, J., Boldú, J. L., Pérez-Ramírez, J. G., Pérez, R., Reyes Gasga, J., Avalos, M., and José-Yacamán, M., (1988). *Journal of Materials Research*, **3**(3), 417.

Crabtree, G. W., Downey, J. W., Flandermeyer, B. K., Jorgensen, J. D., Klippert, T. E., Kupperman, D. S., Kwok, W. K., Lam, D. J., Mitchell, A. W., McKale, A. G., Nevitt, M. V., Nowicki, L. J., Paulikas, A. P., Poeppel, R. B., Rothman, S. J., Routbort, J. L., Singh, J. P., Sowers, C. H., Umezawa, A., and Veal, B. W. (1987). *Adv. Ceram. Mater.* **2**, Spec. Issue, 444.

Dai, Y., Swinnea, J. S., Steinfink, H., Goodenough, J. B., and Campion, A. (1987). *J. Am. Chem. Soc.* **109**, 5291.

Dinger, T. R., Worthington, T. K., Gallagher, W. J., and Sandstrom, R. L. (1987). *Phys. Rev. Lett.* **58**, 2687.

Dixon, M. A., VerNooy, P. D., Stacy, A. M., (1988). *MRS Proc., Ext. Abstr. High-Temp. Supercond., 2nd, 1988* p. 69.

Eaglesham, D. J., Humphreys, C. J., Clegg, W. J., Harmer, M. A., Alford, N. McN., and Birchall, J. D. (1987). *Adv. Ceram. Mater.* **2**, Spec. Issue, 662.

Elam, W. T., Kirkland, J. P., Neiser, R. A., Skelton, E. F., Sampath, S., and Herman, H. (1987). *Adv. Ceram. Mater.* **2**, Spec. Issue, 411.

Emery, V. J. (1987). *Phys. Rev. Lett.* **58**, 2794.

Emery, V. J., (1989)., *Materials Research Bulletin*, **XIV**(1), 67.

Esquinazi, P., Luzariaga, J., Duran, C., Esparza, D. A., and D'Ovidio, C. (1987). *Phys. Rev. B: Condens. Matter* **36**, 2316.

Essmann, U., and Träuble, H. (1971). *Sci. Am.* **224**(3), 74.

Franck, J. P., Jung, J., and Mohamad, M. A.-K. (1987). *Phys. Rev. B: Condens. Matter* **36**, 2308.

Frase, K. G., and Clark, D. R. (1987). *Adv. Ceram. Mater.* **2**, Spec. Issue, 295.

Freltoft, T., Fisher, J. E., Shirane, G., Moncton, D. E., Sinha, S. K., Vaknin, D., Remeika, J. P., Cooper, A. S., and Harshman, D. (1987). *Phys. Rev. B: Condens. Matter* **36**, 826.

Fujita, T., Aoki, Y., Maeno, Y., Sakurai, J., Fukuba, H., and Fujii, H. (1987). *Jpn. J. Appl. Phys.* **26**, Part 2, L368.

Gallagher, P. K. (1987). *Adv. Ceram. Mater.* **2**, Spec. Issue, 632.

Ginley, D. S., Venturini, E. L., Kwak, J. K., Baughman, R. J., Morosin, B., and Schriber, J. E. (1987). *Phys. Rev. B: Condens. Matter* **36**, 829.

Ginzburg, V. L. (1964). *Phys. Lett.* **13**, 101.

Ginzburg, V. L. (1982). In "High Temperature Superconductivity" (V. L. Ginzburg and D. A. Kirzhits, eds.) Consultants Bureau, New York.

Gleick, J. (1987). *N.Y. Times Mag.* August 16, p. 29.

Godwod, K., Gorecka, J., Jasiolek, G., Majewski, J., and Przyslupski, P. (1987). *Z. Phys. B* **67**, 313.

Grader, G. S., and Gallagher, P. K. (1987). *Adv. Ceram. Mater.* **2**, Spec. Issue, 649.

Grant, P. M., Parkin, S. S. P., Lee, V. Y., Engler, E. M., Ramirez, M. L., Vasquez, J. E., Lim, G., Jacowitz, R. D., and Greene, R. L. (1987). *Phys. Rev. Lett.* **58**, 2482.

Gu, H., Li, Q., Zhang, J., Zou, B., and Yin, D. (1988). *MRS Proc., Ext. Abstr. High-Temp. Supercond., 2nd, 1988.* p. 275.

Halley, J. W., Ed., (1988). *Theories of High Temperature Superconductivity*, Addison-Wesley, N.Y.

Hauck, J., Bickmann, K., and Zucht, F. (1987). *Z. Phys. B* **67**, 299.

Hasegawa, T., Kishio, K., Aoki, M., Ooba, N., Kitazawa, K., Fueki, K., Ichida, S., and Tanaka, S. (1987). *Jpn. J. Appl. Phys.* **26**, Part 2, L337.

Hatano, T., Matsushita, A., Nakamura, K., Honda, K., Matsumoto, T., and Ogawa, K. (1987). *Jpn. J. Appl. Phys.* **26**, Part 2, L374.

Hazen, R. M., Finger, L. W., Angel, R. J., Prewitt, C. T., Ross, N. L., Mao, H. K., Hadidiacos, C. G., Hor, P. H., Meng, R. L., and Chu, C. W. (1987). *Phys. Rev. B: Condens. Matter* **35**, 7238.

Hazen, R. M., (1988). *Superconductors: The Breakthrough*, Unwin Hyman, Ltd., London

Hazen, R. M., (1988). *Sci. Amer.*, **258**(6), 74.

Hidaka, Y., Enomoto, M., Suzuki, M., Oda, M., and Murakami, T. (1987). *Jpn. J. Appl. Phys.* **26**, Part 2, L377.

Hirabayashi, M., Ihara, H., Terada, N., Senzaki, K., Hayashi, K., Waki, S., Murata, K., Tokumoto, M., and Kimura, Y. (1987). *Jpn. J. Appl. Phys.* **26**, Part 2, L454.

Hirotsu, Y., Nagakura, S., Murata, Y., Nishihara, T., Takata, M., and Yamashita, T. (1987). *Jpn. J. Appl. Phys.* **26**, Part 2, L380.

Hor, P. H., Meng, R. L., Wang, Y. Q., Gao, L., Huang, Z. J., Betchtold, J., Forster, K., and Chu, C. W. (1987). *Phys. Rev. Lett.* **58**, 1891.

Hummel, R. E. (1985). "Electronic Properties of Materials." Springer-Verlag, New York. (1985).

Ihara, H., Hirabayashi, M., Terada, N., Kimura, Y., Senzaki, K., and Tokumoto, M. (1987). *Jpn. J. Appl. Phys.* **26**, Part 2, L463.

Iqbal, Z., Steinhauser, S. W., Bose, A., Cipollini, N., and Eckhardt, H. (1987). *Phys. Rev. B: Condens. Matter* **36**, 2283.

Iwazumi, T., Yoshizaki, R., Sawada, H., Uwe, H., Sakudo, T., and Matsuura, E. (1987). *Jpn. J. Appl. Phys.* **26**, Part 2, L386.

Jarzebski, Z. M. (1973). "Oxide Semiconductors." Wydawnictwa, Warsaw.

Jorgensen, J. D., Schuttler, H.-B., Hinks, D. G., and Capone, D. W., II (1987). *Phys. Rev. Lett.* **58**, 1024.

Kawasaki, M., Funabashi, M., Nagata, S., Fueki, K., and Koinuma, H. (1987). *Jpn. J. Appl. Phys.* **26**, Part 2, L388.

Khim, Z. G., Lee, S. C., Lee, J. H., Suh, B. J., Park, Y. W., Park, C., Yu, I. S., and Park, J. C. (1987). *Phys. Rev. B: Condens. Matter* **36,** 2305.

Kingery, W. D., Bowen, H. K., and Uhlmann, D. R. (1976). *Introduction to Ceramics,* 2nd ed. Wiley, New York.

Kingon, A. I., Chevacharoenkul, S., Mansfield, J., Brynestad, J., and Haase, D. G. (1987). *Adv. Ceram. Mater.* **2,** Spec. Issue, 678.

Kini, A. M., Geiser, U., Kao, H. C. I., Carson, K. D., Wang, H. H., Monaghan, M. R., and Williams, J. M. (1987). *Inorg. Chem.* **26,** 1834.

Kirschner, I., Bankuti, J., Gal, M., and Torkos, K. (1987). *Phys. Rev. B: Condens. Matter* **36,** 2313.

Kishio, K., Kitazawa, K., Hasegawa, T., Aoki, M., Fueki, K., Ichida, S., and Tanaka, S. (1987). *Jpn. J. Appl. Phys.* **26,** Part 2, L391.

Kishio, K., Sugii, N., Kitazawa, K., and Fueki, K. (1987). *Jpn. J. Appl. Phys.* **26,** Part 2, L466.

Kitano, Y., Kifune, K., Mukouda, I., Kamikura, H., Sakurai, J., Komura, Y., Hoshino, K., Suzuki, A., Minami, A., Maeno, Y., Kato, M., and Fujita, T. (1987). *Jpn. J. Appl. Phys.* **26,** Part 2, L394.

Kittel, C. (1986), *Introduction to Solid State Physics,* 6th Edition, Wiley, New York, Chapter: 8.

Kobayashi, N., Sasaoka, T., Oh-Ishi, K., Sasaki, T., Kikuchi, M., Endo, A., Matsuzaki, K., Inoue, A., Noto, K., Syuono, Y., Saito, Y., Masumoto, T., and Muto, Y. (1987). *Jpn. J. Appl. Phys.* **26,** Part 2, L358.

Kohiki, S., Hamada, T., and Wada, T. (1987). *Phys. Rev. B: Condens. Matter* **36,** 2290.

Koinuma, H., Hashimoto, T., Kawasaki, M., and Fueki, K., (1987). *Jpn. J. Appl. Phys.* **26,** Part 2, L399.

Kramer, S., and Kordas, G. (1988). *MRS Proc., Ext. Abstr. High-Temp. Supercond., 2nd, 1988* p. 67.

Kwak, J. F., Venturini, E. L., Nigrey, P. J., and Ginley, D. S. (1988). *Evidence for Intergranular Weak Links in the High-Tc Oxides,* D. W. Caponell, et al., eds), Materials Research Society, Pittsburgh, Pennsylvania, pp. 164.

Ledbetter, H. M., Kim, S. A., and Capone, D. W., (1988). *MRS Proc., Ext. Abstr. High-Temp. Supercond., 2nd, 1988* p. 293.

Lee, P. A., and Read, N. (1987). *Phys. Rev. Lett.* **58,** 2691.

Liang, J. M., Chang, L., Sung, S. M., Kung, J. H., Wu, P. T., and Chen, L. J. (1988). *MRS Proc., Ext. Abstr. High-Temp. Supercond., 2nd, 1988* p. 327.

Little, W. A. (1964). *Phys. Rev.* **134,** A1416.

Liu, S. H. (1988). *MRS Ext. Proc., Abstr. High-Temp. Supercond., 2nd, 1988* p. 11.

Malozemoff, A. P., and Grant, P. M. (1987). *Z. Phys. BG,* **67,** 275.

Matsuura, T., and Miyake, K. (1987). *Jpn. J. Appl. Phys.* **26,** Part 2, L407.

Matsuzaki, K., Inoue, A., Kimura, H., Moroishi, K., and Masumoto, T. (1987). *Jpn. J. Appl. Phys.* **26,** Part 2, L334.

McCallum, R. W., Verhoeven, J. D., Noack, M. A., Gibson, E. D., Laabs, F. C., Finnemore, D. K., and Moodenbaugh, A. R. (1987). *Adv. Ceram. Mater.* **2,** Spec. Issue, 388.

McKitterick, J., Chen, L.-Q., Sasayama, S., McHenry, M. E., Kalonji, G., and O-Handley, R. C. (1987). *Adv. Ceram. Mater.* **2,** Spec. Issue, 353.

Meng, R. L., Hor, P. H., Wang, Y. Q., Fuang, Z. J., Sun, Y. Y., Gao, L., Bechtold, J., Chu, J. C., and Chu, C. W., (1988).

Mitchell, T. E., Roy, T., Fisk, Z., and Smith, J. L. (1988). *MRS Proc., Ext. Abstr. High-Temp. Supercond., 2nd, 1988* p. 271.

Mueller, F. M. (1987). *J. Met.* **39**(5), 3.

Muromachi, E. T., Uchida, Y., Matsui, Y., and Kato, K. (1987). *Jpn. J. Appl. Phys.* **26,** Part 2, L476.

Murphy, D. W., Sunshine, S., van Dover, R. B., Cava, R. J., Batlogg, B., Zahurak, S. M., and Schneemeyer, L. F. (1987). *Phys. Rev. Lett.* **58,** 1888.

Nagasaka, K., Sato, M., Ihara, H., Tokumoto, M., Hirabayashi, M., Terada, N., Senzaki, K., and Kimura, Y. (1987). *Jpn. J. Appl. Phys.* **26,** Part 2, L479.

Nucker, N., Fink, J., Renker, B., Ewert, D., Politis, C., Weijs, P. J. W., and Fuggle, J. C. (1987). *Z. Phys. B* **67,** 9.

O'Bryan, H. M., and Gallagher, P. K. (1987). *Adv. Ceram. Mater.* **2,** Spec. Issue, 640.

Ogita, N., Ohbayashi, K., Udagawa, M., Aoki, Y., Maeno, Y., and Fujita, T. (1987). *Jpn. J. Appl. Phys.* **26,** Part 2, L415.

Oguchi, T. (1987). *Jpn. J. Appl. Phys.* **26,** Part 2, L417.

Ohbayashi, K., Ogita, N., Udagawa, M., Aoki, Y., Maeno, Y., and Fujita, T. (1987) *Jpn. J. Appl. Phys.* **26,** Part 2, L420.

Ohbayashi, K., Ogita, N., Udagawa, M., Aoki, Y., Maeno, Y., and Fujita, T., (1987). *Jpn. J. Appl. Phys.* **26,** Part 2, L423.

Oh-Ishi, K., Kikuchi, M., Syono, Y., Hiraga, K., and Morioka, Y. (1987). *Jpn. J. Appl. Phys.* **26,** Part 2, L484.

Onoda, M., Shamoto, S., Sato, M., and Hosoya, S. (1987). *Jpn. J. Appl. Phys.* **26,** Part 27 L363.

Ovshinsky, S. R., Young, R. T., Allred, D. D., DeMaggio, G., and Van der Leeden, G. A. (1987). *Phys. Rev. Lett.* **58,** 2579.

Oyanagi, H., Ihara, H., Matsushita, T., Tokumoto, M., Hirabayashi, M., Terada, N., Senzaki, K., Kimura, Y., and Yao, T. (1987). *Jpn. J. Appl. Phys.* **26,** Part 2, L488.

Panson, A. J., Wagner, G. R., Braginski, A. I., Gavaler, J. R., Janocko, M. A., Pohl, H. C., and Talvacchio, J. (1987). *Appl. Phys. Lett.* **50,** 1104.

Paul, D. McK., Balakrisnan, G., Bernhoeft, N. R., David, W. I. F., and Harrison, W. T. A. (1987). *Phys. Rev. Lett.* **58,** 1967.

Pauling, L. (1987). *Phys. Rev. Lett.* **59,** 225.

Peterson, G. G., Weinberger, B. R., Lynds, L. and Krasinski, H. A. (1988). *Journal of Materials Research* **3**(4), 605.

Phillips, J. C. (1987). *Phys. Rev. B: Condens. Matter* **36,** 861.

Pickett, W. E. (1988). *MRS Proc., Ext. Abstr. High-Temp. Supercond., 2nd, 1988.* p. 7.

Politis, C. (1988). *High Temp. Supercond., Late News Symp., MRS Proc., 1988.*

Ramakrishna, B. L., Ong, E. W., Iqbal, Z. (1988). *MRS Proc., Ext. Abstr. High-Temp. Supercond., 2nd, 1988.* p. 51.

Rao, C. N. R., and Ganguly, P. (1987). *Curr. Sci.* **56,** 47.

Reller, A., Bednorz, J. G., and Muller, K. A. (1987). *Z. Phys. B.* **67,** 285.

Rosen, H., Engler, E. M., Strand, T. C., Lee, V. Y., and Bethune, D. (1987). *Phys. Rev. B: Condens. Matter* **36,** 726.

Rosenberg, H. M. (1986). "The Solid State," 2nd ed., p. 88. Oxford Univ. Press, London and New York.

Roth, R. S., Davis, K. L., and Dennis, J. R. (1987). *Adv. Ceram. Mater.* **2,** Spec. Issue, 303.

Saito, Y., Noji, T., Endo, A., Matsuzaki, N., Katsumata, M., and Higushi, N. (1987). *Jpn. J. Appl. Phys.* **26,** Part 2, L366.

Saito, Y., Noji, T., Endo, A., Matsuzaki, N., Katsumata, M., and Higuchi, N. (1987). *Jpn. J. Appl. Phys.* **26,** Part 2, L491.

Sarikaya, M., Stern, E. A., Kikuchi, R., Aksag, I. A. (1988). *MRS Proc., Ext. Abstr. High-Temp. Supercond., 2nd, 1988.* p. 277.

Sawada, H., Asito, Y., Iwazumi, T., Yoshizaki, R., Abe, Y., and Matsuura, E. (1987). *Jpn. J. Appl. Phys.* **26,** Part 2, L426.

Semba, K., Tsurumi, S., Hikita, M., Iwata, T., Noda, J., and Kurihara, S. (1987). *Jpn. J. Appl. Phys.* **26,** Part 2, L394.

Simon, R., and Smith, A., (1988). *Sperconductors: Conquering Technology's New Frontier,* Plenum Press, N.Y. and London.

Sinha, S. K. (1988). *MRS Bull.* **13**(6), 24.

Sishen, X., Cuiying, Y., Xiaojing, W., Guangcan, C., Hanjie, F., Wei, C., Yuqing, Z., Zhongxian, Z., Qiansheng, Y., Genghua, C., Jingkui, L., and Fanghua, L. (9187). *Phys. Rev. B: Condens. Matter* **36,** 2311.

Sleight, A. W. (1988). In *Better Supercond. Chem. MRS Proc., 1988.*

Song, S. N., Hwu, S.-J., Du, F. L., Poeppelmeier, K. R., Mason, T. O., and Ketterson, J. B. (1987). *Adv. Ceram. Mater.* **2,** Spec. Issue, 480.

Stavola, M., Cava, R. J., and Rietmen, E. A. (1987). *Phys. Rev. Lett.* **58,** 1571.

Stavola, M., Krol, D. M., Weber, W., Sunshine, S. A., Jayeraman, A., Kourouklis, G. A., Cava, R. J., and Rietmen, E. A. (1987). *Phys. Rev. B: Condens. Matter.* **36,** 850.

Stone, F. S. (1955). *In* "Chemistry of the Solid Waste" (W. E. Garner, ed.), p. 20. Academic Press, New York.

Sugai, S., Sato, M., and Hosoya, S. (1987). *Jpn. J. Appl. Phys.* **26,** Part 2, L495.

Sugai, S., Uchuda, S., Takagi, H., Kitazawa, K., and Tanaka, S. (1987). *Jpn. J. Appl. Phys.* **26,** Part 2, L879.

Sun, J. Z., Webb, D. J., Naito, M., Char, K., Hahn, M. R., Hsu, J. W. P., Kent, A. D., Mitzi, D. B., Oh, B., Beasley, M. R., Geballe, T. H., Hammond, R. H., and Kapitulnik, A. (1987). *Phys. Rev. Lett.* **58,** 1574.

Sung, C. M., Peng, P., Gorton, A., Chou, Y. T., Jain, H., Smyth, D. M., and Harmer, M. P. (1987). *Adv. Ceram. Mater.* **2,** Spec. Issue, 668.

Suzuki, M., and Murakami, T., (1987). *Jpn. J. Appl. Phys.* **26,** Part 2, L524.

Syono, Y., Kikushi, M., Oh-Ishi, K., Hiraga, K., Arai, H., Matsui, Y., Kobayashi, N., Sasaoka, T., and Muto, Y. (1987). *Jpn. J. Appl. Phys.* **26,** Part 2, L498.

Syono, Y., Kikuchi, M., Oh-Ishi, K., Tokiwa, A., Kajitani, T., and Kobayashi, N. (1988). Crystal in *MRS Proc., Ext. Abstr. High-Temp., Supercond., 2nd, 1988* p. 363.

Tajima, S., Uchida, S., Tanaka, S., Kanbe, S., Kitazawa, K., and Fueki, K. (1987). *Jpn. J. Appl. Phys.* **26,** Part 2, L432.

Takada, Y. (1988). *MRS Proc., Ext. Abstr. High-Tech. Supercond., 2nd, 1988* p. 17.

Takagi, H., Uchida, S., Kitazawa, K., and Tanaka, S. (1987). *Jpn. J. Appl. Phys.* **26,** Part 2, L123.

Takahashi, H., Murayama, C., Yoma, S., Mori, N., Kishio, K., Kitazawa, K., and Fueki, K. (1987). *Jpn. J. Appl. Phys.* **26,** Part 2, L504.

Takegahara, K. (1987). *Jpn. J. Appl. Phys.* **26,** Part 2, L457.

Tanner, D. (1987). *Supercond., 1987.* Gainesville, Florida, Materials Science and Engineering Symposium.

Tao, Y. K., Swinnea, J. S., Manthiram, A., Kim. J. S., Goodenough, J. B., and Steinfink, H., (1988). *Journal of Materials Research* **3**(2), 248.

Tarascon, J. M., McKinnon, W. R., Greene, L. H., Hull, G. W., and Vogel, E. M. (1987). *Phys. Rev. B: Condens. Matter* **36,** 226.

Terada, N., Ihara, H., Hirabayashi, M., Senzaki, K., Kimura, Y., Murata, K., and Tokumoto, M. (1987). *Jpn. J. Appl. Phys.* **26,** Part 2, L508.

Terada, N., Ihara, H., Hirabayashi, M., Senzaki, K., Kimura, Y., Murata, K., Tokumoto, M., Shimomura, O., and Kikegawa, T. (1987). *Jpn. J. Appl. Phys.* **26,** Part 2, L510.

Terakura, K., Ishida, H., Park, K. T., Yanase, A., and Hamada, N. (1987). *Jpn. J. Appl. Phys.* **26,** Part 2, L512.

Thomas, G. A., Millis, A. J., Bhatt, R. N., Cava, R. J., and Rietman, E. A. (1987). *Phys. Rev. B: Condens. Matter* **36,** 736.

Thomas, G. A., Ng, H. K., Millis, A. J., Bhatt, R. N., Cava, R. J., Rietman, E. A., Johnson, D. W. Jr., Spinosa, G. P., and Vandenberg, J. M. (1987). *Phys. Rev. B: Condens. Matter* **36,** 846.

Tien, J. K., Hendrix, B. C., Borofka, J. C., and Abe, T. (1988). *MRS Proc., Ext. Abstr. High-Temp. Supercond., 2nd, 1988* p. 73.

Uchida, S., Takagi, H., Yanagisawa, H., Kishio, K., Kitazawak, K., Fueki, K., and Tanaka, S. (1987). *Jpn. J. Appl. Phys.* **26,** Part 2, L445.

Uchikawa, F., Zheng, H., Chen, K. C., and Mackenzie, J. D. (1988). *MRS Proc., Ext. Abstr. High-Temp. Supercond., 1988* p. 89.

Umarji, A. M., Gopalakrishnan, I. K., Yakhmi, J. V., Gupta, L. C., Vijayaraghavan, R., and Iyer, R. M. (1987). *Curr. Sci.* **56,** 250.

Vaknin, D., Sinha, S. K., Moncton, D. E., Johnston, D. C., Newsam, J. M., Safinya, C. R., and King, H. E., Jr. (1987). *Phys. Rev. Lett.* **58,** 2802.

Venturini, E. L., Schirber, J. E., Morosin, B., Ginley, D. S., and Kwak, J. F. (1988). *MRS Proc., Ext. Abstr. High-Temp. Supercond., 2nd, 1988* p. 47.

Vonsovsky, S. V., Izyumov, Yu. A., and Kurmaev, E. Z. (1982). "Superconductivity of Transition Metals." Springer-Verlag, Berlin.

Wang. G., Hwu, S.-J., Song, S. N., Ketterson, J. B., Marks, L. D., Poeppelmeier, K. R., and Mason, T. O. (1987). *Adv. Ceram. Mater.* **2,** Spec. Issue, 313.

Wang, H. H., Geiser, U., Thorn, R. J., Carson, K. D., Beno, M. A., Monaghan, M. R., Allen, T. J., Proksch, R. B., Stupka, D. L., Kwok, W. K., Crabtree, G. W., and Williams, J. M. (1987). *Inorg. Chem.* **26,** 1190.

Weber, W. J., Pederson, L. R., Prince, J. M., Davis, K. C., Exarhos, G. J., Maupin, G. D., Prater, J. T., Frydrych, W. S., Aksay, I. A., Thiel, B. L., and Sarikaya, M. (1987). *Adv. Ceram. Mater.* **2,** Spec. Issue, 471.

Whangbo, M. H., Evain, M., Beno, M. A., and Williams, J. M. (1987). *Inorg. Chem.* **26,** 1831.

Wu, M. K., Ashburn, J. R., Torng, C. J., Hor, P. H., Meng, R. L., Gao, L., Huang, Z. J., Wang, Y. Q., and Chu, C. W. (1987). *Phys. Rev. Lett.* **58,** 908.

Xu, Y., Moodenbaugh, A. R., Sabatini, R. L., and Sucnaga, M. (1988). *MRS Proc., Ext. Abstr. High-Temp. Supercond., 2nd, 1988* p. 373.

Yamaguchi, Y., Yamauchi, H., Ohashi, M., Yamamoto, H., Shimoda, N., Kiruchi, M., and Syono, Y. (1987). *Jpn. J. Appl. Phys.* **26,** Part 2, L4475.

Yin, D., Han, R., and Lu, G. (1988). *MRS Proc., Ext. Abstr. High-Temp. Supercond., 2nd 1988* p. 21.

Yoshizaki, R., Iwazumi, T., Sawada, H., Ikeda, H., and Matsuura, E. (1987). *Jpn. J. Appl. Phys.* **26,** Part 2.

Yu, J., Freeman, A. J., and Xu, J.-H. (1987). *Phys. Rev. Lett.* **58,** 1035.

Zheng, H., Chen, K. C., and Mackenzie, J. D. (1988). *MRS Proc., Ext. Abstr. High-Temp. Supercond., 2nd, 1988* p. 93.

APPENDIX A

Typical Composition (in Mole %) and Properties of Optical Glasses*

Type	Designation	n_D	v	SiO_2	B_2O_3	Al_2O_3	Na_2O	K_2O	LiO	CaO	BaO	ZnO	PbO	ZrO_2	TiO_2^+	La_2O_3	Sb_2O_3
Borosilicate Crown	BK	1.510	64	71.4	6.5		5.2	13.9		2.0							
Crown	K	1.523	59	73.0	1.0		5.0	10.0		10.0							
Light Barium Crown	BaK	1.541	60	71.0	5.0		3.0	8.0	1.5		8.5	1.0	3.0				
Heavy Crown	SK	1.617	55	57.5	6.0	4.5					25.0	6.0	0.5	0.5			
Very Heavy Crown	SSK	1.656	50	50.5	12.5	1.5					24.0	3.5		0.6		2.0	
Crown Flint	KF	1.529	52	81.0			11.0										1.5
Light Barium Flint	BaLF	1.562	51	70.0	2.5		5.0	5.0			7.0	3.5	3.0				
Barium Flint	BaF	1.605	44	66.0			4.0	3.0			9.0	6.5	4.0				
Heavy Barium Flint	BaSF	1.617	39	69.5			2.5	6.5			4.5	9.0	9.0				
Very Light Flint	LLF	1.558	46	77.0	1.0		4.0	7.0			1.0	1.0	10.0				
Light Flint	LF	1.572	42	76.0			3.0	8.0			1.0		12.0				
Heavy Flint	SF	1.617	37	73.5			2.0	4.5				0.5	19.5				

* Based upon: S. Musikant, "Optical Materials: An Introduction to Selection and Application," Marcel Dekker, Inc., NY (1985). I. Fanderlik, "Optical Properties of Glass," Elsevier, NY (1983). N. P. Bansal and R. H. Doremus, "Handbook of Glass Properties," Academic Press, NY (1986). M. Grayson, "Encyclopedia of Glass, Ceramics and Cement," J. Wiley & Sons, NY (1985).

Typical Fluoride Fiber Optics Glass Compositions

Glass		Composition (wt%)								
		ZrF_4	HfF_4	BaF_2	PbF_2	GdF_3	LaF_3	AlF_3	NaF	InF_3
Z(H)BLAN	Core	53		20			4	3	20	
	Cladding	39.7	13.3	18			4	3	22	
ZB(P)LiAN	Core	54.9		17.7	4.9		3.9	3.7	14.7	0.2
	Cladding	54.9		22.6			3.9	3.7	14.7	0.2
ZBGA	Core	60.5		31.7		3.8	4			
	Cladding	58.6		30.7		3.7	7			

Reference: T. Kanomori "Recent Advances in Fluoride Glass Fiber Optics in Japan," pp. 363–367 *Halide Glasses II*, Materials Science Forum, vol. 19, Trans. Tech. Publications, 1987, Switzerland.

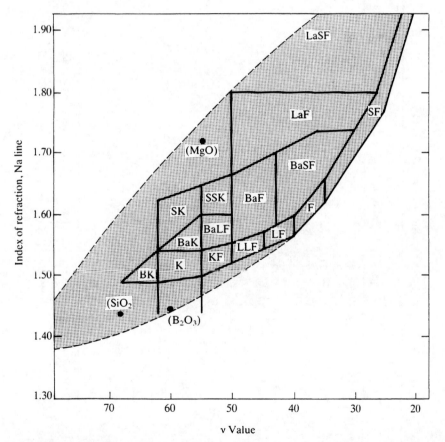

FIGURE A.1. Index of refraction (n_D) and dispersion (v) of various optical glasses described in Appendix A.

INDEX

Absorbed flux, 335
Absorption, 352, 358–385, 475
Activation Energy, 143, 182
Ag, 7
$AgAl_{11}O_{17}$, 143
AgCl, 357
Agglomeration, 1
AgI, 156
Ag_2S, 156
Aircraft, 1
Airy's formulae, 483–484
Al, 7
Alkaline, 20, 147
Alnico, 327
α–SiC, 111, 187
Alumina, Al_2O_3, 3, 12, 27, 143, 149, 155, 187, 223, 355, 468
Amorphous, 19
Amorphous semiconductor(s), 117
Anderson parameter, 122
Angle of polarization, 340
Angular momentum, 51
Anion(s), 4
Anisotropic, structure, 29, 272, 435
Anomalous dispersion, 345
Antibonding molecular orbital, 127, 367
Antiferroelectric, 272–276, 278
Antiferromagnetic, 28, 284, 288, 295
Arrhenius distribution, plot, 217
As, 7
Atom, *see specific element*

Atomic polarization, 189
Au, 7
Automotive, 1
Avalanche breakdown, 222, 229
Avalanche photodiodes, APD, 454
Avogadro's number, 141
Axial ratio, 211

Ba, 7
BaF_2, 357
Ba–La–ZrO_3, 232
Band gap, 70, 89, 91, 114
Band theory, 4, 64
BaO, 30, 187
BaO–P_2O_5, 182
Ba–Pb–BiO_3, 155
Barrier, 45
$BaTiO_2$, 28, 30, 223, 243, 244, 246, 254, 265
BCS Theory, 499, 598
Bell, Alexander, 392
BeO, 143
β–alumina, 156, 159
Bi, 7
Biaxial optical indicatrix, 439
$BiFeO_3$, 278
Binder, 22
Bi_2O_3–Y_2O_3, 156
Birefringence, 272, 434, 446
Bisque ware, 22
BK7, 357
Blue LED, 452

Bohr magneton, 283
Bohr model, 51
Boiling point(s), 17
Boltzmann distribution, 84
Bonding molecular orbital, 367
Bond order matrix, 63
Bonds, 4
Born and Lande theory, 16
Borosilicate, 20
Boundaries, 26, 45
Bound and free electrons, 50, 66
Br, 7
Bragg diffraction, 40
Bragg switch, 41
Breakdown voltage, 223
Brewster angle, 416
Brewster angle windows, 485
Bridging oxygen, 20
Brillouin zones, 67, 113
Brittleness, 25
Bronze, 1
Butadiene, 60

C, carbon, 7
Ca, calcium, 7
CaF_2, 13, 156
Calcite crystal, $CaCO_3$, 436
CaO, 30
$CaO-P_2O_5$, 182
$CaO-SiO_2$, 152
Capacitance, 186
Capacitor(s), 187
Capillary, 26
Carrier concentration, 96
Casting, 22
$CaTiO_3$, 245
Cation(s), 4
Cd, 7
$CdHfO_3$, 278
CdS, 111, 355, 357, 457–459
CdSe, 111, 459
CdTe, 111, 357
$CdTiO_3$, 278
Ce, 7
Central energy, 61
Ceramics, 1, 2
Characterization, 23
Charging constant, 193
Chemicals, 3
Circuits, 191
Cl, chlorine, 7
Classes of magnetic ceramics, 309
Clausius–Mosotti equation, 195, 241, 256
Clay, 1

Co, 7, 307
Coefficient of resistivity, 495
Coercive magnetic field, 293
Coherence length, 501–502, 514
Cole–Cole distribution, plot, 203
Commercial markets, 29
Compensated semiconductor, 96
Compton Effect, 32, 38
Compton scattering, 352
Computers, 35
Conductance, 142, 192
Conduction band, 80, 98
Conduction electron(s), 78
Conductivity, 27, 28, 83
Conjugate, 37
Control(s), 1
Cooper pairs, 499–508
Coordination, 9
Copper oxides, 27, 493–535
Core, 42
Core diameter, 394
Corning 7940, 354
Corrosion, 3, 28, 33
Coulomb, 186
Coulomb force, 15, 51
Coupled oscillator, 299, 499–508
Covalent, 3
Cr, 7, 467–468, 489
Creep, 28
Critical angle, 339
Crown glass, 442
Crystal(s), 2, 5, 6, 150
Crystal classes, 237
Crystal growth, 27
Crystalline structure of superconductors, 521–523
Crystal structure, 6, 19
Cs, 7
CsBr, 357
CsI, 357
CTE, coefficient of thermal expansion, 17, 487
Cu, copper, 7, 27, 495
Cubic, 8, 10, 239, 321
Cubic ferrites, 321
Curie temperature, 242, 267, 292
Curie–Weiss law, 243, 292, 298

de Broglie's wavelength, 38, 502
Debye equation, 202
Debye temperature, 507
Decibel (db), 337
Defect chemistry of superconductors, 523–525
Defect semiconductors, 163, 164, 174
Deformational losses, 213

Degenerate semiconductors, 96
Densification, 26, 427–428
Depletion layer, 99, 102
Depolarizing field, 263
Diamagnetic, 29, 283, 288, 497
Diamond, C, 111, 187, 357
Diatomic linear lattice, 385
Dielectric(s), 29, 185
Dielectric breakdown, 222
Dielectric conductivity, 193
Dielectric constants, 186, 187, 241, 270, 272
Dielectric loss factor, 193
Dielectric waveguides, 419–432
Diffuse reflection, 335
Diode equation, 105–108, 451
Diode(s), 102, 105–108
Dipole moment, 244
Dipole polarization, 189, 257
Disordered, see Glasses; Amorphous
Dispersion, 345, 349, 351
Dispersion-shifted fiber(s), 395
Distributed feedback lasers, 395
Divalent, 147
Domain(s), 237, 259
Domain of closure, 306
Domain formation, 239
Domain motion, 305, 308
Domain wall(s), 263
Donor(s), 94
d orbital(s), 359–376
Drift velocity, 79

E_c, conduction band edge, 94
E_f, Fermi level, 75, 86, 90
E_g, band gap energy, 64
Easy directions of magnetization, 307
Effective mass, 77
Eigenfunction(s), 45, 57
Eigenvalue(s), 45, 48, 57, 500–501
Einstein relation, 138
Electric polarization, 189
Electric susceptibility, 194, 241
Electrode (interfacial) polarization, 144
Electro-mechanical coupling, 237
Electronic polarizability, 342–344
Electronic resonance, 344
Electron mobilities, 83, 111, 116, 138
Electron pairs, 499–508, 516
Electron scattering, 39
Electro-optic(s), 434, 441
Electro-optic devices, 448
Electro-striction, 260
Elliptical polarization, 384
Emission cross section, 476

Energy level splitting, 362–364
Engines, 2
Enthalpy, 144, 167
Entropy, 144
Environmental effects, 2, 266
Equilibrium, 25
Er, 7
Etalon(s), 479–484
Eu, 7
Eu–Mo–O, 252
Eutectic, 24
E_v (valence band edge), 88
Exchange integral, 300
Exchange interaction(s), 297
Excited states, 70
Exciton(s), 96, 455–457
Expansion, 18
Expectation Value(s), 38
Extraordinary ray, 436
Extrinsic semiconductor(s), 93

F, fluorine, 7
Fabricating, see Processing
Fabry–Perot, 395, 479–484
Farad, 186
Faraday effect, 441
Faraday's law, 290
Fast ion conductor(s), 154, 157
FCC, face centered cubic, 14, 245
Fe, 7, 307
$FeAl_2O_4$, 143
$FeCr_2O_4$, 143
Feedback, 464
Feldspar, 1
Fe_2O_3, 313
Fe_3O_4, 143, 312
Fermi–Dirac distribution, 86
Fermi level, energy, E_f, 75, 86, 90
Ferrimagnetic, 284, 288, 305
Ferrites, 311–321
  cubic, 321
  garnet, 327
  hexagonal, 321
Ferroelectric, 1, 237, 238, 240, 252, 259
Ferromagnetic, 28, 284, 288, 292, 305
Ferrospinels, 309
Fiber optics, 30, 392–433
Finesse, 484
Firing, 25
Flint, glass, 1, 350, 442
Fluorescence, 449–459, 469
Fluorescence lifetimes, 488–490
Fluorite, 14
Flux, photon, 20, 335

Fluxoid quantum, 514–515
Focusing, 477
Four-level laser, 470
Fr, 7
Free charges, 188
Free ion transitions, 359
Frenkel defects, 162, 165
Fresnel's equations, 412–415, 427–428
Frohlich high energy criterion, 225

Ga, 7
GaAs, 110, 355, 357, 452, 485–486
Garnet ferrite, 327
GaSb, 111
Gas constant, R, 141
Gas ignitors, 237
Gauss's law, 14
Gd, 7
Ge, 3, 111, 121, 357
Glass conductivity, 130, 142
Glasses, 4, 17, 19, 187, 355
Goos–Haenchen shift, 420
Graded index waveguides, 394, 420–425
Grain boundaries, 1, 27
Grain structure, 1, 27
Green ware, 22

H, hydrogen, 56
Halides, 7
Hall effect, 81
Hamiltonian, 57, 59
Hard magnets, 310
Heating, 189, 190
Heisenberg uncertainty, 44, 490
Henry's law, 167
Herasil, 354
Hertz, 189
Heteroatom parameters, 60
Hexagonal, rhombic, 12, 239, 321, 437
Hexagonal ferrites, 321
Hf, 7
HfO_2, 143
Hg, 7, 495
High-T_c, 123
 superconductor(s), 493, 495, 515–526
Holes, 78–80, 111
HOMO–LUMO gap, 70, 73
Homosil, 354
Hopping, 120, 125
Housewares, 1
Huckel molecular orbital(s), 59
Hydrocarbon(s), 2, 56
Hysteresis, 250

IBM PC, 35
Idler frequency, 448
InAs, 111
Index of refraction, 348, 439
Indirect exchange interactions, 297
Infinite square well, 45, 48
Infrared, IR, 189, 337, 429
Infrasil, 354
Inorganic, 55
InSb, 111
Insulator(s), 28
Integrated optics, 393, 429
Interfacial polarization, 144, 189, 208
Intermediate state, 509
Internal absorptance, 353
Internal magnetic field, 297
Internal transmittance, 353
Interstitial, 167
Intrinsic breakdown, 222, 224
Inverse spinel, 312
Inversion layer, 99
Ionic conductivity, 139
Ionic–covalent, 4
Ionic mobility, 138
Ionic polarization, 385–389
Ionic radii, 7, 147
Ionic refraction, 344–349
Ionic transport, 27
Ionization, 168
Ir, 7
Irtran 4, 357
IR–vitreosil, 354
Isometric, 239
Isometric optical indicatrix, 437
Isotope effect, 505–507
Isotropic, 245, 434
Isotropic losses, 470

K, potassium, 7
KCl, 357
Kerr quadratic effect, 440, 447
KH_2PO_4, KDP, 243, 434
K–Mg–Bi–TiO_2, 156, 278
$KNbO_3$, 243, 252
Kramers–Kronig relation, 129
Kronig–Penny model, 65, 118

La, 7
La–CaO_3, 155
LaF_3, 156
Lambert–Beer law, 353
Λ functions, 248
La–NiO_3, 155

La_2O_3, 156
Laser focusing, 477
Laser mode, selection, 477, 478, 479–484
Lasers, 30, 463–492
Laser wavelengths, 489
La–Sr–CrO_3, 155
Lattice electrons, 67, 77
Lattice interactions, 304, 314
Lattice parameters, 17, 65, 70
Lattice potential energy, 16
LED, light emitting diode, 111, 451–455
Lenses, 30, 477
Lenz's law, 289
Li, 7
LiF, 187, 357
Ligand field theory, 359
$LiNbO_3$, 252
Linear dielectrics, 185, 191
Linear structure, 10, 65, 118
Li_2O–SiO_2, 149, 150, 156, 187, 206, 215
Liquid phase sintering, 1, 25
$LiTaO_3$, 252
Li–V_2O_5, 156
Local field theory, 252
Lodestone, 313
London equations, 510
Lorentz force, 81
Lorentz–Lorentz equation, 346–348
Loss angle, 192
Loss tangent, 193
Loudspeakers, 237
Low energy criterion, 225
Lu, 7

Madelung constant, 15
Magnetic field strength, H, 265, 285
Magnetic moments, 285, 286
Magnetic spin quantum numbers, 283
Magnetizing force, 289
Magneto-optics, 434
Magnetostrictive energy, 306
Markets, 31
Masers, 463
Matrix, 60
Maxwell's equation, 15, 396–398
Maxwell–Wagner–Sillars theory, 210
MD, molecular dynamics, 17
Mean dispersion of optical glasses, 346
Meissner effect, 496–498
Memory effect, 447
Metastable states, 467–468
Mg, magnesium, 4, 7
MgO, 4, 9, 11, 143, 187, 357

MgO–P_2O_5, 182
MgO–SiO_2, 153
Mica, muscovite, 223
Microstructures, 20, 494, 514
Microwave devices, 310
Mie scattering, 352
Migration losses, 190
Military, 30
Miller indices, 11, 307
Mixed alkali effect, 148, 152
Mixed state, 514
Mn, 7
MnO, 304
Mn–Zn–Fe_3O_4, ferrite, 317
Mo, 7
Mobility, see Electron mobilities; Ionic mobility
Mobility gap, 119
Modal dispersion, 410
Mode equation, 417, 486
Modes, 394, 402, 407
Modifiers, 20
Modulator, 430
Molar volume, 349
Molecular orbital, MO, theory, 35, 55, 57, 302, 367
Molten silicates, 150
Monochromatic, 463–464
Monoclinic, 239, 439
Monomode optical fibers, 394
Monovalent, 20
MoO_2, 143
Multimode optical fibers, 394
Multiple reflection, 339
M–W–S heterogeneous model, 210

N, nitrogen, 7
Na, sodium, 7, 149
$NaAl_{11}O_{17}$, 143, 156
Na–Bi–TiO_3, 278
NaCl, 9, 11, 221, 223, 357
$NaNbO_3$, 278
Na_2O–SiO_2, 21, 149, 151, 153, 187, 214
Na–Zr–Si–P_2O_5, 156
Nb, 7
Nd, 7
Nd–YAG, 470–475
Near field, 403, 404
Near infrared, NIR, 337
Network former, 19
Network modifier, 20, 147
Ni, nickel, 3
Nm, 7
Non-bonding molecular orbital, 367

Noncentrosymmetric, 237
Non-linear dielectrics, 237
Non-linear optics, NLO, 460
Nuclear correlation relaxation time, 221
Numerical aperture, 394
Numerical errors, 34
n–type semiconductors, 94

O, oxygen, 7
Occupancy factors, 160
Occupied states, 70
Octahedral, 8, 10, 244
Octahedral ligand field, 362
OH absorption on silica, 354
Operators, 37
Optical activity, 441
Optical axis, 470
Optical cavities, 464
Optical communications, 395
Optical computers, 30
Optical coupled switch, 431
Optical fibers, 392, 393–395
Optical indicatrix, 437
Optical modulator, 430
Optical phase shifter, 430
Optics, 335, 392, 434, 463, 477
Ordinary ray, 436
Organic binders, 22
Organic chemistry, 55
Orthorhombic, 239, 438
Oscillators, 299, 448, 485
Overlap, 55, 62
Oxidation states, 317

Paraelectric, 244, 254
Paramagnetic, 284, 288, 291
Parametric oscillator, amplifier, 448
Particles, 45, 50
Pauli exclusion principle, 15, 86, 285
Pb, lead, 7
PbF_2, 156
Pb–Fe–TaO_3, 252
Pb–Ge–O, 252
Pb–glass, ceramics, 223, 252, 278
PbS, 4, 111, 175
$PbTa_2O_6$, 252
$PbTiO_3$, 243
$PbZrO_3$, 223, 243
$PbZrTiO_6$, 4
Penetration depth, 510–514
Permeability, 285
Permittivity, 15, 186. *See also* Dielectric constants
Perovskite structure, 244, 444, 493–535

Perturbation theory, 34
Phase diagrams, 24
Phase diagrams of superconductors, 518–520
Phase retardation, 446
Phase shifter, 430
Phase velocity, 341
Phonon, coupling, 471, 499–504
Photoconductive diodes, 459
Photodiodes, 112
Photo–electric effect, 72
Photon energy, 38
Photon flux, 335
Photonic ceramics, materials, 30, 335
Pi bonds, 56
Piezoelectric, 1, 238
Planar waveguides, 425–426
Planck's constant, 51
Plastic clad fibers, 394
Plastics, 3
PLZT, Pb–La–Zr–TiO_3, 253–254, 276–278, 442–448
Pm, 7
p–n junction, 104
Po, 7
Pockels linear effect, 440
Point groups, 237, 238
Polar axis, 239
Polarization, 194, 240, 259, 341, 378–389, 446
Polarons, 96
Poling, 274
Polycrystalline, 1, 25
Polymeric clad fibers, 394
Polytypism, 112
Population inversion, 468
p orbitals, 55, 56
Porcelain, 1, 223
Porosity, 1, 27
Potential wells, 45
Pottery, 1
Powders, 2, 26
Power limitation, 487
Pr, 7
Pressing, 22
Principal optical axes, 436
Probability, 42, 47, 49
Processing, 1, 20, 23
Properties, 23
Proton conduction, 180
Pseudoatoms, 36, 69
Pseudopotential, 42
Pt, 7
p–type semiconductors, 94
Pu, 7
Pumped excitation, 469, 488

Pyroelectricity, 238
PZT, Pb–Zr–TiO$_3$, 273, 442–448

Q-switching, 485
Quantum confinement, 457–459
Quantum efficiency, 451, 488
Quantum mechanical repulsion, 15
Quantum mechanics postulates, 36
Quantum numbers, 54
Quantum theory, 33, 45
Quantum well devices, 112, 457–458, 487
Quantum well lasers, 487
Quartz, 223, 355, 357, 442
Quartz indicatrix, 438
Quenching, 27

R, gas constant, 141
Ra, 7
Radiation, 335–337
Radius ratio, 6
Raman spectroscopy, 506
Ramsdell notation, 113
Raoult's law, 167
Rasch–Hinrichsen Law, 142
Rayleigh scattering, 352, 429
Ray path deviations, 422
Rb, 7
RbAg$_4$I$_5$, 156
RC circuit, 204
Reactions, 8, 23
Rectifiers, diodes, 100, 102, 105
Reflected flux, 335
Reflection, 338, 410–412, 418
Refracted flux, 335
Refraction, 410, 435, 439–440
Relative loss factor, 193
Relative permittivity, 186
Relativity, special theory of, 36
Relaxation time, 205
Remanent magnetization, 292
Remanent polarization, 252
Repulsion, 15
Resistance(ivity), 21, 142
Resonance, 465
Resonance energy, 62
Resonating valence bond, RVB, 517
Retardation, 446
Reversible polarization, 240
Rh, 7
Rhombohedral, 247, 273
Rochelle salt, 243, 267
Rock salt structure, 9, 11
Rotational operations, 238
Roundoff errors, 34
Ru, 7

Ruby laser, 467–468
RuO$_2$, 155
Russell–Saunders notation, 359
Rutile, TiO$_2$, 223

S, sulfur, 7
Saturated magnetization, 293
Sb, 7
Sc, 7
Scattered flux, 335
Scattering, 351, 429
Schrodinger equation, 37, 42, 57, 65, 500
SDI, Strategic Defense Initiative, 30
Semiconductor contacts, 99
Semiconductors, 28, 69, 74–135
Semi-empirical calculations, 58
SHO, simple harmonic oscillator, 40
Short range, 185
Shottky defects, 169
Shrinkage, 22
Si, 3, 7, 111, 355, 357
SiC, 3, 69, 187, 452–454
Siemens, 192
Sigma bonds, 55
Signal frequency, 448
Silica, Type I though VI, 3, 142, 187, 354–357
Single mode waveguides, 486
Sintering, 1, 27
Slater–Bethe curve, 300
Sm, 7
Sm(MoO$_4$)$_3$, 252
Sn, 7
Snell's law, 339, 412
Soft magnets, 310
Solarization, 358
Solar system, 33
Sol–gel silica, 216, 354–357
Sol–gel silica waveguides, 426–429
Solid electrolytes, 154
Solid–state continuous wave laser, 474
Solid–state pulsed laser, 473
Solid state sintering, 1, 25
Solidus line, 25
Sparkplugs, 1
Speckle, 463
Spectral transmission, 354–357
Spectrosil, 354
Spectrosil–WF, 354
Specular reflection, 335
Spin bag model, 517
Spinel:
   inverse, 312
   normal, 311
Spinel ferrites, 311, 314, 318

546 INDEX

Spin magnetic moments, 283, 286, 294
Spin order, 303
Spin vectors, 300
Spontaneous magnetization, 305
Spontaneous polarization, 237, 240, 257
Square planar ligand field, 362
Sr, 7
$SrBi_2Ta_2O_9$, 252
$SrTeO_3$, 252
$SrTiO_3$, 155, 223
Step index profiles, 394
Stevels deformation polarization, 190
Stimulated emission, 469–471
Strain birefringence, 487
Structure, *see specific compound*
Substitutional defect, 169
Superconductors, 27, 493–535
Superlattices, 115
Superposition, 43, 75
Suprasil, 354
Suprasil-W, 354
Susceptibility:
  electric, 241
  magnetic, 287
Switches, 431
Symmetry, 237

T_g, glass transition temperature, 17
Ta, 7
Tb, 7
Te, 7
TEM waves, 402
Tensors, 137
Tetragonal, 8, 10, 239, 437
Tetrahedral ligand field, 362
TE waves, 407
Th, 7
Thermal breakdown, 222, 227
Thermal expansion coefficient, CTE, 487
ThO_2, 156
Three-level laser, 468
Threshold of stimulated emission, 488–489
Ti, 7, 245
Tight binding approximation, 69
TiO_2, 155, 357
Tl, 7
Tm, 7
Total loss factor, 193
Transference numbers, 137
Transformers, 317
Transistors, 109
Transition metals, 7, 286

Transmission, 335–337, 442–443, 427–429
Transmitted flux, 335
Transport, 124–129
Trap levels, 225
Triclinic, 239, 438
Trivalent, 7
Tuning, 479–484
Tunneling, 129
Type I superconductors, 508–509
Type I though VI silica, 354–356
Type II superconductors, 513–515

Ultrasonic cleaners, 237
Ultraviolet, UV, 337
Ultraviolet catastrophe, 72
Uniaxial optical indicatrix, 437
Unpoled material, virgin, 293

Vacancy defects, 165–180, 521–522, 523–525
Valence band, 64–71, 80
Valence (electrons), 7, 35
Van der Waals, 16
Verdet constant, 442
Visible, VIS, 337
Vitrification, 1, 25
VLIC, very large integrated circuit, 462
von Neumann bottleneck, 395
V_2O_3–P_2O_3, 155

W, tungsten, 2, 7
Water absorption on silica, 216
Wavefunctions, 36, 57, 367, 500–501
Waveguides, 392–433
Whitewares, 2
Work function, 88, 102
World War II, 1

Y, 7
YAG, 357, 470
Yb, 7
Y–Gd–Fe–Garnet, 330
YIG, 332
Y_2O_3, 143

Zinc–blende lattice, 111
Zinc sulfide, 442
Zn, 7
ZnO, 155
ZnS, 357
ZnSe, 357
Zr, 7
ZrO_2, 143, 156